IFIP Advances in Information and Communication Technology 432

IFIP – The International Federation for Information Processing

IFIP was founded in 1960 under the auspices of UNESCO, following the First World Computer Congress held in Paris the previous year. An umbrella organization for societies working in information processing, IFIP's aim is two-fold: to support information processing within its member countries and to encourage technology transfer to developing nations. As its mission statement clearly states,

> *IFIP's mission is to be the leading, truly international, apolitical organization which encourages and assists in the development, exploitation and application of information technology for the bene t of all people.*

IFIP is a non-profitmaking organization, run almost solely by 2500 volunteers. It operates through a number of technical committees, which organize events and publications. IFIP's events range from an international congress to local seminars, but the most important are:

- The IFIP World Computer Congress, held every second year;
- Open conferences;
- Working conferences.

The flagship event is the IFIP World Computer Congress, at which both invited and contributed papers are presented. Contributed papers are rigorously refereed and the rejection rate is high.

As with the Congress, participation in the open conferences is open to all and papers may be invited or submitted. Again, submitted papers are stringently refereed.

The working conferences are structured differently. They are usually run by a working group and attendance is small and by invitation only. Their purpose is to create an atmosphere conducive to innovation and development. Refereeing is also rigorous and papers are subjected to extensive group discussion.

Publications arising from IFIP events vary. The papers presented at the IFIP World Computer Congress and at open conferences are published as conference proceedings, while the results of the working conferences are often published as collections of selected and edited papers.

Any national society whose primary activity is about information processing may apply to become a full member of IFIP, although full membership is restricted to one society per country. Full members are entitled to vote at the annual General Assembly, National societies preferring a less committed involvement may apply for associate or corresponding membership. Associate members enjoy the same benefits as full members, but without voting rights. Corresponding members are not represented in IFIP bodies. Affiliated membership is open to non-national societies, and individual and honorary membership schemes are also offered.

Zhongzhi Shi   Zhaohui Wu   David Leake
Uli Sattler (Eds.)

Intelligent Information Processing VII

8th IFIP TC 12 International Conference, IIP 2014
Hangzhou, China, October 17-20, 2014
Proceedings

 Springer

Volume Editors

Zhongzhi Shi
Chinese Academy of Sciences
Institute of Computing Technology
Beijing 100190, China
E-mail: shizz@ics.ict.ac.cn

Zhaohui Wu
Zhejiang University
Department of Computer Science
Hangzhou 310027, China
E-mail: wzh@zju.edu.cn

David Leake
Indiana University
Computer Science Department
Bloomington, IN 47405, USA
E-mail: leake@cs.indiana.edu

Uli Sattler
University of Manchester
School of Computer Science
Manchester M13 9PL, UK
E-mail: sattler@cs.man.ac.uk

ISSN 1868-4238 e-ISSN 1868-422X
ISBN 978-3-662-51608-9 e-ISBN 978-3-662-44980-6
DOI 10.1007/978-3-662-44980-6
Springer Heidelberg New York Dordrecht London

Typesetting: Camera-ready by author, data conversion by Scientific Publishing Services, Chennai, India

Printed on acid-free paper

Springer is part of Springer Science+Business Media (www.springer.com)

Preface

This volume comprises the 8th IFIP International Conference on Intelligent Information Processing. As the world proceeds quickly into the Information Age, it encounters both successes and challenges, and it is well recognized nowadays that intelligent information processing provides the key to the Information Age and to mastering many of these challenges. Intelligent information processing supports the most advanced productive tools that are said to be able to change human life and the world itself. However, the path is never a straight one and every new technology brings with it a spate of new research problems to be tackled by researchers; as a result we are not running out of topics; rather the demand is ever increasing. This conference provides a forum for engineers and scientists in academia, university, and industry to present their latest research findings in all aspects of intelligent information processing.

This is the 8th IFIP International Conference on Intelligent Information Processing. We received more than 70 papers, of which 32 papers are included in this program as regular papers and 3 as short papers. We are grateful for the dedicated work of both the authors and the referees, and we hope these proceedings will continue to bear fruit over the years to come. All papers submitted were reviewed by two referees.

A conference such as this cannot succeed without help from many individuals who contributed their valuable time and expertise. We want to express our sincere gratitude to the Program Committee members and referees, who invested many hours for reviews and deliberations. They provided detailed and constructive review reports that significantly improved the papers included in the program.

We are very grateful to have the sponsorship of the following organizations: IFIP TC12, Zhejiang University and Institute of Computing Technology, Chinese Academy of Sciences. Thanks to Dr. Xi Yang for carefully checking the Proceedings.

Finally, we hope you find this volume inspiring and informative. Enjoy your leisurely stay in Hangzhou, China.

August 2014

Zhongzhi Shi
Zhaohui Wu
David Leake
Uli Sattler

Organization

General Chairs

T. Dillon (Australia)
Z. Wu (China)

A. Aamodt (Norway)

Program Chairs

Z. Shi (China)
D. Leake (USA)

U. Sattler (UK)

Program Committee

A. Aamodt (Norway)
B. An (China)
A. Bernardi (Germany)
C. Bryant (UK)
L. Cao (Australia)
E. Chang (Australia)
L. Chang (China)
E. Chen (China)
H. Chen (UK)
F. Coenen (UK)
Z. Cui (China)
S. Dustdar (Austria)
S. Ding (China)
Y. Ding (USA)
Q. Duo (China)
J. Ermine (France)
P. Estraillier (France)
W. Fan (UK)
Y. Gao (China)
L. Hansen (Denmark)
T. Hong (Taiwan)
Q. He (China)
T. Honkela (Finland)
Z. Huang
   (The Netherlands)

G. Kayakutlu (Turkey)
D. Leake (USA)
G. Li (China)
X. Li (Singapore)
J. Liang (China)
Y. Liang (China)
H. Leung (Hong Kong)
S. Matwin (Canada)
E. Mercier-Laurent
   (France)
F. Meziane (UK)
Z. Meng (China)
S. Nefti-Meziani (UK)
G. Osipov (Russia)
M. Owoc (Poland)
G. Pan (China)
A. Rafea (Egypt)
K. Rajkumar (India)
M. Saraee (UK)
F. Segond (France)
Q. Shen (UK)
Z.P. Shi (China)
T. Nishida (Japan)
A. Skowron
   (Poland)

M. Stumptner
   (Australia)
E. Succar (Mexico)
H. Tianfield (UK)
I.J. Timm (Germany)
S. Tsumoto (Japan)
G. Wang (China)
J. Weng (USA)
Z. Wu (China)
S. Vadera (UK)
Y. Xu (Australia)
H. Xiong (USA)
J. Yang (Korea)
X. Yang (China)
Y. Yao (Canada)
J. Yu (China)
C. Zhang (China)
J. Zhang (China)
M. Zhang (Australia)
C. Zhou (China)
J. Zhou (China)
Zhi-Hua. Zhou (China)
J. Zucker (France)

Abstracts of Keynote Presentations

Challenges of Big Data in Scientific Discovery

Benjamin W. Wah

The Chinese University of Hong Kong
Shatin, Hong Kong
bwah@cuhk.edu.hk

Abstract. Big Data is emerging as one of the hottest multi-disciplinary research fields in recent years. Big data innovations are transforming science, engineering, medicine, healthcare, finance, business, and ultimately society itself. In this presentation, we examine the key properties of big data (volume, velocity, variety, and veracity) and their relation to some applications in science and engineering. To truly handle big data, new paradigm shifts (as advocated by the late Dr. Jim Gray) will be necessary. Successful applications in big data will require in situ methods to automatically extracting new knowledge from big data, without requiring the data to be centrally collected and maintained. Traditional theory on algorithmic complexity may no longer hold, since the scale of the data may be too large to be stored or accessed. To address the potential of big data in scientific discovery, challenges on data complexity, computational complexity, and system complexity will need to be solved. We illustrate these challenges by drawing on examples in various applications in science and engineering.

Neuromorphic Computing beyond von Neumann

Karlheinz Meier

Department of Physics and Astronomy
Universität Heidelberg
meierk@kip.uni-heidelberg.de

Abstract. The brain is characterized by extreme power efficiency, fault tolerance, compactness and the ability to develop and to learn. It can make predictions from noisy and unexpected input data. Any artificial system implementing all or some of those features is likely to have a large impact on the way we process information.

With the increasingly detailed data from neuroscience and the availability of advanced VLSI process nodes the dream of building physical models of neural circuits on a meaningful scale of complexity is coming closer to realization. Such models deviate strongly from classical processor-memory based numerical machines as the two functions merge into a massively parallel network of almost identical cells.

The lecture will introduce current projects worldwide and introduce the approach proposed by the EU Human Brain Project to establish a systematic path from biological data, simulations on supercomputers and systematic reduction of cell complexity to derived neuromorphic hardware implementations with a very high degree of configurability.

Ontology-Based Monitoring of Dynamic Systems

Franz Baader

Theoretical Computer Science
TU Dresden, Germany
baader@tcs.inf.tu-dresden.de

Abstract. Our understanding of the notion "dynamic system" is a rather broad one: such a system has states, which can change over time. Ontologies are used to describe the states of the system, possibly in an incomplete way. Monitoring is then concerned with deciding whether some run of the system or all of its runs satisfy a certain property, which can be expressed by a formula of an appropriate temporal logic.

We consider different instances of this broad framework, which can roughly be classified into two cases. In one instance, the system is assumed to be a black box, whose inner working is not known, but whose states can be (partially) observed during a run of the system. In the second instance, one has (partial) knowledge about the inner working of the system, which provides information on which runs of the system are possible. In this talk, we will review some of our recent research that investigates different instances of this general framework of ontology-based monitoring of dynamic systems.

Cyborg Intelligence: Towards the Convergence of Machine and Biological Intelligence

Zhaohui Wu

Zhejiang University
wzh@zju.edu.cn

Abstract. Recent advances in the multidisciplinary fields of brain-machine interfaces, artificial intelligence, computational neuroscience, microelectronics, and neurophysiology signal a growing convergence between machines and living beings. Brain-machine interfaces (BMIs) enable direct communication pathways between the brain and an external device, making it possible to connect organic and computing parts at the signal level. Cyborg means a biological-machine system consisting of both organic and computing components. Cyborg intelligence aims to deeply integrate machine intelligence with biological intelligence by connecting machines and living beings via BMIs, enhancing strengths and compensating for weaknesses by combining the biological cognition capability with the machine computational capability. This talk will introduce the concept, architectures, and applications of cyborg intelligence. It will also discuss issues and challenges.

EEG-Based Visual Brain-Computer Interfaces

Xiaorong Gao

Dept. of Biomedical Engineering
Tsinghua University
gxr-dea@tsinghua.edu.cn

Abstract. Over the past several decades, electroencephalogram (EEG) based brain-computer interfaces (BCIs) have attracted attention from researchers in the field of neuroscience, neural engineering, and clinical rehabilitation. While the performance of BCI systems has improved, they do not yet support widespread usage. Recently, visual BCI systems have become popular because of their high communication speeds, little user training, and low user variation. However, it remains a challenging problem to build robust and practical BCI systems from physiological and technical knowledge of neural modulation of visual brain responses. This talk will review the current state and future challenges of visual BCI systems. And the taxonomy based on the multiple access methods of telecommunication systems is described. Meanwhile, the challenges will be discussed, i.e., how to translate current technology into real-life practices. Specifically, useful guidelines are provided in this talk to help exploring new paradigms and methodologies to improve the current visual BCI technology.

Abstract. ...

Table of Contents

Machine Learning

Semi-paired Probabilistic Canonical Correlation Analysis.............. 1
  Bo Zhang, Jie Hao, Gang Ma, Jinpeng Yue, and Zhongzhi Shi

Using Bat Algorithm with Levy Walk to Solve Directing Orbits of
Chaotic Systems 11
  Xingjuan Cai, Lei Wang, Zhihua Cui, and Qi Kang

Complex Proteomes Analysis Using Label-Free Mass Spectrometry-
Based Quantitative Approach Coupled with Biomedical Knowledge..... 20
  Chao Pan, Wenxian Peng, Huilong Duan, and Ning Deng

Data Mining

Online Migration Solver Based on Instructions Statistics: Algorithm for
Deciding Offload Function Set on Mobile Cloud System 29
  Yuanren Song and Zhijun Li

Improved Hierarchical K-means Clustering Algorithm without Iteration
Based on Distance Measurement 38
  *Wenhua Liu, Yongquan Liang, Jiancong Fan, Zheng Feng, and
  Yuhao Cai*

An Optimized Tag Recommender Algorithm in Folksonomy 47
  *Jie Chen, Baohua Qiang, Yaoguang Wang, Peng Wang, and
  Jun Huang*

Web Mining

Extracting Part-Whole Relations from Online Encyclopedia 57
  Fei Xia and Cungen Cao

Topic Detection and Evolution Analysis on Microblog 67
  Guoyong Cai, Libin Peng, and Yong Wang

A DBN-Based Classifying Approach to Discover the Internet Water
Army... 78
  Weiqiang Sun, Weizhong Zhao, Wenjia Niu, and Liang Chang

An Efficient Microblog Hot Topic Detection Algorithm Based on Two
Stage Clustering 90
  Yuexin Sun, Huifang Ma, Meihuizi Jia, and Wang Peiqing

Collecting Valuable Information from Fast Text Streams 96
   Baoyuan Qi, Gang Ma, Zhongzhi Shi, and Wei Wang

Multi-agent Systems

An AUML State Machine Based Method for Multi-agent Systems
Model Checking 106
   Dapeng Zhang, Xiang Ji, and Xinsheng Wang

Adaptive Mechanism Based on Shared Learning in Multi-agent
System 113
   Qingshan Li, Hua Chu, Liang Diao, and Lu Wang

An Agent-Based Autonomous Management Approach to Dynamic
Services... 122
   Fu Hou, Xinjun Mao, Junwen Yin, and Wei Wu

Research and Application Analysis of Feature Binding Mechanism 133
   Youzhen Han and Shifei Ding

Automatic Reasoning

The Correspondence between Propositional Modal Logic with Axiom
$\Box\phi \leftrightarrow \Diamond\phi$ and the Propositional Logic 141
   Meiying Sun, Shaobo Deng, and Yuefei Sui

A Sound and Complete Axiomatic System
for Modality $\Box\phi \equiv \Box_1\phi \wedge \Box_2\phi$ 152
   Shaobo Deng, Meiying Sun, Cungen Cao, and Yuefei Sui

Verification of Branch-Time Property Based on Dynamic Description
Logic 161
   Yaoguang Wang, Liang Chang, Fengying Li, and Tianlong Gu

Dynamic Description Logic Based on DL-Lite 171
   Na Zhang, Liang Chang, Zhoubo Xu, and Tianlong Gu

Formalizing the Matrix Inversion Based on the Adjugate Matrix in
HOL4 178
   Liming Li, Zhiping Shi, Yong Guan, Jie Zhang, and Hongxing Wei

Decision Algorithms

A Heuristic Approach to Acquisition of Minimum Decision Rule Sets in
Decision Systems 187
   Zuqiang Meng, Liang Jiang, Hongyan Chang, and Yuansheng Zhang

Cooperative Decision Algorithm for Time Critical Assignment without
Explicit Communication 197
    Yulin Zhang, Yang Xu, and Haixiao Hu

Using PDDL to Solve Vehicle Routing Problems 207
    Wenjun Cheng and Yuhui Gao

Automated Localization and Accurate Segmentation of Optic Disc
Based on Intensity within a Minimum Enclosing Circle 216
    Ping Jiang and Quansheng Dou

Multimedia

An Optimization Scheme for SVAC Audio Encoder 221
    Ruo Shu, Shibao Li, and Xin Pan

A Traffic Camera Calibration Method Based on Multi-rectangle:
Calibrating a Camera Using Multi-rectangle Constructed by Mark
Lines in Traffic Road 230
    Liying Lu, Xiaobo Lu, Saiping Ji, and Chen Tong

A Multi-instance Multi-label Learning Framework of Image Retrieval ... 239
    Chaojun Wang, Zhixin Li, and Canlong Zhang

The Retrieval of Shoeprint Images Based on the Integral Histogram of
the Gabor Transform Domain 249
    Xiangyang Li, Minhua Wu, and Zhiping Shi

Scene Classification Using Spatial and Color Features................ 259
    Peilong Zeng, Zhixin Li, and Canlong Zhang

Pattern Recognition

Multipath Convolutional-Recursive Neural Networks for Object
Recognition 269
    Xiangyang Li, Shuqiang Jiang, Xinhang Song, Luis Herranz, and
    Zhiping Shi

Identification of Co-regulated Gene Network by Using Path Consistency
Algorithm Based on Gene Ontology 278
    Hanshi Wang, Chenxiao Wang, Lizhen Liu, Chao Du, and Jingli Lu

Information Security

Case Retrieval for Network Security Emergency Response Based on
Description Logic 284
    Fei Jiang, Tianlong Gu, Liang Chang, and Zhoubo Xu

On the Prevention of Invalid Route Injection Attack................... 294
   Meng Li, Quanliang Jing, Zhongjiang Yao, and Jingang Liu

A Formal Model for Attack Mutation Using Dynamic Description
Logics 303
   *Zhuxiao Wang, Jing Guo, Jin Shi, Hui He, Ying Zhang,
   Hui Peng, and Guanhua Tian*

Efficient Integrity Protection for P2P Streaming 312
   Lingli Deng, Ziyao Xu, Wei Chen, Lu Lu, and Xiaodong Duan

Author Index.. 323

Semi-paired Probabilistic Canonical Correlation Analysis

Bo Zhang[1,2,4], Jie Hao[3], Gang Ma[1,2], Jinpeng Yue[1,2], and Zhongzhi Shi[1]

[1] Chinese Academy of Sciences, Institute of Computing Technology,
The Key Laboratory of Intelligent Information Processing, Beijing, China
{zhangb,mag,yuejp,shizz}@ics.ict.ac.cn
[2] University of Chinese Academy of Sciences, Beijing, China
[3] Xuzhou Medical College School of Medicine Information, Xuzhou, China
haojie@xzmc.edu.cn
[4] China University of Mining and Technology,
School of Computer Science and Technology,
Xuzhou, China

Abstract. CCA is a powerful tool for analyzing paired multi-view data. However, when facing semi-paired multi-view data which widely exist in real-world problems, CCA usually performs poorly due to its requirement of data pairing between different views in nature. To cope with this problem, we propose a semi-paired variant of CCA named SemiPCCA based on the probabilistic model for CCA. Experiments with artificially generated samples demonstrate the effectiveness of the proposed method.

Keywords: canonical correlation analysis, probabilistic canonical correlation analysis, semi-paired multi-view data.

1   Introduction

CCA is a data analysis and dimensionality reduction method similar to PCA. While PCA deals with only one data space, CCA is a technique for joint dimensionality reduction across two spaces that provide heterogeneous representations of the same data. In real world, we often meet with such a case that one object is represented by two or more types of features, e.g., image can be represented by color and texture features, the same person has visual and audio features. Canonical correlation analysis (CCA) is a classical but still powerful method for analyzing these paired multi-view data.

CCA requires the data be rigorously paired or one-to-one correspondence among different views due to its correlation definition. However, such requirement is usually not satisfied in real-world applications due to various reasons, e.g., (1) different sampling frequencies of sensors acquiring data or sensor faulty result in the multi-view data cannot keep one-to-one correspondence any more. (2) we are often given only a few paired and a lot of unpaired multi-view data, because unpaired multi-view data are relatively easier to be collected and pairing them is difficult, time consuming,

Z. Shi et al. (Eds.): IIP 2014, IFIP AICT 432, pp. 1–10, 2014.

even expensive. In literature, such data is referred as semi-paired multi-view data [1], weakly-paired multi-view data [2] or partially-paired multi-view data [3]. To cope with this problem, several extensions of CCA have been proposed to utilize the meaningful prior information hidden in additional unpaired data.

In this paper, we propose a yet another semi-paired variant of CCA called SemiPCCA, which extends the probabilistic CCA model to incorporate unpaired data into the projection. We derive an efficient EM learning algorithm for this model. Experimental results on various learning tasks show promising performance for SemiPCCA model. It is necessary to mention that the actual meaning of "semi-" in SemiPCCA is "semi-paired" rather than "semi-supervised" in popular semi-supervised learning literature.

This paper is organized as follows. After reviewing previous work in Section 2, we formally introduce SemiPCCA model in Section 3 and derive an EM algorithm in Section 4. Finally Section 5 illustrates experiments results and Section 6 concludes the paper.

2   Related Work

In this section, we review canonical correlation analysis and some improved algorithms of CCA that can effectively deal with semi-paired multi-view data.

2.1   CCA: Canonical Correlation Analysis

Let x_1 and x_2 be two set of random variables. Consider the linear combination $u = W_1^T x_1$ and $v = W_2^T x_2$. The problem of canonical correlation analysis reduce to find optimal linear transformation W_1 and W_2, which maximizes the correlation coefficient between u and v in accordance with that between x_1 and x_2. That is:

$$\rho = \max_{W_1, W_2} \frac{W_1^T \Sigma_{12} W_2}{\sqrt{W_1^T \Sigma_{11} W_1 \cdot W_2^T \Sigma_{22} W_2}} \tag{1}$$

Where Σ_{11} and Σ_{22} are the within-set covariance matrix and Σ_{12} is the between-sets covariance matrix. Since the solution of Eq. (1) is not affected by rescaling W_1 and W_2 either together or independently, the optimization of ρ is equivalent to maximizing the numerator subject to $W_1^T \Sigma_{11} W_1 = 1$ and $W_2^T \Sigma_{22} W_2 = 1$. Then with Lagrange multiplier method, we can get

$$\Sigma_{12} \Sigma_{22}^{-1} \Sigma_{21} W_1 = \lambda^2 \Sigma_{11} W_1$$

$$\Sigma_{21} \Sigma_{11}^{-1} \Sigma_{12} W_2 = \lambda^2 \Sigma_{22} W_2$$

Which is a generalized eigenproblem of the form $Ax = \lambda Bx$. A sequence of W_1 and W_2 can be obtained by eigenvectors descending ordered by the corresponding maximal eigenvalues, which indicating the explained correlation. In some literature, CCA is often described as the following:

$$\begin{pmatrix} 0 & \Sigma_{12} \\ \Sigma_{21} & 0 \end{pmatrix} \begin{pmatrix} W_1 \\ W_2 \end{pmatrix} = \lambda \begin{pmatrix} \Sigma_{11} & 0 \\ 0 & \Sigma_{22} \end{pmatrix} \begin{pmatrix} W_1 \\ W_2 \end{pmatrix}$$

2.2   Semi-paired Canonical Correlation Analysis

Recently, some improved algorithms of CCA that can deal with semi-paired multi-view data have emerged. Blaschko et al. [4] proposes semi-supervised Laplacian regularization of kernel canonical correlation (SemiLRKCCA) to find a set of highly correlated directions by exploiting the intrinsic manifold geometry structure of all data (paired and unpaired). SemiCCA [5] resembles the manifold regularization [6], i.e., using the global structure of the whole training data including both paired and unpaired samples to regularize CCA. Consequently, SemiCCA seamlessly bridges CCA and principal component analysis (PCA), and inherits some characteristics of both PCA and CCA. Gu et al. [3] proposed partially paired locality correlation analysis (PPLCA), which effectively deals with the semi-paired scenario of wireless sensor network localization by virtue of the combination of the neighborhood structure information in data. Most recently, Chen et al. [1] presents a general dimensionality reduction framework for semi-paired and semi-supervised multi-view data which naturally generalizes existing related works by using different kinds of prior information. Based on the framework, they develop a novel dimensionality reduction method, termed as semi-paired and semi-supervised generalized correlation analysis (S^2GCA), which exploits a small amount of paired data to perform CCA.

3   The SemiPCCA Model

In this section, we first review a probabilistic model for CCA in section 3.1, and then present our model.

3.1   PCCA: Probabilistic Canonical Correlation Analysis

In [7], Bach and Jordan propose a probabilistic interpretation of CCA. In this model, two random vectors $x_1 \in \mathbb{R}^{m_1}$ and $x_2 \in \mathbb{R}^{m_2}$ are considered generated by the same latent variable $z \in \mathbb{R}^d (\min(m_1, m_2) \geq d \geq 1)$ and thus the "correlated" to each other. The graphical model of the probabilistic CCA model is shown in Figure 1(a).

In this model, the observations of x_1 and x_2 are generated form the same latent variable z (Gaussian distribution with zero mean and unit variance) with unknown linear transformations W_1 and W_2 by adding Gaussian noise ε_1 and ε_2, i.e. ,

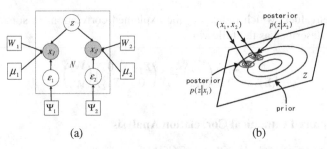

(a)   (b)

Fig. 1. (a) Graphical model for probabilistic CCA. The shaded nodes represent observed variables and unshaded node represents the latent variable. The box denotes a plate comprising a data set of N-independent observations. (b) Illustration of the projection of paired data onto the mean of the posterior distribution in latent space.

$$x_1 = W_1 z + \mu_1 + \varepsilon_1, W_1 \in \mathbb{R}^{m_1 \times d} \tag{2}$$

$$x_2 = W_2 z + \mu_2 + \varepsilon_2, W_2 \in \mathbb{R}^{m_2 \times d} \tag{3}$$

$$P(z) \sim \mathcal{N}(0, I_d), \ P(\varepsilon_1) \sim \mathcal{N}(0, \psi_1), \ P(\varepsilon_2) \sim \mathcal{N}(0, \psi_2)$$

Let $x = \begin{pmatrix} x_1 \\ x_2 \end{pmatrix}$, we have

$$P(x) \sim \mathcal{N}(\mu, \Sigma)$$

where $\mu = \begin{pmatrix} \mu_1 \\ \mu_2 \end{pmatrix}$, $\Sigma = WW^T + \psi$, $W = \begin{pmatrix} W_1 \\ W_2 \end{pmatrix}$ and $\psi = \begin{pmatrix} \psi_1 & 0 \\ 0 & \psi_2 \end{pmatrix}$. From [7], the corresponding maximum-likelihood estimations to the unknown parameters W_1, W_2, μ_1, μ_2, ψ_1 and ψ_2 are

$$\hat{\mu}_1 = \frac{1}{N} \sum_{i=1}^{N} x_1^i, \hat{W}_1 = \widetilde{\Sigma}_{11} U_{1d} M_1, \hat{\psi}_1 = \widetilde{\Sigma}_{11} - \hat{W}_1 \hat{W}_1^T$$

$$\hat{\mu}_2 = \frac{1}{N} \sum_{i=1}^{N} x_2^i, \hat{W}_2 = \widetilde{\Sigma}_{22} U_{2d} M_2, \hat{\psi}_2 = \widetilde{\Sigma}_{22} - \hat{W}_2 \hat{W}_2^T$$

where $\widetilde{\Sigma}_{11}$, $\widetilde{\Sigma}_{22}$ have the same meaning of standard CCA, the columns of U_{1d} and U_{2d} are the first d canonical directions, P_d is the diagonal matrix with its diagonal elements given by the first d canonical correlations and M_1, $M_2 \in \mathbb{R}^{d \times d}$, with spectral norms smaller the one, satisfying $M_1 M_2^T = P_d$. In our expectations, let $M_1 = M_2 = (P_d)^{1/2}$. The posterior expectations of z given x_1 and x_2 are

$$E(z|x_1) = M_1^T U_{1d}^T (x_1 - \hat{\mu}_1), E(z|x_2) = M_2^T U_{2d}^T (x_2 - \hat{\mu}_2) \tag{4}$$

Thus, $E(z|x_1)$ and $E(z|x_2)$ lie in the d dimensional subspace that are identical with those of standard CCA, as illustrate in Figure 1(b).

3.2   SemiPCCA: Semi-paired Probabilistic Canonical Correlation Analysis

Consider a set of paired samples of size N_p, $\mathbf{X}_1^{(P)} = \{(x_1^i)\}_{i=1}^{N_p}$ and $\mathbf{X}_2^{(P)} = \{(x_2^i)\}_{i=1}^{N_p}$, where each sample x_1^i (resp. x_2^i) is represented as a vector with dimension of m_1 (resp. m_2). When the number of paired of samples is small, CCA tends to overfit the given paired samples. Here, let us consider the situation where unpaired samples $\mathbf{X}_1^{(U)} = \{(x_1^j)\}_{j=N_p+1}^{N_1}$ and/or[1] $\mathbf{X}_2^{(U)} = \{(x_2^k)\}_{k=N_p+1}^{N_2}$ are additional provided, where $\mathbf{X}_1^{(U)}$ and $\mathbf{X}_2^{(U)}$ might be independently generated. Since the original CCA and PCCA cannot directly incorporate such unpaired samples, we proposed a novel method named Semi-paired PCCA (SemiPCCA) that can avoid overfitting by utilizing the additional unpaired samples. See Figure 2 for an illustration of the graphical model of the SemiPCCA model.

The whole observation is now $D = \{(x_1^i, x_2^i)\}_{i=1}^{N_p} \cup \{(x_1^j)\}_{j=N_p+1}^{N_1} \cup \{(x_2^k)\}_{k=N_p+1}^{N_2}$. The likelihood, with the independent assumption of all the data points, is calculated as

$$\mathcal{L}(\theta) = \prod_{i=1}^{N_p} P(x_1^i, x_2^i; \theta) \prod_{j=N_p+1}^{N_1} P(x_1^j; \theta) \prod_{k=N_p+1}^{N_2} P(x_2^k; \theta) \tag{5}$$

In SemiPCCA model, for paired samples $\{(x_1^i, x_2^i)\}_{i=1}^{N_p}$, x_1^i and x_2^i are considered generated by the same latent variable z^i and $P(x_1^i, x_2^i; \theta)$ is calculated as in PCCA model, i.e.

$$P(x_1^i, x_2^i; \theta) \sim \mathcal{N}\left(\begin{pmatrix} \mu_1 \\ \mu_2 \end{pmatrix}, \begin{pmatrix} \mathbf{W}_1\mathbf{W}_1^T + \psi_1 & \mathbf{W}_1\mathbf{W}_2^T \\ \mathbf{W}_2\mathbf{W}_1^T & \mathbf{W}_2\mathbf{W}_2^T + \psi_2 \end{pmatrix}\right)$$

Whereas   for   unpaired   observations $\mathbf{X}_1^{(U)} = \{(x_1^j)\}_{j=N_p+1}^{N_1}$ and $\mathbf{X}_2^{(U)} = \{(x_2^k)\}_{k=N_p+1}^{N_2}$, x_1^j and x_2^k are separately generated from the latent variable z_1^j and z_2^k with linear transformations $\mathbf{W}_1$ and $\mathbf{W}_2$ by adding Gaussian noise ε_1 and ε_2. From Eq. (2) and Eq. (3),

$$P(x_1^j; \theta) = \int P(x_1^j|z_1^j)P(z_1^j) \, dz_1^j \sim \mathcal{N}(\mu_1, \mathbf{W}_1\mathbf{W}_1^T + \psi_1)$$

$$P(x_2^k; \theta) = \int P(x_2^k|z_2^k)P(z_2^k) \, dz_2^k \sim \mathcal{N}(\mu_2, \mathbf{W}_2\mathbf{W}_2^T + \psi_2)$$

---

[1] In the context of automatic image annotation, $\mathbf{X}_1^{(U)}$ only exists, whereas $\mathbf{X}_2^{(U)}$ is empty.

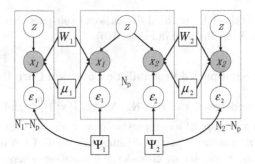

Fig. 2. Graphical model for probabilistic CCA. The box denotes a plate comprising a data set of N_p paired observations, and additional unpaired samples.

3.3   Projections in SemiPCCA Model

Analogous to the PCCA model, in SemiPCCA model the projection of paired obser-vation (x_1^i, x_2^i) is directly given by Eq. (4).

Although this result looks similar as that in PCCA model, the learning of $\mathbf{W}_1$ and $\mathbf{W}_2$ are influenced by those unpaired samples. Unpaired samples reveal the global structure of whole the samples in each domain. Note once a basis in one sample space is rectified, the corresponding bases in the other sample space is also rectified so that correlations between two bases are maximized.

4   EM learning for SemiPCCA

The log likelihood of the observations in $\mathcal{L}(\theta)$ is a sum of three parts. Therefore in E-step we need to deal with them differently. For each paired observation i in $\{(x_1^i, x_2^i)\}_{i=1}^{N_p}$, we estimate the posterior distribution of z^i given (x_1^i, x_2^i). This is done using

$$P(z^i|x_1^i, x_2^i; \theta) \sim \mathcal{N}\left(\mathbf{W}^\mathbf{T}(\mathbf{W}\mathbf{W}^\mathbf{T} + \boldsymbol{\psi})^{-1}\left(\begin{pmatrix} x_1^i \\ x_2^i \end{pmatrix} - \mu\right), \mathbf{I} - \mathbf{W}^\mathbf{T}(\mathbf{W}\mathbf{W}^\mathbf{T} + \boldsymbol{\psi})^{-1}\mathbf{W}\right)$$

and we calculate the expectation with respect to the posterior distribution $P(z^i|x_1^i, x_2^i; \theta)$ as

$$\langle z^i \rangle = \mathbf{W}^\mathbf{T}(\mathbf{W}\mathbf{W}^\mathbf{T} + \boldsymbol{\psi})^{-1}\left(\begin{pmatrix} x_1^i \\ x_2^i \end{pmatrix} - \mu\right) \tag{6}$$

$$\langle z^i {z^i}^\mathbf{T} \rangle = \langle z^i \rangle \langle z^i \rangle^\mathbf{T} + \mathbf{I} - \mathbf{W}^\mathbf{T}(\mathbf{W}\mathbf{W}^\mathbf{T} + \boldsymbol{\psi})^{-1}\mathbf{W} \tag{7}$$

For unpaired points $\{(x_1^j)\}_{j=N_p+1}^{N_1}$, latent variable z_1^j is only conditioned on x_1^j, which can be calculated posterior distribution via

$$P\big(z_1^j\big|x_1^j;\theta\big)\sim\mathcal{N}\left(W_1^T(W_1W_1^T+\psi_1)^{-1}(x_1^j-\mu_1),\,I-W_1^T(W_1W_1^T+\psi_1)^{-1}W_1\right)$$

and we calculate the expectation with respect to the posterior distribution $P\big(z_1^j\big|x_1^j;\theta\big)$ as

$$\langle z_1^j\rangle = W_1^T(W_1W_1^T+\psi_1)^{-1}(x_1^j-\mu_1) \tag{8}$$

$$\langle z_1^j z_1^{jT}\rangle = \langle z_1^j\rangle\langle z_1^j\rangle^T + I - W_1^T(W_1W_1^T+\psi_1)^{-1}W_1 \tag{9}$$

For unpaired points $\big\{(x_2^k)\big\}_{k=N_p+1}^{N_2}$, latent variable z_2^k is only conditioned on x_2^k, which can be calculated posterior distribution via

$$P\big(z_2^k\big|x_2^k;\theta\big)\sim\mathcal{N}\left(W_2^T(W_2W_2^T+\psi_2)^{-1}(x_2^k-\mu_2),\,I-W_2^T(W_2W_2^T+\psi_2)^{-1}W_2\right)$$

and we calculate the expectation with respect to the posterior distribution $P\big(z_2^k\big|x_2^k;\theta\big)$ as

$$\langle z_2^k\rangle = W_2^T(W_2W_2^T+\psi_2)^{-1}(x_2^j-\mu_2) \tag{10}$$

$$\langle z_2^k z_2^{kT}\rangle = \langle z_2^k\rangle\langle z_2^k\rangle^T + I - W_2^T(W_2W_2^T+\psi_2)^{-1}W_2 \tag{11}$$

In the M-step, we maximize the complete log-likelihood $\mathcal{L}(\theta)$ by setting the partial derivatives of the complete log likelihood with respect to each parameter to zero, holding $P\big(z^i\big|x_1^i,x_2^i;\theta\big)$, $P\big(z_1^j\big|x_1^j;\theta\big)$ and $P\big(z_2^k\big|x_2^k;\theta\big)$ fixed from the E-step.

For means of x_1 and x_2 we have

$$\hat{\mu}_1 = \tilde{\mu}_1 = \frac{1}{N_1}\sum_{i=1}^{N_1}x_1^i,\ \hat{\mu}_2 = \tilde{\mu}_2 = \frac{1}{N_2}\sum_{i=1}^{N_2}x_2^i \tag{12}$$

Which are just the sample means. Since they are always the same in all EM iterations, we can centre the data $X_1^{(P)}\cup X_1^{(U)}$, $X_2^{(P)}\cup X_2^{(U)}$ by subtracting these means in the beginning and ignore these parameters in the learning process. So for simplicity we change the notation $x_1^i, x_2^i,\ x_1^j$ and x_2^k to be the centred vectors in the following.

For the two mapping matrices, we have the updates

$$\widehat{W}_1 = \left[\sum_{i=1}^{N_p}x_1^i\langle z^i\rangle^T + \sum_{j=N_p+1}^{N_1}x_1^j\langle z_1^j\rangle^T\right]\left[\sum_{i=1}^{N_p}\langle z^i z^{iT}\rangle + \sum_{j=N_p+1}^{N_1}\langle z_1^j z_1^{jT}\rangle\right]^{-1} \tag{13}$$

$$\widehat{W}_2 = \left[\sum_{i=1}^{N_p}x_2^i\langle z^i\rangle^T + \sum_{k=N_p+1}^{N_2}x_2^k\langle z_2^k\rangle^T\right]\left[\sum_{i=1}^{N_p}\langle z^i z^{iT}\rangle + \sum_{k=N_p+1}^{N_2}\langle z_2^k z_2^{kT}\rangle\right]^{-1} \tag{14}$$

Finally the noise levels are updated as

$$\widehat{\Psi}_1 = \frac{1}{N_1}\left\{ \sum_{i=1}^{N_p}(x_1^i - \widehat{W}_1\langle z^i\rangle)(x_1^i - \widehat{W}_1\langle z^i\rangle)^T + \sum_{j=N_p+1}^{N_1}(x_1^j - \widehat{W}_1\langle z_1^j\rangle)(x_1^j - \widehat{W}_1\langle z_1^j\rangle)^T \right\} \tag{15}$$

$$\widehat{\Psi}_2 = \frac{1}{N_2}\left\{ \sum_{i=1}^{N_p}(x_2^i - \widehat{W}_2\langle z^i\rangle)(x_2^i - \widehat{W}_2\langle z^i\rangle)^T + \sum_{k=N_p+1}^{N_2}(x_2^k - \widehat{W}_2\langle z_2^k\rangle)(x_2^k - \widehat{W}_2\langle z_2^k\rangle)^T \right\} \tag{16}$$

The whole algorithm is summarized in Table 1.

Table 1. Algorithm of learning in SemiPCCA Model

Input: Paired observations $\{(x_1^i, x_2^i)\}_{i=1}^{N_p}$. Unpaired observations $\{(x_1^j)\}_{j=N_p+1}^{N_1}$ and $\{(x_2^k)\}_{k=N_p+1}^{N_2}$. A desired dimension d.

1: Initialize model parameters $\theta = \{W_1, W_2, \Psi_1, \Psi_2\}$.
2: Calculate the sample means (12) and centre the data $X_1^{(P)} \cup X_1^{(U)}$, $X_2^{(P)} \cup X_2^{(U)}$.
3: **repeat**
    {E-step}
4:    **for** i = 1 to N_p **do**
5:        Calculate Eq. (6) and Eq. (7) for paired data (x_1^i, x_2^i);
6:    **end for**
7:    **for** j = N_p+1 to N_1 **do**
8:        Calculate Eq. (8) and Eq. (9) for unpaired data (x_1^j);
9:    **end for**
10:   **for** k = N_p+1 to N_2 **do**
11:      Calculate Eq. (10) and Eq. (11) for unpaired data (x_2^k);
12:   **end for**
    {M-step}
13:   Update W_1 and W_2 via Eq. (13) and Eq. (14);
14:   Update Ψ_1 and Ψ_2 via Eq. (15) and Eq. (16);
15: **until** the change of θ is smaller than a threshold.
Output: θ and projection vectors $\langle z^i\rangle (i = 1 \dots N_p)$ which are obtained from E-step.

---

5   Experiments

The performance of the proposed method is evaluated using the artificial data set created as follows: we drew samples $\{z^i\}_{i=1}^N$ from $\mathcal{N}(0, I_d)$ of dimension d = 2 and number of samples N = 300. Then the complete paired samples $\{(x_1^i, x_2^i)\}_{i=1}^N$ were created as

$$x_1 = T_1 z + \varepsilon_1, T_1 \in \mathbb{R}^{m_1 \times d}$$

$$x_2 = T_2 z + \varepsilon_2, T_2 \in \mathbb{R}^{m_2 \times d}$$

Where $P(\varepsilon_1) \sim \mathcal{N}\left(0, \begin{bmatrix} 0.75 & 0.5 \\ 0.5 & 0.75 \end{bmatrix}\right)$, $P(\varepsilon_2) \sim \mathcal{N}\left(0, \begin{bmatrix} 1 & 1 \\ 1 & 1 \end{bmatrix}\right)$, $T_1 = \begin{bmatrix} 0.6 & -1/\sqrt{2} \\ 0.8 & -1/\sqrt{2} \end{bmatrix}$, $T_2 = \begin{bmatrix} 0.3 & -0.7 \\ 0.4 & 0.7 \end{bmatrix}$. The dimension of samples are set to $m_1 = 2$ and $m_2 = 2$.

We removed several samples from $\{x_2^i\}_{i=1}^N$ by a simple linear discrimination. As a discriminant function, we used $f(x_2) = a^T x_2 - \theta$ where $a = (a_1, \ldots, a_{m_2})^T$, and θ is the discrimination threshold such that the larger θ we set, the more samples removed. A sample (x_1^i, x_2^i) was kept paired if $f(x_2^i) > 0$, and (x_1^i, x_2^i) was removed otherwise. Then, we compare the proposed SemiPCCA with the original CCA and PCCA. We evaluated the performance of (Semi)CCA by the weighted sum of cosine distances defined as follows:

$$C(W_x, W_x^*, \Lambda^*) = \sum_{i=1}^d \lambda_i^* \frac{w_{x,i}^T w_{x,i}^*}{\|w_{x,i}\| \cdot \|w_{x,i}^*\|}$$

Where $W_x^* = (w_{x,1}^*, w_{x,2}^*, \cdots, w_{x,d}^*)^T$ and $\Lambda^* = diag(\lambda_1^*, \lambda_2^*, , \lambda_d^*)$ are the "true" first d canonical correlation directions and coefficients of fully paired samples. [5]

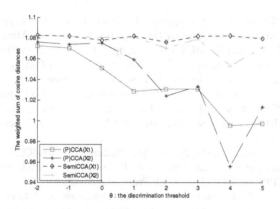

Fig. 3. Average cosine distances for artificial data

Figure 3 shows the weighted sum of cosine distances averaged over 1000 independent trials for different discrimination thresholds θ from -2 to 5. The results indicate that SemiPCCA tends to outperform the ordinary (P)CCA; it is noteworthy that even when the number of unpaired samples is not so large, SemiPCCA performs better than the original (P)CCA.

6   Conclusions

In this paper, we proposed a new semi-paired variant of CCA that we named SemiPCCA. Unlike the previous semi-paired CCA, our model is based on the probabilistic model for CCA and also intuitively comprehensive. We evaluated its performance by using artificially generated samples, and SemiPCCA performs better than the original (P)CCA.

Our future work includes some comparison of SemiPCCA with other semi-paired variants of CCA, evaluated its performance by using other data, such as PASCAL Visual Object Challenge (VOC) data sets.

Acknowledgements. This work is supported by the National Program on Key Basic Research Project (973 Program) (No. 2013CB329502), National Natural Science Foundation of China (No.61035003, No.61202212, No.61072085, No.60933004), National High-tech R&D Program of China (863 Program) (No.2012AA011003), National Science and Technology Support Program (2012BA107B02).

References

1. Chen, X., Chen, S., Xue, H., Zhou, X.: A unified dimensionality reduction framework for semi-paired and semi-supervised multi-view data. Pattern Recognition 45(5), 2005–2018 (2012)
2. Lampert, C.H., Krömer, O.: Weakly-paired Maximum Covariance Analysis for Multimodal Dimensionality Reduction and Transfer Learning. In: Daniilidis, K., Maragos, P., Paragios, N. (eds.) ECCV 2010, Part II. LNCS, vol. 6312, pp. 566–579. Springer, Heidelberg (2010)
3. Gu, J., Chen, S., Sun, T.: Localization with Incompletely Paired Data in Complex Wireless Sensor Network. IEEE Transactions on Wireless Communications 10(9), 2841–2849 (2011)
4. Blaschko, M., Lampert, C., Gretton, A.: Semi-supervised Laplacian Regularization of Kernel Canonical Correlation Analysis. In: Daelemans, W., Goethals, B., Morik, K. (eds.) ECML PKDD 2008, Part I. LNCS (LNAI), vol. 5211, pp. 133–145. Springer, Heidelberg (2008)
5. Kimura, A., Kameoka, H., Sugiyama, M., Nakano, T.: SemiCCA: Efficient Semi-supervised Learning of Canonical Correlations. In: 20th International Conference on Pattern Recognition (ICPR), pp. 2933–2936. IEEE Press, Istanbul (2010)
6. Belkin, M., Niyogi, P., Sindhwani, V.: Manifold Regularization: A Geometric Framework for Learning from Labeled and Unlabeled Examples. The Journal of Machine Learning 7, 2399–2434 (2006)
7. Bach, F.R., Jordan, M.I.: A Probability Interpretation of Canonical Correlation Analysis. Technical Report 688, Department of Statistics, Universityof California, Berkeley (2005)

Using Bat Algorithm with Levy Walk to Solve Directing Orbits of Chaotic Systems

Xingjuan Cai[1,2], Lei Wang[1], Zhihua Cui[2], and Qi Kang[1]

[1] Department of Control Science and Engineering, Tongji University, 201804, China
[2] Complex System and Computational Intelligence Laboratory,
Taiyuan University of Science and Technology, 030024, China
xingjuancai@gmail.com

Abstract. Bat algorithm is a novel swarm intelligent version by simulating the bat seeking behavior. However, due to the poor exploitation, the performance is not well. Therefore, a new variant named bat algorithm with Gaussian walk is proposed in which the original uniform sample manner is replaced by one Gaussian sample. In this paper, we want to further investigate the performance with Levy sampling manner. To test its performance, three other variants of bat algorithms are employed to compare, simulation results show our modification is superior to other algorithms.

Keywords: Bat algorithm, Levy walk, chaotic system.

1 Introduction

Bat algorithm (BA)[1] is novel swarm intelligent algorithm [2] inspired by the echolocation behavior of bats with varying pulse rates of emission and loudness. Due to the simple concepts and easy implementation, it has been applied to many areas successfully. Tsai et al. [3] proposed an improved EBA to solve numerical optimization problems. G. Wang et al. [4] incorporated the mutation operator into the methodology of BA. Bora et al. [5] applied BA to solve the brushless DC wheel motor problem. Recently, Gandomi and Yang introduced chaos into BA so as to increase its global search mobility for robust global optimization [6].

However, due to the poor exploitation, BA can not achieve good performance when dealing with multi-modal problems. To solve this problem, a novel variant which is called bat algorithm with Gaussian walk (BAGW)[7] is designed, simulation results show it is effective. In this paper, we provide one attempt to replace Gaussian walk with Levy walk because Levy sample can escape from local optima with more probability, furthermore, we apply this algorithm to solve the directing orbits of chaotic systems.

The rest of this paper was organized as follows: Section 2 provides a short introduction for the bat algorithm with Levy walk, as well as in section 3, our experiments are provided.

Z. Shi et al. (Eds.): IIP 2014, IFIP AICT 432, pp. 11–19, 2014.

2   Bat Algorithm with Levy Walk

In BA with Levy walk, each bat is defined by its position $\overrightarrow{x_i}(t)$, velocity $\overrightarrow{v_i}(t)$, frequency $\overrightarrow{f_i}$, loudness $A_i(t)$, and the emission pulse rate $r_i(t)$ in a $D-$dimensional search space. The new solutions $\overrightarrow{x_i}(t+1)$ at time $t+1$ is given by

$$x_{ik}(t+1) = x_{ik}(t) + v_{ik}(t+1) \tag{1}$$

and the velocity $v_{ik}(t+1)$ is updated by

$$v_{ik}(t+1) = v_{ik}(t) + (x_{ik}(t) - p_k(t)) \cdot f_{ik} \tag{2}$$

where $\overrightarrow{p}(t)$ is the current global best location found by all bats in the past generations. Frequency f_{ik} is dominated by:

$$f_{ik} = f_{min} + (f_{max} - f_{min}) \cdot \beta \tag{3}$$

where $\beta \in [0, 1]$ is a random number drawn from a uniform distribution, two parameters f_{min} and f_{max} are assigned to 0.0 and 2.0 in practical implementation, respectively.

To improve the exploitation capability, there exists one local search strategy for each bat, once a solution (e.g. $\overrightarrow{x_j}(t+1)$) is selected, it will be changed as with a random walk:

$$\overrightarrow{x_j}(t+1) = \overrightarrow{p}(t) + \eta\mu \tag{4}$$

where μ is a random number sampled with Levy distribution, η is a parameter to control the size of the random influence. Generally, in the first period, η assigns a large number to enhance the escaping probability from local optimum. as well as η is a small value in the later period to improve the local search probability. From our experiments, the parameter η is decreased from 0.8 to 0.001.

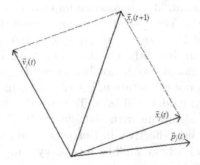

Fig. 1. Problem Illustration

For those bats whose positions are changed with Eq.(4) in iteration t, then in iteration $t+1$, the position of them will be updated with Eq.(2). In this velocity update equation, $x_{ik}(t) - p_k(t)$ is closely to zero because $\vec{x_i}(t)$ is located in the neighbor of $\vec{p}(t)$, therefore, if $\vec{v_i}(t)$ is large, then this local search pattern will be destroyed (please refer to Fig.1). Thus, in our modification, if the position $\vec{x_i}(t)$ is updated with Eq.(4), then the velocity update equation of bat i is changed as follows:

$$v_{ik}(t+1) = (x_{ik}(t) - p_k(t)) \cdot f_{ik} \tag{5}$$

while for other bats, the velocities are still updated by Eq.(2).

For bat j, the loudness $A_i(t+1)$ and the rate $r_i(t+1)$ of pulse emission are updated as follows:

$$A_i(t+1) = \alpha A_i(t) \tag{6}$$

$$r_i(t+1) = r_i(0)[1 - exp(-\gamma t)] \tag{7}$$

In this paper, $r_i(0) = 0.3$, $\gamma = 0.9$, $\alpha = 0.95$, and for each bat i, $A_i(0) = 0.9$. All of them are coming from experiments.

Algorithm 1. Bat Algorithm with Gaussian Walk

Objective function $f(\vec{x})$ $\vec{x} = (x_1, x_2, ..., x_D)^T$;
Initialize the parameters for each bat: position $\vec{x_i}(0)$, velocity $\vec{v_i}(0)$, loudness $\vec{A_i}(0)$, rate $\vec{r_i}(0)$ and repulse frequency $\vec{f_i}(0)$;
t=0;
while t<Largest iterations **do**
    Update the velocity for each bat with Eq.(2) or Eq.(5);
    Update the position for each bat with Eq.(1);
    if $rand < \vec{r_i}(t)$ **then**
        Re-update the position and velocity of bat j around the selected best solution $\vec{p}(t)$ with Eq.(4);
        Accept the new solution $\vec{x_i}(t+1)$ and velocity $\vec{v_i}(t)$;
    end if
    if $rand < \vec{A_i}(t) \& f(\vec{x_i}(t)) < f(\vec{p}(t))$ **then**
        Increase $\vec{r_i}(t)$ with Eq.(7) and reduce $\vec{A_i}(t)$ with Eq.(6);
    end if
    Rank the bats and find the current best $\vec{p}(t)$;
    t=t+1;
end while
Output the best solution $\vec{p}(t)$;

---

3   Application to Direct Orbits of Chaotic Systems

Directing orbits of chaotic systems is a multi-modal numerical optimization problem [8][9]. Consider the following discrete chaotic dynamical system:

$$\vec{x}(t+1) = \vec{f}(\vec{x}(t)), \quad t = 1, 2, ..., N \tag{8}$$

where state $\vec{x}(t) \in R^n$, $\vec{f} : R^n \to R^n$ is continuously differentiable.

Let $\vec{x}_0 \in R^n$ be an initial state of the system. If small perturbation $\vec{u}(t) \in R^n$ is added to the chaotic system, then

$$\vec{x}(t+1) = \vec{f}(\vec{x}(t)) + \vec{u}(t), \quad t = 0, 2, ..., N - 1 \tag{9}$$

where $\|\vec{u}(t)\| \le \mu$, μ is a positive real constant.

The goal is to determine suitable $\vec{u}(t)$ so as to make $\vec{x}(N)$ in the $\epsilon-$neighborhood of the target $\vec{x}(t)$, i.e., $\|\vec{x}(N) - \vec{x}(t)\| < \epsilon$, where a local controller is effective for chaos control.

Generally, assuming that $\vec{u}(t)$ acts only on the first component of $\vec{f}$, then the problem can be re-formulated as follows:

min $\|\vec{x}(N) - \vec{x}(t)\|$ by choosing suitable $\vec{u}(t)$,   t=0,2,...,N-1
S.t.

$$\begin{cases} x_1(t+1) = f_1(\vec{x}(t)) + \vec{u}(t) \\ x_j(t+1) = \vec{f}_j(\vec{x}(t)) \qquad \text{j=2,3,...,n} \end{cases} \tag{10}$$

$$|u(t)| \le \mu \tag{11}$$

$$\vec{x}(0) = \vec{x}_0 \tag{12}$$

As a typical discrete chaotic system, Hénon Map is employed as an example in this paper. Hénon Map can be described as follows:

$$\begin{cases} x_1(t+1) = -px_1^2(t) + x_2(t) + 1 \\ x_2(t+1) = qx_1(t) \end{cases} \tag{13}$$

where $p = 1.4$, $q = 0.3$.

The target $\vec{x}(t)$ is set to be a fixed point of the system $(0.63135, 0.18941)^T$, $\vec{x}_0 = (0,0)^T$, and $\vec{u}(t)$ is only added to $\vec{x}_1$ with the bound $\mu = 0.01, 0.02$ and 0.03. The population is 20, and the largest generation is 1000.

To test the performance of BAGW, the standard version of bat algorithm (SBA)[1], bat algorithm with chaotic frequency (CBA)[6], bat algorithm with mutation operator (BAM)[4] and our modified bat algorithm (MBA) are used to compare.

Under the different values of N, Table 1-3 and Figure 2-10 list the comparison of SBA, CBA, BAM and MBA. It is obviously MBA is better than other two algorithms significantly.

Table 1. Statistics Performance with μ=0.01

N	Algorithm	Mean Value	Standard Deviation
	SBA	1.3507e-02	2.4871e-04
7	CBA	1.3574e-02	2.2013e-04
	BAM	1.3500e-02	1.7893e-04
	MBA	1.3432e-02	1.6884e-04
	SBA	1.4627e-03	2.4469e-04
8	CBA	1.4435e-03	3.9140e-04
	BAM	1.2050e-03	3.4073e-04
	MBA	1.2487e-03	2.3760e-04
	SBA	9.1685e-04	4.4396e-04
9	CBA	7.3079e-04	3.3920e-04
	BAM	6.8183e-04	3.5905e-04
	MBA	7.0345e-04	5.1549e-04

Table 2. Statistics Performance with μ=0.02

N	Algorithm	Mean Value	Standard Deviation
	SBA	1.1719e-02	3.0371e-04
7	CBA	1.1785e-02	3.3121e-04
	BAM	1.1674e-02	3.0044e-04
	MBA	1.1618e-02	3.4935e-04
	SBA	6.7194e-04	3.7888e-04
8	CBA	5.3853e-04	2.1505e-04
	BAM	4.3811e-04	3.2369e-04
	MBA	4.2412e-04	2.8488e-04
	SBA	7.8986e-04	3.0203e-04
9	CBA	6.9046e-04	3.2074e-04
	BAM	6.2232e-04	3.4635e-04
	MBA	5.4411e-04	3.8880e-04

Table 3. Statistics Performance with μ=0.03

N	Algorithm	Mean Value	Standard Deviation
	SBA	1.0274e-02	5.4125e-04
7	CBA	1.0065e-02	4.3152e-04
	BAM	1.0060e-02	3.3370e-04
	MBA	1.0016e-02	4.0748e-04
	SBA	8.0668e-04	4.1823e-04
8	CBA	6.7150e-04	3.1692e-04
	BAM	6.3732e-04	3.8370e-04
	MBA	5.3046e-04	3.4043e-04
	SBA	8.4733e-04	4.9613e-04
9	CBA	1.0249e-03	4.8139e-04
	BAM	8.6568e-04	5.3757e-04
	MBA	7.1709e-04	3.9151e-04

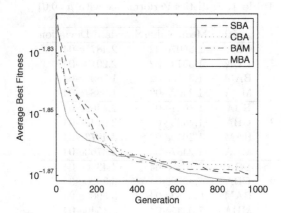

Fig. 2. Dynamic Performance with Error=0.01 and N=7

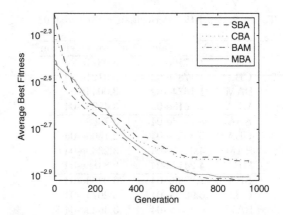

Fig. 3. Dynamic Performance with Error=0.01 and N=8

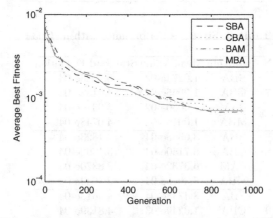

Fig. 4. Dynamic Performance with Error=0.01 and N=9

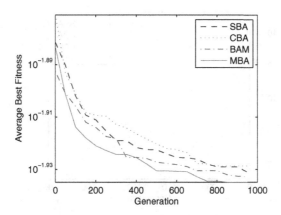

Fig. 5. Dynamic Performance with Error=0.02 and N=7

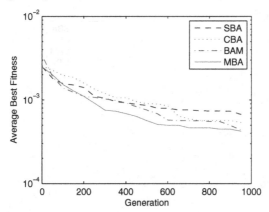

Fig. 6. Dynamic Performance with Error=0.02 and N=8

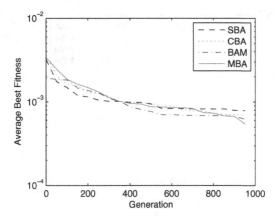

Fig. 7. Dynamic Performance with Error=0.02 and N=9

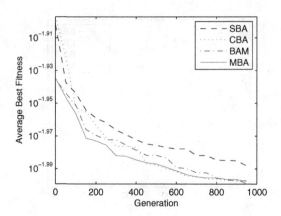

Fig. 8. Dynamic Performance with Error=0.03 and N=7

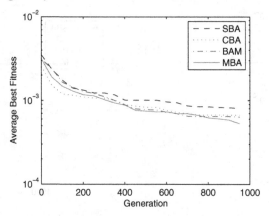

Fig. 9. Dynamic Performance with Error=0.03 and N=8

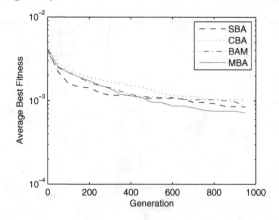

Fig. 10. Dynamic Performance with Error=0.03 and N=9

4   Conclusion

In this paper, a new variant of BA, bat algorithm with Levy walk, is applied to solve the directing orbits of chaotic systems. Simulation results show it is effective. Furthermore research topics include other applications.

Acknowledgment. This paper is supported by Natural Science Foundation of China (61003053, 71371142, 61005090), the Fundamental Research Funds for the Central Universities, the Research Fund of State Key Lab. of Management and Control for Complex Systems and the Program for the Innovative Talents of Higher Learning Institutions of Shanxi.

References

1. Yang, X., Gandomi, A.H.: Bat algorithm: a novel approach for global engineering optimization. Engineering Computation 29, 464–483 (2012)
2. Parpinelli, R.S., Lopes, H.S.: New inspirations in swarm intelligence: a survey. International Journal of Bio-inspired Computation 3(1), 1–16 (2011)
3. Tsai, P.W., Pan, J.S., Liao, B.Y., Tsai, M.J., Istanda, V.: Bat algorithm inspired algorithm for solving numerical optimization problems. Applied Mechanics and Materials 148-149, 134–137 (2011)
4. Wang, G., Guo, L., Duan, H., Liu, L., Wang, H.: A bat algorithm with mutation for UCAV path planning. The Sciencific World Journal, Article ID 418946, 15 pages (2012), doi:10.1100/2012/418946
5. Bora, T.C., Coelho, L.S., Lebensztajn, L.: Bat-inspired optimization approach for the brushless DC wheel motor problem. IEEE Transactions on Magnetics 48(2), 947–950 (2012)
6. Gandomi, A.H., Yang, X.S.: Chaotic bat algorithm. Journal of Computational Science 5, 224–232 (2014)
7. Cai, X.J., Wang, L., Kang, Q., Wu, Q.D.: Bat algorithm with Gaussian walk. International Journal of Bio-inspired Computation 6(3), 166–174 (2014)
8. Liu, B., Wang, L., Jin, Y.H.: Directing orbits of chaotic systems by particle swarm optimization. Chaos Solitons & Fractals 29, 454–461 (2006)
9. Wang, L., Li, L.L., Tang, F.: Directing orbits of chaotic dynamical systems using a hybrid optimization strategy. Physical Letters A 324, 22–25 (2004)

Complex Proteomes Analysis Using Label-Free Mass Spectrometry-Based Quantitative Approach Coupled with Biomedical Knowledge

Chao Pan[1], Wenxian Peng[2], Huilong Duan[1], and Ning Deng[1,*]

[1] College of Biomedical Engineering and Instrument Science,
Key Laboratory of Biomedical Engineering of Ministry of Education of China,
Zhejiang University, Hangzhou, China
[2] Department of Radiology, Zhejiang Medical College, Hangzhou, China
zju.dengning@gmail.com

Abstract. Label-free quantitative proteomics based on mass spectrometry plays an essential role in large-scale analysis of complex proteomes. Meanwhile, quantitative proteomics is not only a way for data processing, but also an important approach for exploring protein functions and interactions in a large-scale manner. An effective method combining quantitation and qualification should be built. To systematically overcome this challenge, we proposed a new label-free quantitative method using spectral counting in the proposed method, the count of shared peptides was considered as an optimized factor to accurately appraise abundance of Isoforms for complex proteomes. Large-scale functional annotations for complex proteomes were extracted by g:Profiler and were assigned to functional clusters. To test the effect of the methods, three groups of mitochondrial proteins including mouse heart mitochondrial dataset, mouse liver mitochondrial dataset and human heart mitochondrial dataset were selected for analysis. According to the biochemical properties of mitochondrial proteins, all functional annotations were assigned to various signalling pathway or functional clusters. We came to draw a conclusion that the strategy with shared peptides overcame inaccurate and overestimated results for low-abundant isoforms to improve accuracy, and quantitative proteomics coupled with biomedical knowledge can thoroughly comprehend functions and relationships for complex proteomes, and contribute to providing a new method for large-scale comparative or diseased proteomics.

Keywords: Complex Proteomes, Label-free Quantitation, Mass Spectrometry, Biomedical Knowledge.

1 Introduction

Relative quantitative proteomics is aiming at quantifying and detecting differential protein expression between various biological samples of interest, such as biomarkers discovery, signalling pathway or drug discovery [1-3]. Generally, quantitative

---

* Corresponding author.

Z. Shi et al. (Eds.): IIP 2014, IFIP AICT 432, pp. 20–28, 2014.

proteomics can be separated into two major approaches: the use of stable isotope labelling and label-free techniques[4]. Particularly, label-free approaches, which directly use MS feature of abundance such as spectral or peptide count or chromatographic peak area, is a reliable, versatile, and cost effective alternative to labelled quantitation [5]. However, issues arise with peptides that are shared between multiple proteins[6]. Which protein did they originate from and how should these shared peptides be used in a quantitative proteomics workflow[6]? In addition, post-translational modifications, isoforms, and splice variants are not captured by the mere analysis of transcript abundances. Protein mixtures today can routinely be characterized in terms of proteins present in the sample, but in order to allow biological interpretation, quantitative analyses are necessary[4].

In this paper, we used Normalized Spectral Abundance Factor (NSAF) based peptide count as starting point for our analysis[7] and proposed a new method for shared peptides to accurately evaluate abundance of Isoforms[6]. Label-free quantitative approaches can accurately describe abundance of complex proteins, and quantitative proteomics is not only a way for data processing, but a method for comprehending and explaining functions and relationships of proteins. Therefore, large-scale functional annotations were extracted from biomedical knowledge by g:Profiler[9] and were assigned to 12 functional clusters due to the biochemical properties of mitochondrial proteins. We found that the new strategy with shared peptides overcame inaccurate and overestimated results for low-abundant isoforms to improve accuracy, and analysis of biomedical knowledge based quantitative proteomics contributed to discovering biomarkers and targets.

2   Data and Methods

2.1   Data Acquisition

All MS/MS spectral we used were from the preliminary work of authors[10-12]. Mitochondria were treated with 0.5% DDM to extract membrane proteins, separated by SDS-PAGE followed by CBB G250 staining. Bands were sequentially cut from the continuum of the gel lane and were labelled to obtain much more accurate results in the peptide shared quantitation. Proteins were digested with trypsin, and peptides were analysed by LTQ-Orbitrap.

2.2   Data Preparation

All MS/MS spectra including mouse heart mitochondrial dataset, mouse liver mitochondrial dataset and human heart mitochondrial dataset were searched against the IPI mouse database (version 3.47) and   IPI human database (version 3.68)[8] using the pFind software kit (version 2.6)[13]. Detailed search parameters were performed using as follows: partial tryptic digest allowing two missed cleavages; fixed modification of cysteine with carbamidomethylation (57.021 Da) and   variable modification of methionine with oxidation (15.995 Da), the precursor and fragment mass tolerances were set up at 1.5 and 0.5 Da, respectively. Peptides matching the following criteria were used for protein identification: DeltaCN$\geqslant$0.1; FDR$\leqslant$1.0%; peptide mass was 600.0~6000.0; peptide length was 6~60.

2.3   Label Free Quantitative Algorithm

NSAF which was described by Old *et al*[7] gained popularity because it used protein length to rectify spectral counts to improve accuracy. We used the normalized spectral abundance factor (NSAF) based peptide count for quantitative proteomics. All peptide spectral counts were summed for each identified protein, and then divided by protein length, generating the values of Spectral Abundance Factor (SAF); the SAF value of each identified subunit was then normalized against the sum of all SAFs within an individual biological sample, resulting in the Normalized SAF (NSAF) value; all NSAF values were then calculated separately for all biological samples. The average value of NSAF for each identified protein was used for further quantitative and biological analyses. The normalization process, as a routine operation to eliminate systematic errors, can only be applied in some certain circumstances, for instance, when comparing relative changes between two complex mixture samples[14]. Meanwhile, we proposed a new method with shared peptides to explore how to accurately estimate abundance of isoforms for complex proteomes. We used distinct peptides as a proportional factor and allocated shared peptides to isoforms. Corresponding with NSAF, we similarly used protein length to rectify distinct peptides and obtained the proportional factor by normalizing distinct peptides, then allocated shared peptides to isoforms to obtain the final spectral count.

$$\left(NSAF_s\right)_J = \frac{\left(Sc/L\right)_J}{\sum_{i=1}^{N}\left(Sc/L\right)_i} \tag{1}$$

$$Sc = \sum_{band=1}^{B}\left(Scu_{band} + Scs_{band} \times R_{band}\right), \text{ and } R_{band} = \frac{Scu/L}{\sum_{k=1}^{n}\left(Scu_k/L_k\right)} \tag{2}$$

In first equation, where Sc was the spectral count for protein J and L was the length of protein J, N was the total proteins; in the second equation, B was the total bands, Scu was the count of distinct peptide, Scs was the count of shared peptide, R was the proportional factor.

2.4   Functional Analysis

g:Profiler is a web-based toolset for functional profiling of gene lists from large-scale experiments[9]. It adopts the Benjamin-Hochberg statistic method to control false discovery rate (FDR)[15] to improve accuracy. According to these properties, g:Profiler was used to obtain functional annotation of complex proteomes in large-scale experiments.

We extracted gene name of each protein from IPI fasta database, and analysed these files including gene names by g:Profiler, then outcomes were analysed by in-house software toolkit to obtain functional annotation of complex proteomes. In this paper, we used pValue as a key factor to filter functional annotation. For the protein with multiple functions, the functional annotation corresponding to the smallest pValue was filtered. According to the biochemical properties of mitochondrial proteins, all mitochondrial proteins were assigned to 12 functional clusters.

3   Results and Discussion

In order to completely test the above methods, we selected mitochondrial proteins for analysis. Mitochondria have received extensive attention due to their importance in cellular function and known causative role in diseases. Mammalian mitochondria are double-membrane organelles, serving as the metabolic power houses of eukaryotic cells[16-18]. In this paper, mouse heart mitochondrial dataset, mouse liver mitochondrial dataset and human heart mitochondrial dataset were selected to obtain all MS/MS spectra, then each group was analysed by NSAF and its optimization algorithm. We found that the new strategy with shared peptides overcame inaccurate and overestimated results for low-abundant isoforms. Meanwhile, functional annotations of mitochondrial proteins in large-scale experiments were assigned to 12 functional clusters due to the biochemical properties of mitochondrial proteins. The work flow was shown as Figure 1.

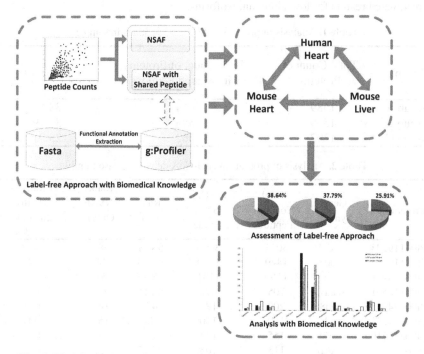

Fig. 1. Workflow of label free quantitative proteomics with biomedical knowledge

3.1 Assessment of Label -Free Quantitative Algorithm

Spectral count, defined as the total number of spectra identified for a protein, has recently gained acceptance, therefore we evaluated the label-free approach based spectral count, especially the new strategy with shared peptides. We selected a group of mouse heart mitochondrial proteins to obtain MS/MS spectra by LTQ-Oribitrap, and this group of proteins were repeated three times, named as Group A, Group B, Group C. All proteins were searched by pFind toolkit to identify total counts of proteins and the counts of proteins with shared peptides. Then each group was quantified by NSAF and the new method, and quantitative results were sorted in descending order. We found that proteins with shared peptide were accounted twenty five to forty percent of total proteins, as table 1 shows; especially those proteins which rank have dramatic changes nearly had shared peptides. Importantly, we found that such identified proteins even came from a same family, such as proteins belonging to acyl-CoA dehydrogenase family. Proteins in the family play an important role in life event due to their biochemical properties of fatty acid metabolism and lipid metabolism. As table 2 shows, proteins with shared peptides reached 90%. Ranking of these proteins in the family generally ascended after approaching by the new strategy. Additionally, if all peptides of a protein were shared, the quantitative results were extremely different, such as IPI00331251. Therefore, we concluded that normalized processes we designed eliminated systematic errors and should be considered when dealing with MS/MS spectra. Simultaneously, new strategy with shared peptide overcame inaccurate and overestimated results for low-abundant isoforms.

Table 1. Analysis of proteins with shared peptides in sample

Sample	Total Count of Proteins	Total Count of Proteins with Shared Peptides	Rate (%)
Group A	1589	614	38.64
Group B	1569	593	37.79
Group C	1397	362	25.91

Table 2. Analysis of proteins in acyl-CoA dehydrogenase family

Protein ID	Gene Symbol	Total Count of Peptides	Total Count of Shared Peptides	Rate (%)	Rank (NSAF Only)	Rank (with Shared Peptides)
IPI00119203	Acadvl	3033	3000	98.91	8	8
IPI00119114	Acadl	1449	1419	97.93	11	12
IPI00134961	Acadm	1150	1131	98.35	51	39
IPI00116591	Acads	705	692	98.16	63	63
IPI00274222	Acad8	200	188	94.00	140	100
IPI00331251	Acads	180	180	100	157	1444
IPI00331710	Acad9	155	144	92.90	182	111
IPI00119842	Acadsb	113	105	92.92	228	165
IPI00170013	Acad10	85	80	94.12	395	289

3.2   Functional Annotation of Identified Proteins

According to the biochemical properties of mitochondrial proteins, all mitochondrial proteins were assigned to 12 functional clusters including apoptosis, DNA/RNA/protein synthesis, metabolism, oxidative phosphorylation, protein binding/folding, proteolysis, redox, signal transduction, structure, transport, cell adhesion and cell cycle.   As table 3 shows, metabolic proteins have highest abundance in mouse liver mitochondrial dataset, while oxidative phosphorylation proteins show highest abundance in cardiac mitochondrial dataset. This explains that liver is important in metabolic process including nutrients synthesis, transformation and decomposition, however, heart promotes blood flowing to provide adequate blood to the organs or tissues, supplies oxygen or various nutrients and takes metabolic products away. Functional clustering for complex proteomes contributes to comprehending physiological and pathological characteristics of mitochondrial proteins.

Table 3. Analysis of functional clustering for mitochondrial proteins

Functional Clusters	% of Total Abundance (Mouse Liver)	% of Total Abundance (Mouse Heart)	% of Total Abundance (Human Heart)
OXPHOS	19.13	37.20	28.62
Metabolism	46.27	34.86	36.35
Transport	7.66	8.39	6.79
Apoptosis	1.82	2.56	5.70
Redox	6.82	1.94	3.94
Binding	4.20	3.47	7.40
Signaling	1.83	2.95	1.82
Biosynthesis	4.35	2.91	3.44
Structure	0.73	0.31	3.49
Proteolysis	1.22	0.74	0.37
Cell Adhesion	0.17	0.03	0.00
Cell Cycle	0.03	0.06	0.01
Unknown	5.76	2.40	1.89

3.3   Biochemical Properties of Identified Proteins

We analysed the biochemical proprieties of identified proteins including: MW (in kDa) and IEF point (pI) based NSAF value to discovery some new regularity using heat map. The heat map generated from software which was designed by us, as figure 2 shows. We found that *three* groups of proteins did not express much more differences, and most of abundant proteins with low molecular weight fall into the area of high pI value.

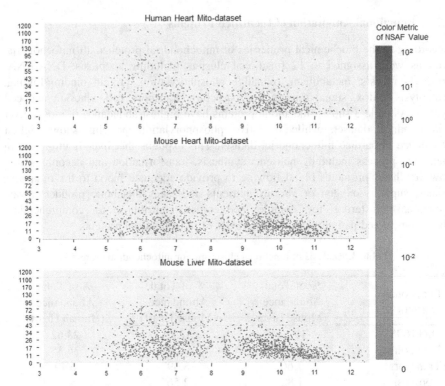

Fig. 2. Individual Heatmaps Show Protein Distribution. X-axis represents *p*I and Y-axis represents molecular weight. Point color shows the abundance of proteins.

4   Conclusions

The method using peptide count can accurately describe abundance of proteins, in addition, the strategy dealing with shared peptides can estimate relative abundance of isoforms for complex proteomes and overcame inaccurate and overestimated results for low-abundant isoforms. According to the biochemical properties of mitochondrial proteins, large-scale functional annotations which were extracted from biomedical knowledge were assigned to 12 functional clusters. We provided a new method based on quantitative analysis to explain functions and relationships of complex proteomes and contribute to bioinformatics research including quantitative expression, difference comparison and diseased proteomics. Even though, NSAF could achieve the best precision using spectral as abundant features of isoforms, it seriously underestimate the actual fold change[14]. Therefore, in order to precisely estimate quantitative results of low-abundant isoforms and further explore the deep relationship between peptides and MS/MS spectral, developing a new method becomes extremely important for complex proteomes.

Acknowledgements. This work was supported by the National High Technology Research and Development Programs of China (863 Programs, no. 2012AA02A601, no. 2012AA02A602, no. 2012AA020201), the National Science and Technology Major Project of China (No. 2013ZX03005012), and the National Natural Science Foundation of China, no. 31100592.

References

[1] Zhao, Y., Lee, W.N.P., Xiao, G.G.: Quantitative proteomics and biomarker discovery in human cancer. Expert Rev. Proteomics 6(2), 115–118 (2009)

[2] Dong, M.Q., Venable, J.D., Au, N., et al.: Quantitative mass spectrometry identifies insulin signaling targets in C. elegans. Science 317(5838), 660–663 (2007)

[3] Lill, J.: Proteomic tools for quantitation by mass spectrometry. Mass Spectrometry Reviews 22(3), 182–194 (2003)

[4] Schulze, W.X., Usadel, B.: Quantitation in mass-spectrometry-based proteomics. Annual Review of Plant Biology 61, 491–516 (2010)

[5] Zhu, W., Smith, J.W., Huang, C.M.: Mass spectrometry-based label-free quantitative proteomics. Journal of Biomedicine and Biotechnology (2010)

[6] Zhang, Y., Wen, Z., Washburn, M.P., et al.: Refinements to label free proteome quantitation: how to deal with peptides shared by multiple proteins. Analytical Chemistry 82(6), 2272–2281 (2010)

[7] Zybailov, B., Mosley, A.L., Sardiu, M.E., et al.: Statistical Analysis of Membrane Proteome Expression Changes in Saccharomyces c erevisiae. Journal of Proteome Research 5(9), 2339–2347 (2006)

[8] Kersey, P.J., Duarte, J., Williams, A., et al.: The International Protein Index: an integrated database for proteomics experiments. Proteomics 4(7), 1985–1988 (2004)

[9] Reimand, J., Kull, M., Peterson, H., et al.: g: Profiler—a web-based toolset for functional profiling of gene lists from large-scale experiments. Nucleic Acids Research 35(suppl. 2), W193–W200 (2007)

[10] Zhang, J., Li, X., Mueller, M., et al.: Systematic characterization of the murine mitochondrial proteome using functionally validated cardiac mitochondria. Proteomics 8(8), 1564–1575 (2008)

[11] Zhang, J., Liem, D.A., Mueller, M., et al.: Altered proteome biology of cardiac mitochondria under stress conditions. The Journal of Proteome Research 7(6), 2204–2214 (2008)

[12] Zhang, J., Lin, A., Powers, J., et al.: Mitochondrial proteome design: From molecular identity to pathophysiological regulation. The Journal of General Physiology 139(6), 395–406 (2012)

[13] Wang, L., Li, D.Q., Fu, Y., et al.: pFind 2.0: a software package for peptide and protein identification via tandem mass spectrometry. Rapid Communications in Mass Spectrometry 21(18), 2985–2991 (2007)

[14] Wu, Q., Zhao, Q., Liang, Z., et al.: NSI and NSMT: usages of MS/MS fragmention intensity for sensitive differential proteome detection and accurate protein fold change calculation in relative label-free proteome quantification. Analyst 137(13), 3146–3153 (2012)

[15] Benjamini, Y., Yekutieli, D.: The control of the false discovery rate in multiple testing under dependency. Annals of Statistics, 1165–1188 (2001)

[16] McDonald, T.G., Van Eyk, J.E.: Mitochondrial proteomics. Basic Research in Cardiology 98(4), 219–227 (2003)

[17] Weiss, J.N., Korge, P., Honda, H.M., et al.: Role of the mitochondrial permeability transition in myocardial disease. Circulation Research 93(4), 292–301 (2003)

[18] Honda, H.M., Korge, P., Weiss, J.N.: Mitochondria and ischemia/reperfusion injury. Annals of the New York Academy of Sciences 1047(1), 248–258 (2005)

Online Migration Solver Based on Instructions Statistics

Algorithm for Deciding Offload Function Set on Mobile Cloud System

Yuanren Song[1] and Zhijun Li[2]

[1] Harbin Institute of Technology, China
windstonedream@gmail.com

Abstract. With the popularization of cloud computing, solving the power shortage problem of mobile devices in a cloud way comes into the eyes. And migrating the whole virtual machine which can save cost for both developers and users has been realized. This paper redefine the function migration problem, and use a dynamic programming algorithm to solve it. This paper releases all functions based on COMET. 3 groups of experiment are conducted, and give speed-up of 7.11x, 23.23x, 8.24x.

Keywords: Mobile Cloud, migrate, offload.

1   Introduction

With the development of smartphone and the growing number of the users. Lots of sensors are deployed on the mobile devices. Of course, more applications will acquire the data, analyse the data and use the data. But it will cost much computing resources which are tight on the mobile devices too. Another problem is power shortage. In MAUI system, the battery dried up in 2 hours by downloading 100kb file repeatedly.

There are several systems based on migrating the whole virtual machine. MAUI is based on Microsoft .Net Framework. The system propose a method to partition the program by functions. Then profile the energy consumption for each function. At last, it uses a solver to decide which part of functions needs to be offloaded to Server. The system is released on CLR, which can save 90% energy when running face recognition programme, and has a speed up of 9.5x. Comparing to MAUI, Clonecloud system doesn't need any annotations when programming. It have a speed-up of 14.05x when running virus scan program.

Joule Benchmark style system model the energy consumption. The model indicates energy consumption is in proportional to the CPU cycles. So in the paper, we use CPU cycles to calculate the running time, and use running time as the optimization objective. We propose an algorithm to solve such a problem.

In Section 2, we give the definition of the problem and give an algorithm to solve the problem. In Section 3, we provides some details when we implement the system. In Section 4, three groups of experiments are conducted to evaluate the system performance.

Z. Shi et al. (Eds.): IIP 2014, IFIP AICT 432, pp. 29–37, 2014.

2   Function Migration Problem

2.1   Definition

The problem is based on a FIT (Function Invoking Tree). FIT is used to reflect the construction of the program. Each function can form a node in FIT. When Function A calls Function B, B will be one of A's children. When a function is invoked by different function, there will be several node in FIT to present the function in different context. Here we import (V, E) to present the FIT. V means the nodes for function, and E= {u, v} means Function u called Function v.

   We also needs some other symbols to complete the model. Ii presents where should the Function i be run. Ii =0 means it will be run on mobile device, and Ii =1 means it will be run on cloud. The instruction length of Function i is presents by Fi. The number of called times is presents by Ri. And SL and SR represents the CPU speed of mobile device and cloud server. Δi presents the time used to synchronize the memory between the mobile device and the cloud. The variables Fi, Ri, SL, SR, Δi, all can be measured or counted by statistics. Then we use the running time migration can save as the optimization objective, and the problem is as above:

$$\text{maximize} \quad \sum_{i \in N} I_i \left(\frac{R_i \cdot F_i}{S_L} - \frac{R_i \cdot F_i}{S_R} \right) - \sum_{i \in V} \sum_{(i,j) \in E} |I_i - I_j| R_j \Delta_j \tag{1}$$

$$\text{subject to} \quad I_i \in \{0, 1\}$$

2.2   Algorithm

Comparing to former problem, energy is replaced by time, so there's no constraint in this problem. So it isn't a 0-1 ILP strictly, and it's a P problem. Here we propose a dynamic programming algorithm, which will cost $O(|V|)$ time to calculate the optimal solution.

   In the algorithm, there's a weight w(T, IT) for each node T. It means the time migration can save when node T chooses IT as the schedule. The weight equals the optimization objective, and the proof will be shown in next chapter.

   The algorithm start from the root node of the FIT. And the algorithm can be divided into 2 parts: calculate weight for each node, and decide whether to offload for each node. The algorithm is as below:

```
program CalcWeight(w(T, I))
  {Assuming Input the Node T and Decision I};
  begin
    If w(T, I) is calculated then
      Return w(T, I)
    Else
      If T is leaf node then
        w(T, 0) := 0
      Else
```

$$w(T,0) = \sum_{(T,i)\in E} \frac{R_i}{R_T} \cdot \max\left(CalcWeight(i,0), CalcWeight(i,1)-\Delta_i\right) \tag{2}$$

```
End if
If node T is native function then
   w(T, 1) := -∞
Else
```

$$w(T,1) := \sum_{(T,i)\in E} \frac{R_i}{R_T} \cdot \max\left(w(i,0)-\Delta_i, w(i,1)\right)+F_T\left(\frac{1}{S_L}-\frac{1}{S_R}\right) \tag{3}$$

```
   End if
   End if
end
```

2.3   Proof

Theory 1. The root node weight calculated by the algorithm is equal to the optimization objective.

Proof: The root node weight is decided by max function in the algorithm. Then the max function can be replaced by I and 1-I. So the root weight w can be written:

$$w = I_{root} w(root,1) - I_{root}\Delta_{root} + (1-I_{root}) w(root,0)$$

$$= I_{root} \sum_{(root,i)\in E} \frac{R_i}{R_{root}}\left(I_i w(i,1)+(1-I_i)w(i,0)-(1-I_i)\Delta_i\right)+I_{root}F_{root}\left(\frac{1}{S_L}-\frac{1}{S_R}\right)$$

$$+(1-I_{root}) \sum_{(root,i)\in E} \frac{R_i}{R_{root}}\left(I_i w(i,1)-(I_i)\Delta_i+(1-I_i)w(i,0)\right)$$

$$= \frac{R_j}{R_{root}} \sum_{(root,i)\in E}\sum_{(i,j)\in E} I_{root}I_i I_j w(i,1)+(1-I_{root})I_i I_j(i,1)+I_{root}I_i\left(1-I_j\right)w(i,0)+$$

$$+ \left(1-I_{root}\right) I_i\left(1-I_j\right)w(i,0) + I_{root}\left(1-I_i\right) I_j w(i,1) + \left(1-I_{root}\right)\left(1-I_i\right) I_j w(i,1) \tag{4}$$

$$I_{root}\left(1-I_i\right)\left(1-I_j\right)w(i,0) + \left(1-I_{root}\right)\left(1-I_i\right)\left(1-I_j\right)w(i,0) + I_{root}F_{root}\left(\frac{1}{S_L}-\frac{1}{S_R}\right)$$

$$-I_{root}I_i\left(1-I_j\right)\Delta_j-I_{root}\left(1-I_i\right)I_j\Delta_j-\left(1-I_{root}\right)\left(1-I_i\right)I_j\Delta_j-\left(1-I_{root}\right)I_i\left(1-I_j\right)\Delta_j$$

$$+I_{root} \sum_{(root,i)\in E} \frac{R_i}{R_{root}}I_i F_i\left(\frac{1}{S_L}-\frac{1}{S_R}\right)+(1-I_{root}) \sum_{(root,i)\in E} \frac{R_i}{R_{root}}I_i F_i\left(\frac{1}{S_L}-\frac{1}{S_R}\right)$$

$$= \frac{R_j}{R_{root}} \sum_{(root,i)\in E}\sum_{(i,j)\in E} I_j w(i,1)+(1-I_j)w(i,0)-\left|I_i-I_j\right|\Delta_j$$

$$+I_{root}F_{root}\left(\frac{1}{S_L}-\frac{1}{S_R}\right)+ \sum_{(root,i)\in E} \frac{R_i}{R_{root}}I_i F_i\left(\frac{1}{S_L}-\frac{1}{S_R}\right)$$

Expand all the node in FIT and let $R_{root} = 1$, we'll get:

$$w = \sum_{i \in V} I_i R_i F_i \left(\frac{1}{S_L} - \frac{1}{S_R} \right) - \sum_{i \in V} \sum_{(i,j) \in E} |I_i - I_j| \Delta_j \tag{5}$$

Theory 2. This problem satisfies dynamic programming property.

Proof: For one node i and one of its son Node j. Let the weight for i be w_i which includes largest weight w_j, but w_i isn't the largest weight for i. Let the largest weight for i be w_i', and it includes Node j's solution w_j'. Then we have $f_i' = f_i - f_j + f_j' < f_i$, which is contradict with that fi' is the largest weight for i.

3   System Implements

The system is realized based on COMET which is published by 2013. Comparing to the Clonecloud system, COMET uses DSM (Distributed Shared Memory) instead of migrating whole stack. So all the threads can access the same block memory simultaneous in a synchronized way. And more than one threads can be migrated to the cloud server at the same time.

COMET is an open source project which modifies the Dalvik Virtual Machine used in Android system. We save the part of network transportation and DSM synchronization, and add our features on it.

First we run a group of test program to get the real instructions in the program and measure the running time to model the CPU speed on both mobile devices and cloud server.

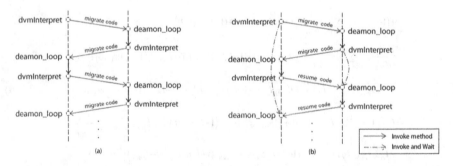

Fig. 1. (a) COMET migration procedure (b) Our system migration procedure

Then we start to run our algorithm to calculate the set of function to offload. When calculating the function set, we need to know the length of instructions F for each function. We just count the running time of the function, then use the mobile speed to evaluate the length of instructions. Another variables Δ should be measured during migration. For each function we set Δ be the RTT between mobile devices and cloud server. As the migration done, we update Δ for each migrated function.

For native method, we use the dictionary provided by COMET to judge which to offload or not. The offloadable methods include some math library, thread library and so on. So this part of method can be run on the cloud server, which can save lots of work for the mobile device.

In COMET, it uses migration code to trigger migration. The code is handled by a method, and when the thread is migrated back it calls another such method to handle the migration code. So the original method will never return (Fig 1a). With more times of migration, the stack grows. We add the function of return, in order to solve the problem and we can add some other work after the migration (Fig 1b).

4   Experiments and Results

4.1   CPU Speed Measure

We program the test program with basic arithmetic instructions, control instructions and loop instructions. The loop is used to control the whole length of the instructions. In order not to increase any extra waste of CPU cycles, we counts length of instructions when a jump instruction comes. The measure of the CPU speed is shown as Fig 2.

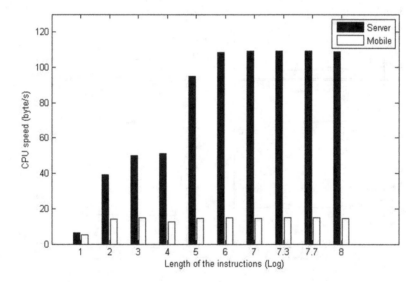

Fig. 2. The CPU speed becomes stable with the growth of the instruction length

The result means we can't use linear model for the functions that have few instructions. But when instructions' length grows more than 106 bytes, the CPU speed becomes stable. And the mobile device and server speed SL, SR is:

$$S_L = 109.207MB / s$$
$$S_R = 14.729MB / s \qquad (6)$$

Further, when the length is less than 106 bytes, the running time is less than 106/SL=68ms. So the functions runs less than 68ms on mobile are ignored. And when the length is less than 106 bytes, the migration saves time less than 106 / SL - 106 / SR = 59ms. In ordinary wireless channel, the RTT time usually more than 59ms. So the ignoring is reasonable.

4.2   Matrix Multiplication

Matrix Multiplication experiment is designed for situation that we want to run specific length of instructions as soon as possible. The experiment use with native algorithm time cost of $O(N^3)$. Multiply the 200*200 matrixes 20, 50, 100, 200 times. The result is shown in Fig 3.

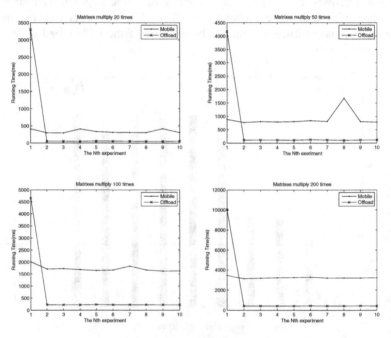

Fig. 3. The time cost of matrix multiplication 20, 50, 100, 200 times on mobile and server

As shown above, the first migration cost much time, because the memory data needs to be synchronized is accumulated. After the first migration, the memory modification is much less, so the running time on server falls into an acceptable threshold. Another question is why it can still offload again after the first migration cost much of the time. Because it first chooses thread creating function as the migration function, then it chooses matrix multiplication function. And we have an average speed-up of 7.11x.

4.3   Seeking Primes

Seeking Primes experiment is designed for situation that we want to get the result as better as possible. We modify sieve algorithm to satisfy the requirement. We divide the positive integers into intervals, and each interval have 8192 integers. The program seeks an interval and test the time then. The length of the interval sought is used as the result. The result is shown in Fig 4.

When the RTT is more than 200ms, when we ask the program run 200ms, it will not be offloaded. Until RTT is less than 150ms, it can be offloaded. The speed-up increases obviously, and average speed-up is 23.23x.

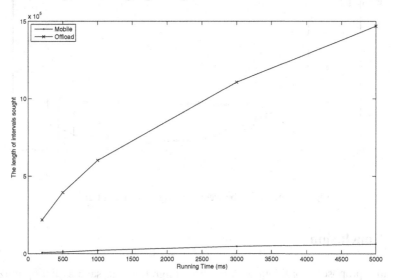

Fig. 4. The relation between the running time and length of interval

4.4   IO Test

IO Test experiment is designed for situation that we process the file on the mobile device. We modify the text statistics program provided by Bell Lab. The input file is a 64K english essay. We test for different length of buffer size, and the result is shown in Fig 5.

The running time on mobile is always 6s. When it comes to offloading, it will be stable when the buffer size is 10000 bytes. In this situation, the time for network communication is minimum. When the buffer size is too small, server will require the file for many times, which cost much time for network communication. When the buffer size is too large, it will need more memory, which will causes more errors during synchronization. So the running time become larger and more unstable. The average speed-up is 8.24x, according to the 7 experiments in the middle.

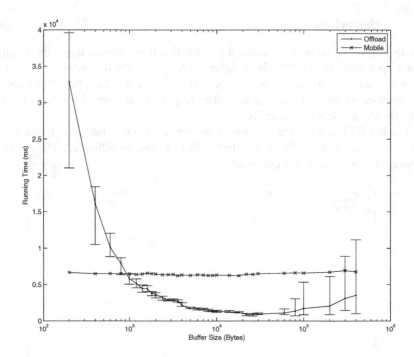

Fig. 5. The relation between running time and buffer size

5   Conclusion

The paper proposes a problem with the time migration can save as the optimization objective. And a dynamic programming algorithm solves the problem. Then we implement the algorithm on COMET system. 3 groups of experiment are conducted, which shows speed-up of 7.11x, 23.23x, 8.24x.

References

1. Siegele, L.: Let it rise: a special report on corporate it (2008),
   http://www.economist.com/node/12411882
2. Cuervo, E., Balasubramanian, A., Cho, D.-K., Wolman, A., Saroiu, S., Chandra, R., Bahl, P.: Maui: making smartphones last longer with code offload. In: Proceedings of the 8th International Conference on Mobile Systems, Applications, and Services, MobiSys 2010, pp. 49–62. ACM, New York (2010)
3. Chun, B.-G., Ihm, S., Maniatis, P., Naik, M., Patti, A.: Clonecloud: elastic execution between mobile device and cloud. In: Proceedings of the Sixth Conference on Computer Systems, EuroSys 2011, pp. 301–314. ACM, New York (2011)

4. Huang, D., Zhang, X., Kang, M., Luo, J.: Mobicloud: building secure cloud framework for mobile computing and communication. In: Proceedings of the Fifth IEEE International Symposium on Service Oriented System Engineering, SOSE, pp. 27–34
5. Gordon, M.S., et al.: COMET: Code Offload by Migrating Execution Transparently. In: 10th USENIX Symposium on Operating Systems Design and Implementation, pp. 93–106 (2012)
6. Kansal, A., Zhao, F.: Fine-grained energy profiling for power-aware application design. SIGMETRICS Performance Evaluation Review 36, 26–31 (2008)

Improved Hierarchical K-means Clustering Algorithm without Iteration Based on Distance Measurement

Wenhua Liu, Yongquan Liang[*], Jiancong Fan, Zheng Feng, and Yuhao Cai

College of Information Science and Engineering,
Shandong University of Science and Technology,
Qingdao City, 266590, China
lyq@sdust.edu.cn

Abstract. Hierarchical K-means has got rapid development and wide application because of combining the advantage of high accuracy of hierarchical algorithm and fast convergence of K-means in recent years. Traditional HK clustering algorithm first determines to the initial cluster centers and the number of clusters by agglomerative algorithm, but agglomerative algorithm merges two data objects of minimum distance in dataset every time. Hence, its time complexity can not be acceptable for analyzing huge dataset. In view of the above problem of the traditional HK, this paper proposes a new clustering algorithm iHK. Its basic idea is that the each layer of the N data objects constructs $\left\lceil \dfrac{N}{2} \right\rceil$ clusters by running K-means algorithm, and the mean vector of each cluster is used as the input of the next layer. iHK algorithm is tested on many different types of dataset and excellent experimental results are got.

Keywords: basic K-means, traditional HK, iHK, Clustering Algorithm.

1   Introduction

Traditional HK algorithm has the advantage of simple and easy to convergence, but also it has some obvious deficiencies. For instance, HK would have a high computational complexity when the k value is uncertain. Agglomerative algorithm only mergers two clusters having minimum distance every time, which leads to have a higher time complexity at high dimension and big data. To overcome the shortcoming of traditional HK, some researchers have done different degrees of improvement for HK algorithm [1-4]. But most of researchers have modified HK algorithm for their specific research fields [5-13].

Now society is rapidly being from information era to age of data, it is significant to accurately grasp the valuable information in dataset. Therefore clustering analysis has become a hot research field chased by researchers. To precisely and quickly analyze the data information carried by the dataset, this paper presents a new clustering algorithm iHK, which is a easily convergent and quite accurate clustering algorithm by integrating with the feature of the K-means and hierarchical algorithm. Moreover, iHK

---

[*] Corresponding author.

Z. Shi et al. (Eds.): IIP 2014, IFIP AICT 432, pp. 38–46, 2014.

algorithm is not limit to a particular research field. In essence, iHK clustering algorithm is an improved HK.

Section 1 describes the summary of iHK algorithm. Section 2 briefly introduces some improved HK algorithm in recent years. Section 3 details iHK algorithm proposed by this paper. And Section 4 presents the experimental results . Section 5 makes a conclusion.

2   Related Work

The training set is divided into two parts, a part of the dataset uses hierarchical algorithm to obtain distribution information of data, then runs K-means at another part. This hybrid algorithm was put forward for the first time by Bernard Chen et al.[14] at 2005. Due to its accuracy, simplicity and convergence, the method attracted wide attention after it has been proposed. He Ying et al.[15] presented HK based on PCA, the general idea of algorithm is that on the whole dataset (rather than two parts), it first makes use of PCA technology to reduce dimension of the dataset and then determines the initial cluster center by executing agglomerative algorithm. Finally it gets clustering results by using K-means. Improved HK based on PCA has more accuracy than traditional HK. To overcome the limitation of a binary tree constructed by hierarchical algorithm, a kind of divisive clustering algorithm was come up with by Lamrous S, Taileb M et al.[16]. The algorithm generates an non-binary tree where each node can split more than two branches by employing K-means, where the k value of K-means is determined by Silhouette index. Kohei Arai et al.[2] studied on an integrated HK algorithm to cluster high-dimensional data set. An improved HK algorithm was put forward by Yongxin Liu et al.[5] to solve the problem of document clustering possessing big data and high-dimension. Li Zhang et al.[17] combined divisive algorithm with agglomerative algorithm to address irreversibility of HK. Divisive algorithm gets several clusters by executing K-means at each layer, then utilizes agglomerative algorithm to merger clusters. Bernard Chen et al.[18] added fuzzy theory into traditional HK for boosting the precision of finding protein sequences theme and reducing the time complexity of HK algorithm.

According to the above related work, some ideas about significantly improving efficiency and accuracy are drawn which they can be applied to iHK. Such as, when using agglomerative algorithm no longer relies on binary tree rules and similarity measure between clusters would adopt mean distance rather than minimum distance to avoid serious impact of noise data on precision of algorithm. In iHK algorithm, the number of clusters produced by K-means always varies with the change of layer (data size). So cluster center can present the distribution of data as far as possible.

3   iHK Clustering Algorithm

3.1   Normalization of Data

Paper gives a new algorithm iHK on the basis of learning a large number of improved HK clustering algorithms. iHK firstly standardizes attributes by formula (1), because diverse attributes may adopt different units of measurement in some dataset.

$$Z_{if} = \frac{x_{if} - m_f}{S_f} \tag{1.1}$$

In (1.1), S_f and m_f is given in detail in the following (1.2) and (1.3).

$$S_f = \frac{1}{n}\left(\left|x_{1f} - m_f\right| + \left|x_{2f} - m_f\right| + \ldots + \left|x_{nf} - m_f\right|\right) \tag{1.2}$$

where $x_{1f}, x_{2f}, \ldots\ldots, x_{nf}$ is n values of the f attribute and m_f is average of f attribute.

$$m_f = \frac{x_{1f} + x_{2f} + \ldots\ldots + x_{nf}}{n} \tag{1.3}$$

Then data objects obtained by above-mentioned steps are used as clustering data. iHK algorithm greatly promotes execution efficiency by combining the idea of 2-way merge, because running algorithm can reduce half the number of clusters at every time. To improve traditional HK algorithm would use K-means algorithm to clustering data at each layer rather than simply merger two clusters by adopting minimum distance between clusters. Next, traditional Hierarchical clustering algorithm and iHK algorithm are simply introduced.

3.2   Traditional Hierarchical Algorithm

Assuming that dataset is D and the total number of data object is N. At first, treating every data as a single cluster, hence there are N clusters. Similarity measure between cluster C_i and C_j is shown by (2).

$$dist_{min} = (C_i, C_j) = \min_{p \in C_i, q \in C_j} \{\|p - q\|\} \tag{2}$$

The number of clusters just reduces one at a time through merging two clusters by (2). This process is being performed continually until meeting the given threshold or all data are in one cluster. The time complexity of hierarchical algorithm is $O(N^2)$, thus it is not suitable for processing huge dataset.

3.3   iHK Clustering Algorithm

iHK overcomes limit of merely aggregating two clusters one time at hierarchical algorithm so that traditional hierarchical algorithm can be better applied to clustering big data. In addition, iHK no longer merges clusters in a minimum distance manner at each layer, but it makes use of K-means algorithm based on mean distance measurement. Supposing that hth layer has L data objects, the process of iHK algorithm based above assumption is that it selects cluster center by employing (3) before L data objects are divided into $\left[\frac{L}{2}\right]$ clusters through K-means algorithm.

$$\begin{cases} 1,3,5,\ldots . i, i+2, \ldots . L, & L \text{ is odd number} \\ 1,3,5,\ldots . i, i+2, \ldots . L-1, L \text{ is even numbe} \end{cases} \tag{3}$$

where $1, 3, \ldots\ldots i$ is a sequence of data objects.

By comparing the distance between the remaining data and cluster centers, data is divided into the closest cluster center. However, attributes may be mixed, therefore it can't adopt simple distance metric value as a criterion. New standard is defined by (4.1).

$$dist\ (x,\ y) = \frac{\sum\limits_{i=1}^{n} d^{i}_{x_i,y_i}}{n} \qquad (4.1)$$

where x is a data object with n attributes, namely $x = (x_1, x_2, \ldots\ldots, x_n)$, and y is a cluster center with n attributes, namely $y = (y_1, y_2, \ldots\ldots, y_n)$. If the ith attribute of x is discrete data type, then $d^{i}_{x_i,y_i} = 0$ iff $x_i = y_i$. Otherwise $d^{i}_{x_i,y_i} = 1$. If this attribute is numeric type, then $d^{i}_{x_iy_i}$ is calculated by (4.2).

$$d^{i}_{x_i,y_i} = |x_i - y_i| \qquad (4.2)$$

After merging, altogether forms $\left[\frac{L}{2}\right]$ clusters. Mean vector $\overline{x} = (\overline{x_1}, \overline{x_2}, \ldots\ldots, \overline{x_n})$ is used to represent cluster and the calculation formula of mean vector is given by (5).

$$\begin{cases} \overline{x_i} = \dfrac{1}{count\ (C_k)} \sum\limits_{x_{ij} \in C_k} x_{ij}, & x_i \text{ is a numeric type attribute} \\[3mm] \overline{x_i} = \max(\ x_{ij} \in C_k | count\ (x_{ij})), & x_i \text{ is a discrete type attriute} \end{cases} \qquad (5)$$

In formula (5), $count(C_k)$ is applied to count the number of data object in C_k cluster and $count(x_{ij})$ is used to compute how many values equal x_{ij}. If x_i is numeric type, $\overline{x_i}$ is a average of x_i in C_k. If x_i is a discrete type, then $\overline{x_i}$ equals to the value of attribute appearing most times in dataset.

$\left[\frac{L}{2}\right]$ clusters are produced after the aforementioned process has been executed, then mean vectors obtained at h th layer are used as the new input of $h+1th$ layer. iHK algorithm is being executed until satisfying given condition or threshold. iHK algorithm performs $\log_2 N$ times K-means in total, in which K-means just compares half of the input data in this layer. Through the above details, the overall description of the iHK algorithm is shown in Figure 1.

In Figure 1, Step1 is to standardize the numeric type data so that between attributes of different measure units can make the correct comparison. Distance calculation between data and cluster center is completed by Step3-4. Mean vectors generated by (5) are used to represent clusters and use them as input of next circulation.

The pseudocode of the algorithm is described by Figure 2.

Input: data set D , the total number of data object n .

Output: Data object are divided into different clusters.

Step1:preprocessing attributes in data set D by (1).

Step2:employing step length 2 to select cluster center by according to the order of the data storage.

Step3:computing distance by (4.2) if attribute is numeric type.

Step4:comparing discrete attribute values whether or not they are the same. Distance is 0,iff their value is the same. Otherwise, the value of distance is 1.

Step5:Performing K-means on the basis of step3-4.

Step6: Treating the results of K-means clustering as new dataset.

Step7:Implementing (5) to process dataset formed by Step6.

Step8:Repeat Step2-7 until meeting the given conditions.

Fig. 1. A complete description of the iHK clustering algorithm

4   Experimental Results

This paper tests the performance of iHK at some typical dataset of diverse type. Test set contain Abalone, Iris, Seeds_dataset, Wine, Credit Approval, Haberman's Survival, Teaching Assistant Evaluation, Ecoli, MONK'S Problem, Balance scale, balloons and so on. They are from UCI database. (http://archive.ics.uci.edu/ml/)

In iHK algorithm, K-means based on new distance metric is executed at each layer. Among, the value of k is different, which is half size of each layer input data. Accuracy rate changes with different k values, which is shown in Figure 3.

The accuracy of iHK algorithm on five large dataset is shown at top of the Figure 3 and Accuracy on relatively small dataset is displayed at the bottom.

Input: dataset D , the total number of data object n .

Output: several clusters.

for each numeric attribute f

do

for i←0 to n

standardizing attribute value;

do

for i←0 to n/2

datacenter[i][t]=datacenter[i*2][t];

exdata[i][t]=datacenter[(i*2)+1][t];

end;

dividing exdata[i] into appropriate cluster by defined distance formula;

$if\,(x_i$ is a numeric type attribute)

then $\overline{x}_i = \dfrac{1}{count(C_k)} \sum_{x_{ij} \in C_k} x_{ij};$

$else \quad \overline{x}_i = \max(x_{ij} \in C_k | count(x_{ij}));$

forming new dataset by the mean vector of each cluster;

until meeting the given conditions.

Fig. 2. The pseudocode of iHK algorithm

The overall trend of all line chart is that they begin to rise gradually and reach maximum at some value, then decrease. Besides, accuracy obviously varies with the change of k values on some dataset, as Figure 3 expresses.

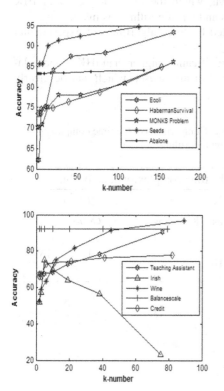

Table 1. Accuracy of three algorithms

Data Sets	basic k-mean	HK	iHK
Abalone	57.4%	50.8%	83.4%
Wine	68.1%	68.5%	75.3%
Seeds	75.7%	82.8%	85.7%
Iris	70.2%	88.5%	75.3%
Credit Approval	54.7%	54.6%	72.9%
Haberman Survival	73%	73.5%	75.2%
Teaching Assistant Evaluation	65.2%	40%	67.6%
Ecoli	75%	78.3%	76%
MONK Problem	67.6%	66.7%	70.2%
Balance scale	90%	62.2%	92.2%
balloons	65.4%	80%	80%

Fig. 3. The line chart of clustering accuracy rate

To make experimental results can show the superiority of iHK clustering algorithm, this paper compares iHK algorithm with basic K-means and traditional HK algorithm at some evaluation indicators of the performance of algorithm. Accuracy rate is frequently used as an important indicator. The accuracy rate of three algorithms at some dataset are shown in Table 1.

Experimental results in Table 1 show that iHK clustering algorithm has higher accuracy than the basic K-means and HK algorithm at most dataset.

Now most of the data come from Web and Web data is mainly big data. When these data are clustered, the time complexity of the algorithm is also considered as a significant indicator of performance of prediction algorithm.

In Table 2, time complexity of HK algorithm, basic K-means, iHK algorithm are compared.

The time complexity of HK algorithm should be the sum of the time complexity of Hierarchical algorithm and K-means', it is greater than $O(N^2)$.

The time complexity of basic K-means is linear, where m is the number of iteration. iHK is similar to K-means. It is also linear, where N is the total number of data objects. By comparison, this conclusion can be drawn, which the time complexity of iHK algorithm is minimum, basic K-means follows and HK algorithm's is maximum .

The different time complexity are generated by three algorithms at experimental data set, which use Figure 4 to show.

The time complexity of HK algorithm is significantly higher than iHK at most dataset. iHK and Basic K-means are about equal at all dataset, which can be clearly observed from Figure 4.

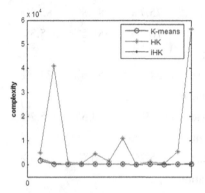

Fig. 4. The time complexity of three algorithms at some data set

Table 2. Comparisons of the time complexity of different algorithms

The time complexity of algorithms	
HK	$O(N^2)$
Basic K-means	$O(mN)$
iHK	$O((1-\dfrac{1}{2}^{\log_2 N})N$

5   Conclusion

HK algorithm has been widely used, due to its superiority in the clustering analysis. But the HK algorithm also has shortcoming, such as high time complexity. So, it can not be applied to clustering big data. Therefore, some improved HK algorithms have been studied by some researchers for resolving the problem of specific application areas. Some improved HK algorithm have good clustering outcome, but these clustering results are not good in certain application domain. iHK algorithm has good generalization in that it is not limited to a particular field. And it can be used to clustering big data since it has low time complexity.

From accuracy, efficiency and the time complexity acquired by comparing these three algorithms, a conclusion can be drawn that iHK has the advantage of high accuracy and easy to convergence. In addition, its performance is distinctly superior to other algorithms, but time complexity is similar to basic K-means's. The important point is that iHK is not based on any special application areas and easy to integrate to other clustering algorithm. However, iHK algorithm still can not solve irreversibility of HK algorithm. The next mainly task is that iHK algorithm is further improved.

Acknowledgment. This paper is supported by State Key Laboratory of Mining Disaster Prevention and Control Co-founded by Shandong Province and the Ministry of Science and Technology, Shandong University of Science and Technology, Leading talent development program of Shandong University of Science and Technology, National Natural Science Foundation of China under Grant 61203305 and Natural Science Foundation of Shandong Province of China under Grant ZR2012FM003.

References

[1] Wang, Y.C.F., Casasent, D.: Hierarchical k-means clustering using new support vector machines for multi-class classification. In: International Joint Conference on IEEE Neural Networks, IJCNN 2006, pp. 3457–3464 (2006)

[2] Arai, K., Barakbah, A.R.: Hierarchical K-means: an algorithm for centroids initialization for K-means. Reports of the Faculty of Science and Engineering 36(1), 25–31 (2007)

[3] Lu, J.F., Tang, J.B., Tang, Z.M., et al.: Hierarchical initialization approach for K-Means clustering. Pattern Recognition Letters 29(6), 787–795 (2008)

[4] Celebi, M.E., Kingravi, H.A.: Deterministic initialization of the k-means algorithm using hierarchical clustering. International Journal of Pattern Recognition and Artificial Intelligence 26(07) (2012)

[5] Archetti, F., Campanelli, P., Fersini, E., et al.: A Hierarchical document clustering environment bases on the induced bisecting k-means. Flexible Query Answering System, pp. 257–269. Springer, Heidelberg (2006)

[6] Liu, Y., Liu, Z.: An improved hierarchical K-means algorithm for web document clustering. In: International Conference on Computer Science and Information Technology, ICCSIT 2008, pp. 606–610. IEEE (2008)

[7] Murthy, V., Vamsidhar, E., Rao, P.S., et al.: Application of hierarchical and K-means techniques in Content based image retrieval. International Journal of Engineering Science and Technology 2(5), 749–755 (2010)

[8] Chen, T.W., Chien, S.Y.: Flexible hardware architecture of hierarchical K-means clustering for large cluster number. IEEE Transactions on Very Large Scale Integration (VLSI) Systems 19(8), 1336–1345 (2011)

[9] Hu, X., Qi, P., Zhang, B.: Hierarchical K-means algorithm for modeling visual area V2 neurons. Neural Information Processing, pp. 373–381. Springer, Heidelberg (2012)

[10] Mantena, G., Anguera, X.: Speed improvements to information retrieval-based dynamic time warping using hierarchical k-means clustering. In: 2013 IEEE International Conference on Acoustics, Speech and Signal Processing (ICASSP), pp. 8515–8519. IEEE (2013)

[11] Chen, B., He, J., Pellicer, S., et al.: Using Hybrid Hierarchical K-means (HHK) clustering algorithm for protein sequence motif Super-Rule-Tree (SRT) structure construction. International Journal of Data Mining and Bioin-formatics 4(3), 316–330 (2010)

[12] Ghwanmeh, S.H.: Applying Clustering of Hierarchical K-means-like Algorithm on Arabic Language. International Journal of Information Technology 3(3) (2007)

[13] Chehata, N., David, N., Bretar, F.: LIDAR data classification using hierarchical K-means clustering. In: ISPRS Congress, Beijing, vol. 37, pp. 325–330 (2008)

[14] Chen, B., Tai, P.C., Harrison, R., et al.: Novel hybrid hierarchical-K-means clustering method (HK-means) for microarray analysis. In: Computational Systems Bioinformatics Conferences, Workshops and Poster Abstracts, pp. 105–108. IEEE (2005)

[15] Ying, H., Qin, L.X.: Study on PCA based Hierarchical K-means Clustering Algorithm. Control and Automation 6, 68 (2012)

[16] Lamrous, S., Taileb, M.: Divisive hierarchical k-means. In: 2006 and International Conference on Intelligent Agents, Web Technologies and Internet Commerce, International Conference on Computational Intelligence for Modelling, Control and Automation, p. 18. IEEE (2006)

[17] Zhang, L., Cui, W.D., et al.: Hybird clustering algorithm based on partitioning and hierarchical method. Computer Engineering and Applications 46(16), 127–129 (2010)

[18] Chen, B., He, J., Pellicer, S., et al.: Protein Sequence Motif Super-Rule-Tree (SRT) Structure Constructed by Hybrid Hierarchical K-means Clustering Algorithm. In: IEEE International Conference on Bioinformatics and Biomedicine, BIBM 2008, pp. 98–103. IEEE (2008)

An Optimized Tag Recommender Algorithm in Folksonomy

Jie Chen[1], Baohua Qiang[1,2,*], Yaoguang Wang[1], Peng Wang[1], and Jun Huang[1]

[1] Guilin University of Electronic Technology, Guilin 541004, China
`cj134cj@163.com`
[2] Guangxi Key Laboratory of Trusted Software, Guilin University of Electronic Technology, Guilin 541004, China
`qiangbh@yahoo.com.cn`

Abstract. In the existing folksonomy system, users can be allowed to add any social tags to the resources, but tags are fuzzy and redundancy in semantic, which make it hard to obtain the required information for users. An optimized tag recommender algorithm is proposed to solve the problem in this paper. First, based on the motivation theory, the recommender system uses the model given to calculate the user retrieval motivation before searching information. Second, we use the results in first step to distinguish the user's type and then cluster the resources tagged according to users who have the similar retrieval motivation with k-means++ algorithm and recommend the most relevant resources to users. The experimental results show that our proposed algorithm with user retrieval motivation can have higher accuracy and stability than traditional retrieval algorithms in folksonomy system.

Keywords: Folksonomy, tag recommender system, collaborative filtering, user retrieval motivation, k-means++.

1   Introduction

Nowadays, with the development of web 2.0, a large amount of digital resources appear and rise. Accord to them, a new network information classification system folksonomy appears. The term folksonomy is generally attributed to Thomas Vander Wal [1], which is a portmanteau of folk and taxonomy. In folksonomy, the tags and resources are created by users, and users can tag the information according to their own needs and preferences in order to make others to retrieve and share the resources. As shown in Figure 1, the resource "Baidu.com" is tagged by user "Brown" with the tag "Search". In folksonomy, users can easily find other users with similar preferences, resources and tags which have been used.

In the existing folksonomy system, most of tags are lack of semantic precision and not standard, which affect the use of it to a certain degree. In order to better organize the information resources, the collaborative filtering [2] recommender system was

---

* Corresponding author.

Z. Shi et al. (Eds.): IIP 2014, IFIP AICT 432, pp. 47–56, 2014.

introduced to folksonomy. But, the clustering result is not as good as we imagine. According to this idea, we put forward an optimized tag recommender algorithm with the users' retrieval motivation.

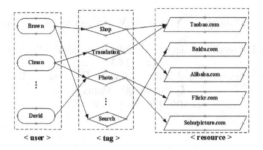

Fig. 1. The structure of a folksonomy system

First, we get a certain amount of datasets from the network, and use the stemmer technology [3] to get more effective and useful text content of tags in the datasets. Second, according to the motivation theory proposed by Strohmaier [4], we get two types of users, Categorizers and Describers [5], which have different retrieval motivations. Third, according to the new user's retrieval motivation, we judge user is a Categorizer or Describer, and then cluster the tags and resources that types of users have tagged with k-means++ algorithm [6]. Finally, according to the similarity, the algorithm give the recommended information which is more accuracy and relevant to users. The experimental results show that the optimized algorithm with user retrieval motivation can have higher accuracy and stability than traditional tag recommender algorithm in folksonomy system.

2   Preliminaries

2.1   Recommender Algorithm

Recommender system is a special system that it is always used to deal with some problems, such as information filtering. It actively provided the options to users that the options are identified as the most suitable results by recommender algorithm. And all we known, the most important part of the recommender system is recommender algorithm, the algorithm can determine the way to work and the recommended strategy. Generally speaking, recommender algorithm has three main types: Content-Based Filtering, Collaborative Filtering and Hybrid Recommender. In this paper, the optimized tag recommender algorithm belongs to the kind of collaborative filtering. The optimized tag recommender algorithm is based on the behavior or interests of the user groups. Firstly, the recommender system will find the neighbor users, according to the historical data of the target users. Secondly, it can get the object value from the neighbor users' evaluations. Finally, according to the previous data, the recommender system will provide the personalized recommender to the target user.

The traditional collaborative recommender algorithm usually has some problems, such as concept drift and sparse, because the limitation of a algorithm is a fact. In this paper, an optimized tag recommender algorithm is proposed to improve these disadvantages based on the user's retrieval motivation. At the same time, this optimized tag recommender algorithm is proved to feasible after the experiment.

2.2   The User's Tagging Motivation

In this paper, we add the users' motivation theory to folksonomy system and use the theory proposed by Strohmaier. In this theory, there are two kinds of users in the datasets, and they are called categorizers and describers. Categorizer is the users who are motivated by categorization and view tagging as a means to categorize resources according to some high-level characteristics. For example, when a popular music is tagged by categorizers, they will use tag 'music' rather the tag 'song', 'tune' even if they have the similar meaning. Describers are the users who are motivated by description view tagging as a means to accurately and precisely describe resources. For example, when a popular music is tagged by describers, the tag "music", "popular", and "favorite" can be used to describe the resource. We consider the tags tagged respectively rather than the tags that can't be classified, which can give the benefit to us and improve the quality of result. In next section, we will give the model to calculate the users' motivation.

2.2.1   Measures

In this part, we use four indicators to measure the users' tagging motivation. The four measures are respectively Tags per Post (TPP), Tag Resource Ratio (TRR), Low-frequency tagging ratio (LFTR) and Interrogative adverbs tagging ratio (IATR). According to the results obtained by the above measures, we can get the type of the users, a categorizer or a describer. The detail description of the measures can be found in literature [5].

- Tags per Post (TPP)

$$TPP(u) = \frac{\sum_{i=1}^{r}|T_{ui}|}{R_u} \tag{1}$$

Where R_u is the number of resources tagged by the user u, T_{ui} is the numbers of tags annotated by user u on resources i, r is the total number of resources. This measure relies on the verbosity of users. So, if the TPP reflects in a higher score, the user is more likely a Describer.

- Tag Resource Ratio (TRR)

$$TRR(u) = \frac{T_u}{R_u} \tag{2}$$

Where T_u is the number of tags annotated by the users, R_u is the number of resources annotated by the users. Because a typical Categorizer would apply only a small of tags to his resources and score a low number on this measure.

- Low-frequency tagging ratio (LFTR)

$$LFTR(u) = \frac{|T_u^0|}{|T_u|} \;,\; T_u^0 = \{t||R(t \le n)|\} \;,\; n = \left\lceil \frac{|R(t_{max})|}{100} \right\rceil \tag{3}$$

Equation 3 shows the calculation of the final measure where are seldom used tags. T_u are all tags of the given user. t_{max} denotes the tag which was tagging the most by the user. n means the critical value. If the LFTR reflects in a lower score, the user is more likely a Categorizer.

- Interrogative adverbs tagging ratio (IATR)

$$IAIR(u) = \frac{Card(t \in T_{str})}{|T_u|} \tag{4}$$

Where $T_{str} = \{$ what, who, when, where, ...$\}$ is a set of Interrogative adverbs, $Card(t \in T_{str})$ is the number of interrogative adverbs tags annotated by user, T_u is the number of tags annotated by the users. Obviously, $IATR(u) \in (0,1)$, and if the IATR reflects in a lower score, the user is more likely a Categorizer.

2.2.2 An Evaluation Model for User's Motivation

According to the four indicators above to distinguish the users' type, categorizers or describers, we can construct a reasonable evaluation model [7] to calculate the orientation of users' motivation below. M=a*TPP(u)+b*TRR(u)+c*LFTR(u)+d*IA--TR(u),Where, a,b,c,d $\in$ (0, 1). According to the results of the experiment, we find that each datasets has a different optimal coefficient. So, we choose the average score of M and use M' to denote the value of it.

$$M' = (TPP(u) + TRR(u) + LFTR(u) + IATR(u))/4 \tag{5}$$

According to the formula 6, we find that M' is a monotonic function and it also has a threshold M_t. If M' is larger than M_t, we consider the user a Describer; On the contrary, the user is a Categorizer. If both are equal, he has not a special motivation and we treat the user either Describer or Categorizer; if the user who have little tagging behaviors and it is hard to get the users' retrieval motivation, we call them the users without motivation.

2.3   Porter Stemmer and K-means++

In the existing folksonomy system, most of tags and resources are created by users without restriction, so they are short of semantic precision and standardization, which affect the use of it to a certain degree. In order to reduce these interference factors, we use porter stemmer to deal with these tags.

Porter stemmer has five steps and each step defines a set of rules. To stem a word, the rules are tested sequentially, if one of these rules matched the current word, then the conditions attached to that rule are tested. Once a rule is accepted; the suffix is removed and the control moves to the next step. If the rule is not accepted then the next rule in the same step will be tested, until either a rule from that step is accepted or there are no more rules in that step. And the control passes to the next step. In the last step, the resultant stem is returned by the stemmer. After that, the irregular tags reduce the interference factors to understand and they can help us to improve the quality of retrieving information.

At the same time, according to the model of the users' motivation, we can get the types of users. Then, we need to deal with the resources tagged by users who have the similar motivation and find the similar resources which have been tagged. In this paper, we choose the k-means++ clustering [8].

The k-means++ is an algorithm for choosing the initial values for the k-means clustering algorithm. It was proposed in 2007 by David Arthur, as an approximation algorithm for the NP-hard k-means problem—a way of avoiding the sometimes poor clustering found by the standard k-means algorithm. And the k-means problem is to find cluster centers that minimize the intra-class variance, the sum of squared distances from each data point being clustered to its cluster center.

3   The Optimized Algorithm

According to the knowledge given above, we propose an optimized algorithm with users' retrieval motivation in folksonomy system. The detail procedure of the algorithm is shown in Figure 2, and the detail algorithm is shown in Table 1.

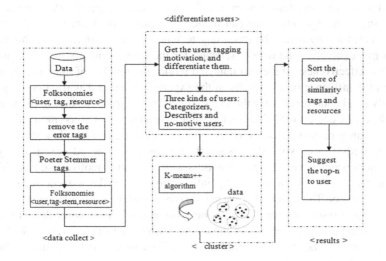

Fig. 2. The procedure of the algorithm

Table 1. The optimized algorithm

An optimized algorithm with users' retrieval motivation:

Input ： a folksonomy datasets，the cluster center k，a retrieved tag t_u by user u.

Output： some tags(to user u).

```
1)      get T(t₁, t₂, ..., tₙ) from a folksonomy datasets;
2)          for each tᵢ (tᵢ ∈ T) ；
3)              if  {   tᵢ has some problem;
4)                      remove;            }
5)              else {   porter stemmer;
6)                      get the stem tⱼ of tᵢ;
7)                          Q←tⱼ;              }
8)          end;
9)          for each tⱼ, tⱼ ∈ Q;
10)             Switch(the tⱼ(userⱼ) tagging motivation Mᵤⱼ')
11)                 {   case 'categoritizers':  k-means++(Q'(tⱼ∈ categoritizers));
12)                                     find(sim max(Qᵤⱼ',tⱼ))→L;
13)                                     tf-idf L, return the top-n to u ;
14)                     case 'describers':    k-means++(Q'(tⱼ∈ describers));
15)                                     find(sim max(Qᵤⱼ',tⱼ))→L;
16)                                     tf-idf L, return the top-n to u ;
17)                     case 'no-motive users': k-means++(Q'(tⱼ∈ Q));
18)                                     find(sim max(Qᵤⱼ',tⱼ))→L;
19)                                     tf-idf L, return the top-n to u;
20)                     deefault:       system.out.print("Absence recommender.")
21)     end;
```

There are three modules in this procedure. First, in the module of data collection, we get the data from the internet and some preprocessing will be done to get the more useful data which have less interference factors. When we get the data, we need to remove the error tags. Then, we can reduce the number of the tags which have the similar semantics by using the stemming technique that we introduced above. And then, some relatively good data is selected for the next module, with which we can ensure the effectiveness of the text data and reduce the number of the text data which has the little influence on the results in the experiment. Second part, after we get the relatively good data, we classify the users into different types. According to the model given above, we can get the M' score, the value of orientation of users' motivation, with which we can get a ranked lists of users, where Categorizers rank high, and Describers rank low. With the two types of users and the tags they have tagged, we can easily find the resources which have tagged by the users with the similar motivation. And this can improve the efficiency of retrieving information when the system recommends resources to users. Third part, after we get the types of users, we need to get the smaller range of search resources, we use the k-means++ algorithm to cluster these resources tagged by users who have the similar motivation. After the data cluster, we calculate the similarity between the retrieved information and the clustered data and then return the resources which have the high similarity to the user. But, for users who have little tagging behaviors and it is hard for us to get the users' retrieval motivation, we only use TF-IDF algorithm to find the similar resources, and then recommend them to the user.

4   Experiments

According to the optimized algorithm proposed above, we need to collect the datasets first. We use the datasets downloaded from http://www.flickr.com in our experiments. This datasets has 4 classifications and 3000 documents. The classifications contain the following types: Resources, Tags, users, Messages of the photo.

In this experiment, we just use three classifications which are Resources, Tags and User. After data processing in the module of data collection, we get the 89 resources, 2537 users and 8478 tags.

4.1   Evaluation Measures

To evaluate the proposed approach, we introduce three measures, precision, recall and F-measure to evaluate the quality of retrieval by different methods, which are defined as follows.

Precision. In the field domain of information retrieval, precision is the fraction of retrieved instances that are relevant, and it is also used with recall. The definition is:

$$\text{precision} = \frac{|\{\text{relevant documents}\} \cap \{\text{retrieved documents}\}|}{|\text{retrieved documents}|} \tag{6}$$

Recall. Recall in information retrieval is the fraction of the documents that are relevant to the query that are successfully retrieved. The definition is:

$$\text{recall} = \frac{|\{\text{relevant documents}\} \cap \{\text{retrieved documents}\}|}{|\text{relevant documents}|} \tag{7}$$

F-measure. Generally speaking, a good retrieval algorithm should be able to have a higher score in precision and recall. Therefore, we use the compromise value to measure, such as F-measure($\beta=1$), that it indicates the precision and recall are equally important now. The formula is shown:

$$F_{(\beta=1)} = \frac{2 * \text{precision} * \text{recall}}{\text{precision} + \text{recall}} \tag{8}$$

If F-measure score is higher, it means that this retrieval algorithm is better than the other.

4.2 Results and Discussion

Compare the F-measure score of three different algorithms to measure them. In Algorithm 1, we use the traditional collaborative recommender algorithm. In Algorithm 2, we add the user motivation theory to the algorithm and use the value of orientation of users' motivation to distinguish the users' type, and then cluster the resources tagged by the users who have the similar retrieval motivation with k-means++ algorithm, and then use the same way as Algorithm 1 to get the results f-measure. In Algorithm 3, different with Algorithm 2, the value of orientation of users' motivation is not used for distinguishing the users' type, but it is used as an auxiliary value of the calculation of the similarity between relevant the tags and the tags which will be retrieved. Finally, we use the same way as Algorithm 1to get the f-measure of result.

The three algorithms can be regarded as the three models of the tag recommender systems. Then we choose the three types of users, Describers, Categorizers and users with no motivation. Let the three models recommend the required source retrieved by three types of users respectively. For the results the models recommend to the users, we take the average of the results after five times' test in order to reduce the experimental error. We can see the results from Figure3, Figure4 and Figure 5. In Figure 3, the users are all the Describers. At the same time, we can see that the F-measure value in Algorithm 2 and 3 have the higher score than the Algorithm 1, which means that for the user of Describers, the algorithms with motivation theory have the better results. In Figure 4, the users are all the Categorizers Meanwhile, and the F-measure value in Algorithm 2 and 3 also have the higher score than the Algorithm 1, which means that the algorithm with users' motivation have a better result too. In Figure 5, the users are all those without retrieval motivation. So, we cannot use the users' motivation as an auxiliary value in Algorithm 2 and 3. For this reason, the three algorithms have the similar results the recommender system gives to the users and the value of F-measure stay the same level intuitively.

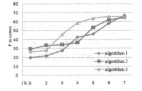 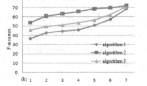 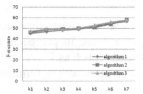

Fig. 3. Compared the results based on Describers

Fig. 4. Compared the results based on Categorizers

Fig. 5. Compared the results based on the sparse users

And then, we random chose a few users to test the optimized algorithm and no longer deliberately distinguish the type of the users. Then, we can see the results from table 2.

Table 1. The F-measure score(%) results at random

The F-measure score(%)	User1(k=260)		User2(k=260)		User3(k=260)	
	Tag1	Tag2	Tag3	Tag4	Tag5	Tag6
Algorithm 1	56.2	57.0	62.4	71.6	46.2	58.2
Algorithm 2	56.3	57.0	69.1	79.4	53.3	62.3
Algorithm 3	56.3	57.1	66.1	77.7	65.7	65.5

From the analysis above, we can conclude that in tag recommender system, the optimized algorithm with users' retrieval motivation can have the better results than the traditional algorithm without users' retrieval motivation when recommend the resource to the users.

5   Conclusions

In folksonomy system, the traditional tag recommender algorithm [9] need to deal with the resources tagged by all the users. For improving the efficiency of the recommender, the users' motivation is added to our tag recommender algorithm. When users retrieved the information, their motivations were given to the model. Then, we just need to deal with the resources tagged by the users who has the similar motivation, and this need less time just to deal with the more relevant resources, meanwhile, the accuracy of resources recommend was improved. The experimental results show that this algorithm with user retrieval motivation can have higher accuracy and stability than traditional retrieval algorithms when recommend the resources to users.

Acknowledgement. This work is supported by National Natural Science Foundation of China (grant 61163057), Guangxi Nature Science Foundation (grant 2012jjAAG0063), Open Fund of Guangxi Key Laboratory of Trusted Software (kx201308). The authors would also like to express their gratitude to the anonymous reviewers for providing helpful suggestions.

References

1. Thoms, V.W.: Folksonomy Coinage and Definition (2013)(retrieved March 2013)
2. Konstan, J.A., Riedl, J.: Recommended for you. IEEE Spectrum 49(10), 54–61 (2012)
3. Karaa, W.B.A.: A new stemmer to improve information retrieval. International Journal of Network Security (IJMIT) (4) (July 2013)
4. Strohmaier, M., Körner, C., Kern, R.: Why do users tag? Detecting users' motivation for tagging in social tagging systems. In: International AAAI Conference on Web blogs and Social Media (ICWSM 2010), Washington, DC, USA (May 2010)
5. Zubiaga, A., Körner, C., Strohmaier, M.: Tags vs Shelves: From Social Tagging to Social Classification. In: Proceedings of the 22nd ACM Conference on Hypertext and Hypermedia, pp. 93–102. ACM, New York (2011)

6. Gemmell, J., Ramezanim, M., et al.: The impact of ambiguity and redundancy on tag recommender in folksonomies. In: Proceedings of the 2009 ACM Conference on Recommender Systems, pp. 23–25. ACM Press, New York (2009)
7. Trattner, C., Körner, C., Helic, D.: Enhancing the Navigability of Social Tagging Systems with Tag Taxonomies. In: Proceedings of the 11th International Conference on Knowledge Management and Knowledge Technologies. ACM, New York (2011)
8. Li, M.: Immune Network Based Text Clustering Algorithm. In: 2012 13th ACIS International Conference on Software Engineering, Artificial Intelligence, Networking and Parallel & Distributed Computing (SNPD) (2012)
9. Harvey, M., Baillie, M., Ruthven, I., Carman, M.J.: Tripartite hidden topic models for personalized tag suggestion. In: Gurrin, C., He, Y., Kazai, G., Kruschwitz, U., Little, S., Roelleke, T., Rüger, S., van Rijsbergen, K. (eds.) ECIR 2010. LNCS, vol. 5993, pp. 432–443. Springer, Heidelberg (2010)

Extracting Part-Whole Relations
from Online Encyclopedia

Fei Xia[1,2] and Cungen Cao[1]

[1] Key Laboratory of Intelligent Information Processing, Institute of Computer Technology,
Chinese Academy of Sciences, Beijing, China
[2] University of Chinese Academy of Sciences, Beijing, China
`xiafei.1986@163.com, cgcao@ict.ac.cn`

Abstract. Automatic discovery of part-whole relations is a fundamental problem in the area of information extraction. In this paper, we present an unsupervised approach to learning lexical patterns from online encyclopedia for extracting part-whole relations. The only input is a few part-whole instances. To tackle the term recognition problem, terms from the domain of the seeds are extracted taking use of the semantic information contained in the online encyclopedia. Instead of collecting sentences that contain relation instances from the seeds, we introduce a novel process to select sentences that may indicate part-whole relations. Patterns are produced from these sentences with terms replaced by *Part* and *Whole* tags. A similarity measurement based on a new edit distance is used and an algorithm is described to cluster similar patterns. We rank the pattern clusters according to their frequencies, and patterns from the top-k clusters are chosen to be applied to identify the new part-whole relations. Experimental results show that our method can extract abundant part-whole relations and achieve a preferable precision compared to the other state-of-the-art approaches.

Keywords: part-whole relations, lexico-syntactical patterns, online encyclopedia, edit distance, clustering.

1 Introduction

Part-whole relations, also known as meronymy, are fundamental semantic relations that exist in many semantic networks, such as WordNet and HowNet. Those semantic networks play a key role in many natural language processing (NLP) systems like information retrieval, and automatic question answering. One of the traditional approaches to automatically extracting part-whole relations is pattern-based, which identifies the relation from patterns like "Y consists of X" that indicates Part-Whole(X, Y) (means X is a part of Y). Those patterns are either designed manually by experts [1], which is very low efficient, or learned from sentences that contain the given part-whole relation instances [2], which needs massive corpora such as the Web.

In recent years, online encyclopedias have drawn attention to many researchers and become an ideal source for semantic information extraction, since they have both broad coverage and high accuracy. BaiduBaike is one of the largest Chinese Wikipedia-like online encyclopedias, and it contains more than 7 million entries by March 2014. Each entry has a corresponding web page which describes that entry in detail (Fig 1a). More importantly, tags (also called folksonomy) labeled by editors and the related entries are listed in the bottom of that entry's page (Fig 1b), which are of great help for semantic relation extraction since they contain semantic information of that entry.

(a) *The top of the page of "banana"*   (b) *Tags and related entries of "banana"*

Fig. 1. The corresponding web page of the entry "banana"

This paper presents an approach to learning lexical patterns from BaiduBaike to extract part-whole relations. The only input are some part-whole instance pairs, and we firstly extract domain terms of both parts and wholes using their related entries and tags. Sentences from pages of domain terms are labeled and those contain both part-terms and whole-terms are selected and transferred into patterns. A similarity formula based on edit distance is used to cluster patterns. Pattern clusters are ranked according to their frequencies, and patterns from the top-k clusters are chosen to be applied to identify the new part-whole relations.

The paper is organized as follows. Section 2 presents an overview of some related work. The details of our procedure are described in section 3. In section 4, the experimental design and results are presented. Finally, section 5 summarizes the work, and proposes future work.

2   Related Work

Several taxonomies of part-whole relations exist since researchers believe part-whole relations "should be treated as a collection of relations, not as a single relation" [3]. One of the most widely accepted taxonomies is developed by Winston et al. [4], and they mentioned six types of part-whole relations based on the way that the parts contribute to the structure of the wholes: component-integral, member-collection, stuff-object, place-area, portion-mass, and feature-activity.

Automatic acquisition of part-whole relations from unlabeled corpora was first presented in [5]. A few part-whole instance pairs as initial seeds were used to learn lexical patterns, and then they extracted the corresponding "parts" of six "whole" instances by these patterns from the North American News Corpus (NANC), which achieves an accuracy of 55%. The low accuracy is due to the noisy patterns they used that tend to indicate both part-whole and non-part-whole relations. Patterns extracted in [2] are more reliable since they were learned from the web by using 503 part-whole seeds derived from a special thesaurus. The Espresso algorithm in [6] uses a novel measurement of pattern reliability based on point-wise mutual information to rank all patterns and keep the top-k ones. Both of the latter two approaches achieve higher precision.

Instead of generating more reliable patterns, some researchers spent more effort on verifying the new extracted part-whole instances. Algorithms developed in [7] improved the performance by using the part-whole relations from WordNet to train a decision-tree classifier which was used to predict the previously unseen instance pairs, and they provided a superior precision rate (83%) and recall rate (98%). However, the supervised approach they used requires extensive manual work and relies heavily on external tools like word-sense disambiguation. Other verification methods include heuristic rules [8] and graph models [9].

With the rapid growth of online encyclopedias, more and more work emerged to extract semantic relations from them. [10] first introduced an algorithm to extract part-whole relationships from the Wikipedia. Sentences containing two terms that share part-whole relation in WordNet were collected and transferred to lexical patterns. Similar patterns were generalized using the edit-distance algorithm and were used to extract new part-whole relations from the Wikipedia. The precision they reported is around 60% with different thresholds in the generalization step.

3   Our Method

Our method to learn part-whole relations from online encyclopedia consists of three steps:
(1) Domain Terms Extraction: terms from domains of part and whole instances are extracted using the semantic information in online encyclopedia.
(2) Pattern Learning: sentences from web pages in the online encyclopedia are analyzed and transferred to patterns if both part-terms and whole-terms are found. Patterns are clustered if they are similar.
(3) Identification of New Part-Whole Relations: those patterns with high frequency learned in step 2 are used to discover new part-whole relation instances.
The details of all the steps are described in the following sections.

3.1   Domain Terms Extraction

One of the critical problems in relation extraction is term recognition, which lies in both pattern learning and new relation identification phrases. Some approaches recognize terms with the help of specialized thesauri [2] or online dictionaries [7, 10, 11].

An online encyclopedia can be treated as a dictionary since it consists of tremendous entries and most of them can be seen as terms. The aim of this step is to extract as many domain terms as possible by giving some entries as seeds. We'll go through the related entries of seeds and pick all the relevant ones. To do this, several relevance measurements are defined as follows.

Tags-Based Relevance

The tags are a collection of keywords that refer to the category of that entry or other useful information. For example, the tags of the entry 香蕉 (banana) consists of 农产品 (agricultural product), 水果 (fruit), 食品 (food), 绿色植物 (greenery) and so on. The more tags that two entries have in common, the more relevant they are considered to be. Based on this hypothesis, we use a tags-based relevance measurement as follows.

$$relevance_1\left(e_i, e_j\right) = \frac{|Tags(e_i) \cap Tags(e_j)|}{|Tags(e_i) \cup Tags(e_j)|} \tag{1}$$

Where $Tags(e_i)$ means all the tags of the entry e_i.

Related-Entries-Based Relevance

As depicted in Fig 1b, many related entries are listed in the bottom of an entry's page. We use $related(e_i{\rightarrow}e_j)$ to denote that e_j is one of the related entries of e_i. Clearly, two entries e_i and e_j are relevant if $related(e_i{\rightarrow}e_j)$ or $related(e_j{\rightarrow}e_i)$.

$$relevance_2\left(e_i, e_j\right) = \begin{cases} 1, if \; related\left(e_i \rightarrow e_j\right) and \; related\left(e_j \rightarrow e_i\right) \\ 0.5, if \; related\left(e_i \rightarrow e_j\right) or \; related\left(e_j \rightarrow e_i\right) \\ 0, otherwise \end{cases} \tag{2}$$

On the other side, the more related entries that two entries have in common, the more relevant they are considered to be. So, another relevance measurement based on related entries is defined.

$$relevance_3\left(e_i, e_j\right) = \frac{|RelatedEntries(e_i) \cap RelatedEntries(e_j)|}{|RelatedEntries(e_i) \cup RelatedEntries(e_j)|} \tag{3}$$

Where $RelatedEntries(e_i)$ means all the related entries of the entry e_i.

Finally, we developed a relevance measurement using the above three measurements.

$$relevance\left(e_i, e_j\right) = \sum_{k=1}^{3} \theta_k \times relevance_k\left(e_i, e_j\right) \tag{4}$$

Where θ_k are tunable parameters, and $\theta_1 + \theta_2 + \theta_3 = 1$.

Giving a set of entries E from a certain domain $Domain(E)$, the probability of a new entry e belongs to $Domain(E)$ is estimated by the following formula:

$$prob\left(e \in Domain(E)\right) = \frac{\sum_{e_i \in E} relevance(e, e_i)}{|E|} \tag{5}$$

Therefore, the iterative algorithm to extract domain terms with a set of seeds E is described in Algorithm 1:

Algorithm 1. Extracting domain terms

Input: domain entries seeds E, threshold λ_1
Output: domain terms V
  1.   add all entries in E to V
  2. $S=E$
  3. $S'=$NULL
  4.   Do Begin
  5.   foreach entry $e \in$ RelatedEntries(S)
  6.   if $\text{prob}(e \in \text{Domain}(S)) \geq \lambda_1$
  7.   add e to S'
  8.   add e to V
  9. $S=S'$
10. $S'=$NULL
11. Repeat (4)~(10) Until S is empty
12. End

We use this algorithm to extract two set of terms, one of them from the domain of part instances in the seeds, and the other from the domain of wholes. We name them $DV(p)$ and $DV(w)$, respectively. So the output of this step is $DV(p)$ and $DV(w)$.

3.2   Pattern Learning

The goal of this step is to extract patterns that indicate a part-whole relation. Traditionally, patterns are learned from those sentences that contain both part and whole instances from the seeds. However, this needs a massive corpus that contains lots of duplicated pieces of text that provides cues for relations between two terms. Compared to the whole Web, an online encyclopedia is a small corpus and the content is not so abundant. So with a small set of seeds, only a few sentences can be extracted from the online encyclopedia.

We collect the sentences that imply part-whole patterns in a different way. After the previous step, two set of terms are build. Sentences that contain terms from $DV(p)$ and terms from $DV(w)$ are selected and patterns are produced from them. The sentence selection process works in the following way:

(1) Go through all sentences from the web pages of entries from $DV(p)$ and $DV(w)$.
(2) Sentences are segmented and part-of-speech tagged using ictclas [12].
(3) If an entry from $DV(p)$ appears in the sentence, replace it by the tag *Part*. A list of *Part* tags appearing in a coordinate structure will be replaced with only one *Part*.
(4) If an entry from $DV(w)$ appears in the sentence, replace it by the tag *Whole*. A list of *Whole* tags appearing in a coordinate structure will be replaced with only one *Whole*.
(5) Select all sentences that contain both *Part* and *Whole* tags.

After this process, the selected sentences are transferred into patterns; for example, sentence (s1c) contains an entry 苹果 (apple) from $DV(w)$, and two entries 维生素C (vitamin C) and 果胶 (pectin) in a coordinate structure from $DV(p)$:

(s1c) 苹果/n 富含/v 维生素/n C/x 和/c 果胶

(s1e) Apples are high in vitamin C and pectin

Therefor pattern (p1c) produced from (s1c) is:

(p1c) **_Whole_** 富含/v **_Part_**

(p1e) **_Whole_** are high in **_Part_**.

Usually, a reliability measurement will be chosen to rank those produced patterns, and the top-k most reliable patterns are selected. Instead of measuring a single pattern, we firstly cluster similar patterns and then pattern-clusters are ranked according to their frequency, and then the patterns in the top-k most frequent clusters are selected.

We use a similarity measurement based on edit distance to judge whether two patterns are similar:

$$\text{Sim}(p_i, p_j) = 1 - \frac{\text{EditDist}(p_i, p_j)}{\max(|p_i|, |p_j|)} \tag{6}$$

Where EditDist(p_i, p_j) is the edit distance between two patterns p_i and p_j, | p_i | is the length of p_i which is defined as the number of the words it contains. A pattern p with length m can be expressed as $p(1...m)$, and the k-th word in p is expressed as $p[k]$. The edit distance between two patterns p_i and p_j is defined as the minimum number of changes (word insertion, deletion and replacement) to transfer p_i to p_j:

$$\text{EditDist}(p_i, p_j) = \text{EditDist}\left(p_i(1 \dots m), p_j(1 \dots n)\right) =$$

$$\min \begin{pmatrix} \text{EditDist}\left(p_i(1 \dots m-1), p_j(1 \dots n)\right) + 1, \\ \text{EditDist}\left(p_i(1 \dots m), p_j(1 \dots n-1)\right) + 1, \\ \text{EditDist}(p_i(1 \dots m-1), p_j(1 \dots n-1)) + diff(p_i[m], p_j[n]), \end{pmatrix} \tag{7}$$

Where m is the length of p_i and n is the length of p_j, and

$$\text{diff}(p_i[m], p_j[n]) = \begin{cases} 0, \text{if } p_i[m] = p_j[n] \text{ or they are synonymous} \\ 0.5, \text{if } \text{POS}(p_i[m]) = \text{POS}(p_j[n]) \\ 1, \text{otherwise} \end{cases} \tag{8}$$

Where POS($p_i[m]$) means the part-of-speech tag of p_i, and we use a Chinese synonymy dictionary "TongYiCiCiLin" (Extension Version) [13] to judge whether two words are synonymous or not.

The algorithm to cluster similar patterns is described in Algorithm 2.

Algorithm 2. Clustering patterns

Input: set of patterns P={p_1, p_2, ..., p_n}, threshold λ_2
Output: clusters of patterns C={c_1, c_2, ...}
   1.   C is initialized as {c_1, c_2, ..., c_n} where c_i={p_i}
   2.   Do Begin
   3.   calculate the similarities of each pair of clusters
   4.   take the two clusters with the biggest similarity, c_i and c_j, the max similarity is *max*
   5.   if max < λ_2
   6.   return
   7.   else
   8.   merge c_i and c_j to a new cluster c'
   9.   recalculate the similarities between c' and all the other clusters
  10. Repeat (2)~(9)
  11. End

The similarity between two clusters is calculated using the average-linkage method; that is, their similarity is computed as the average similarity between every pair of patterns from the two clusters:

$$\text{Sim}(c_1, c_2) = \frac{\sum_{\substack{p_i \in c_1 \\ p_j \in c_2}} \text{Sim}(p_i, p_j)}{|c_1| \times |c_2|} \tag{9}$$

We rank those generalized patterns according to their frequencies which are computed as the sum of the frequencies of the patterns before generalization. Finally, patterns in the top-k clusters are chosen as the output of this step.

3.3   Identification of New Part-Whole Relations

We apply the patterns learned in the previous step to the online encyclopedia corpus to identify new part-whole relations. The corpus is split into individual sentences and preprocessed with word segmentation and part-of-speech tagging. Once a sentence matches a pattern, substrings corresponding to the *Part* and *Whole* tags are extracted, and then terms are recognized using *DV(p)* and *DV(w)*.

Since part instances are always showed in coordinate structures, one sentence may contain several pairs of part-whole relations. The output of this step is a list of pairs of candidate part-whole relations.

4   Experimental Results

We evaluate the performance of our method in extracting stuff-object relation from the BaiduBaike corpus. Our method also can be applied to other types of part-whole relations and we plan to do it in the future work. The five pairs of part-whole relation seeds used in the experiment are listed in Table 1. They are also entries in the online encyclopedia.

Table 1. Part-whole relation seeds

Whole	Part
苹果 (apple)	维生素C (vitamin c)
香蕉 (banana)	钾 (potassium)
橘子 (orange)	柠檬酸 (citric acid)
栗子 (chestnut)	蛋白质 (protein)
龙眼 (longan)	葡萄糖 (glucose)

Two set of domain terms are extracted using Algorithm 1 with $\theta_1 = 0.5, \theta_2 = 0.3, \theta_3 = 0.2, \lambda_1 = 0.25$. The size of $DV(p)$ is 2070, and the size of $DV(w)$ is 2556. We randomly chose 200 terms from each set, and the precisions all exceed 95%, which shows the efficiency of our domain terms extraction algorithm.

After the sentence selection process, 11003 sentences that contain both terms from $DV(p)$ and $DV(w)$ are extracted and 9235 distinct patterns are produced, which indicates that online encyclopedia really contain few duplicated sentences. Similar patterns are clustered by Algorithm 2 with $\lambda_2 = 0.5$, and 330 patterns from top-15 clusters are chosen to be applied to extract new part-whole relations. Table 2 shows some of the patterns learned by our approach.

Table 2. Samples of part-whole patterns

Part-whole pattern	Example
Whole 含有 *Part* *Whole* contains *Part*	桑枝 含有 鞣质 ramulus mori contains tannin
Whole 所含的 *Part* *Part* contained in *Whole*	葱 所含的 大蒜素 allicin contained in oniions
Whole 中的 Part 含量 *Part* content of *Whole*	玉米 中的 蛋白质 含量 protein content of corns
Whole 含 Part 量 *Part* in *Whole*	樱桃 含 铁 量 iron in cherrys
Part 的 主要 来源 是 *Whole* the main source of *Part* is *Whole*	维生素B6 的 主要 来源 是 瘦肉 the main source of vitamin B6 is lean
食用 *Whole* 可以 补充 *Part* eating *Whole* can supplement *Part*	食用 芒果 可以 补充 维生素C eating mangoes can supplement vitamin C

In the new-relation identification step, 5184 sentences match at least one of the patterns, 9868 pairs of candidate part-whole relations are extracted. We calculate the precision of each pattern manually according to the following equation:

$$P = \frac{Cnt(correct-extracted)}{Cnt(all-extracted)} \times 100\% \tag{10}$$

Where Cnt(all-extracted) means the number of all the part-whole relations extracted by that pattern and Cnt(correct-extracted) means the number of the correct ones among them. Table 3 shows the results of those patterns listed in the above table:

Table 3. Relation extraction results

Part-whole pattern	Cnt(all-extracted)	precision
Whole 含有 *Part*	6776	82.1%
Whole 所含的 *Part*	788	84.6%
Whole 中的 Part 含量	698	87.7%
Whole 含 Part 量	169	95.9%
Part 的 主要 来源 是 *Whole*	35	91.4%
食用 *Whole* 可以 补充 *Part*	13	100%

Over 85.8% relations are extracted by the above patterns and 80% of them are identified by the first one. We have got a preferable precision compared to other reported algorithms, such as Girju's [7] 83%.

Two types of common errors are identified in our experimental results:

1) Unable to recognize some of the multiword terms. For example, the part-whole pair extracted from the following sentence (s2c) is Part-Whole(酸, 黄瓜) (Part-Whole(acid, cucumber)), while the correct part instance should be 丙醇二酸 (tartronic acid). This accounts for over 50% of the errors.

   (s2c) 黄瓜/n 中/f 含有/v 丙醇/n 二/m 酸/a

   (s2e) Cucumbers contain tartronic acid

2) Some errors are due to the lack of anaphora resolution algorithm. For example, sentence (s3c) returns the relation Part-Whole(胡萝卜素, 蔬菜) (Part-Whole(carotene, vegetable)), but the whole instance has to be the specific vegetable it referred to, which may be contained in the previous sentence.

   (s3c) 这些/rz 蔬菜/n 富含/v 胡萝卜素/n

   (s3e) These vegetables are high in carotene

To improve the performance of our approach, external tools like term recognition and anaphora resolution will be used in the future work.

5   Conclusions and Future Work

In this paper, we present an unsupervised approach to learning lexical patterns from online encyclopedia to extract part-whole relations. The only input is 5 pairs of part-whole instances. The major contributions of this paper include:

1) An algorithm to extract domain terms taking use of the semantic information contained in online encyclopedia is proposed, these terms are of great help for term recognition in relation identification;

2) A novel method to collect sentences that may indicate part-whole relation is described, compared to those approaches used in previous work, which set a high requirement for the corpus;

3) A new method to select reliable patterns. Similar patterns are clustered and pattern clusters are ranked according to their frequencies, those patterns from the top-k clusters are chose to be applied to identify the new part-whole relations.

Experimental results show that our method can extract abundant part-whole relations and achieve a preferable precision compared to the other approaches.

Future work focuses on improving the accuracy of our approach and applying the method to other relations.

Acknowledgments. This work is supported by the National Natural Science Foundation of China under grant No.91224006, 61173063, 61035004, 61203284, 30973716 and National Social Science Foundation of China under grant No.10AYY003.

References

1. Hearst, M.A.: Automatic acquisition of hyponyms from large text corpora. In: Proceedings of the 14th Conference on Computational Linguistics, vol. 2, Association for Computational Linguistics (1992)
2. Van Hage, W.R., Kolb, H., Schreiber, G.: A method for learning part-whole relations. In: Cruz, I., Decker, S., Allemang, D., Preist, C., Schwabe, D., Mika, P., Uschold, M., Aroyo, L.M. (eds.) ISWC 2006. LNCS, vol. 4273, pp. 723–735. Springer, Heidelberg (2006)
3. Iris, M.A.: Problems of the part-whole relation. In: Relational Models of the Lexicon. Cambridge University Press (1989)
4. Winston, M.E., Chaffin, R., Herrmann, D.: A taxonomy of part-whole relations. Cognitive Science 11(4), 417–444 (1987)
5. Berland, M., Charniak, E.: Finding parts in very large corpora. In: Proceedings of the 37th Annual Meeting of the Association for Computational Linguistics on Computational Linguistics. Association for Computational Linguistics (1999)
6. Pantel, P., Pennacchiotti, M.: Espresso: Leveraging generic patterns for automatically harvesting semantic relations. In: Proceedings of the 21st International Conference on Computational Linguistics and the 44th Annual Meeting of the Association for Computational Linguistics. Association for Computational Linguistics (2006)
7. Girju, R., Badulescu, A., Moldovan, D.: Automatic discovery of part-whole relations. Computational Linguistics 32(1), 83–135 (2006)
8. Cao, X., et al.: Extracting Part-Whole Relations from Unstructured Chinese Corpus. In: Fifth International Conference on Fuzzy Systems and Knowledge Discovery, FSKD 2008. IEEE (2008)
9. Wu, J., Luo, B., Cao, C.: Acquisition and verification of mereological knowledge from Web page texts. Journal-East China University of Science and Technology 32(11), 1310 (2006)
10. Ruiz-Casado, M., Alfonseca, E., Castells, P.: Automatising the learning of lexical patterns: An application to the enrichment of wordnet by extracting semantic relationships from wikipedia. Data & Knowledge Engineering 61(3), 484–499 (2007)
11. Hearst, M.A.: Automated discovery of WordNet relations. WordNet: an electronic lexical database, pp. 131–151 (1998)
12. Zhang, H.P., Liu, Q.: ICTCLAS. Institute of Computing Technology. Chinese Academy of Sciences (2002), http://www.ict.ac.cn/freeware/003_ictclas.asp
13. Mei, J.: Chinese Synonym Thesaurus. Shanghai Lexicology Press, Shanghai (1983)

Topic Detection and Evolution Analysis on Microblog[*]

Guoyong Cai, Libin Peng, and Yong Wang

Guilin University of Electronic Technology, Guangxi Key Lab of Trusted Software,
541004 China
ccgycai@guet.edu.cn, ccsupeng@163.com, hellowy@126.com

Abstract. The study on topic evolution can help people to understand the ins and outs of topics. Traditional study on topic evolution is based on LDA model, but for microblog data, the effect of this model is not significant. An MLDA model is proposed in this paper, which takes microblog document relation, topic tag and authors relations into consideration. Then, the topic evolution in content and intensity is analyzed. The experiments on microblog data have shown the effectiveness and efficiency of the proposed approach to topic detection and evolution analysis on Microblog.

Keywords: Microblog, LDA model, Topic Evolution, Topic Detection.

1 Introduction

Microblog is the shortened form of micro blog (MicroBlog). It is a type of platform which allows users to disseminate/obtain/share information based on relationship among users. Users can set up personal community through WEB, WAP and other client terminals, update the information with no more than 140 words of text, and achieve instantly information sharing. However, the content of microblog is diverse and changing rapidly. It is a challenging problem of how to discover automatically an effective topic and to analyze the evolution of the discovered topic.

Topic is defined as a number of related events caused by a set of seed in Topic Detection and Tracking[1]. The topic model represented by LDA (Latent Dirichlet Allocation) [2] is an important technology in the field of text mining in recent years. LDA model has a good ability of dimension reduction and scalability. It has achieved great success in mining traditional network news topic. Topic evolution is referred to the migration of topic content and intensity over times[3]. Researches on topic evolution in general are based on LDA model to extract topics, and then the topic evolution of content and intensity are analyzed. However, compared with the traditional network text, Microblog has distinct characteristics, such as a short text (usually less than 140 characters), sparse data, noise data, mixed and disorder content, etc. Besides, there also exist social relations, structural social network information. Therefore, it is not effective to use LDA model for microblog text. An MLDA model is proposed in this paper, which takes microblog document relation, topic tag and authors' relationships into consideration. The topic evolution in content and intensity are analyzed based on MLDA model.

---

[*] This work is supported by the NSFC(#61063039), Guangxi Key Lab of Trusted Software (#kx201202).

Z. Shi et al. (Eds.): IIP 2014, IFIP AICT 432, pp. 67–77, 2014.

2   Related Works

Due to its good scalability, LDA model, which proposed by Blei in 2003[2], is extended by many scholars. Researches on topic evolution based on the scalable LDA model are divided into the following three categories:

(1) Continuous time models. The information of time is taken into topic model as a variable to study the topic evolution. For example, a TOT (Topic Over Time) model is proposed by Wang [4]. In TOT model, each document is a mixture of topics that is influenced jointly by its time stamp. The main disadvantage of TOT is that it uses the beta distribution to model the topic development trends. Therefore, for documents that don't release time, it will predict the release date of the document. For documents with given released date, it can predict the document distribution.

(2) Pre-discretization methods. In this type of method, documents are divided into some parts according to the time windows before modeling topics, and then documents are processed and the topic distribution is generated in each time window. Song and etc.[5] proposed a ILDA model (Incremental Latent Dirichlet Allocation) to study the content evolution of a topic and solve ILDA model with Gibbs sampling approach[6]. Online LDA model is proposed by AlSumait[7], history data are used as the prior distribution of the proposed model and LDA model are used to study topic evolution for the arrived data in each time interval. Hu Yanli and etc. [9] have also implemented the online topic evolution analysis based on online LDA model, but they consider the inter-connection of topic distribution in each time slice.

(3) Post-discretization methods. In this type of method, the affect of time is not considered first. After topics are extracted based on LDA in the document set, the topics are divided into time slice according to its time stamp. Grillffiths and Steyvers[8] have proposed a post-discretization method based on LDA model, they use the intensity of topics in each time slice to indicate the topic trends.

3   Topic Evolution Analysis on Microblog

3.1   LDA Model

LDA model is a three-layer probability model. It consists of word, topic and document. The key idea behind the LDA model is to assume that the words in each document were generated by a mixture of topics, where a topic is represented as a multinomial probability distribution over words[9]. Each document has a specific mixture of topics, which generated from the Dirichlet distribution. The specific idea behind LDA is to assume that each document correspond to the multinomial distribution θ of T topics, each topic correspond with the multinomial distribution φ of N_d words, θ is the prior of Dirichlet distribution with parameters α, φ is the prior of Dirichlet distribution with parameters β. For each word in a document d, topic z is extracted from the distribution θ of the document, and the word w is extracted from the distribution φ of topic z. This process will be repeated N_d times, then the document d is generated. The generative process of LDA is shown in figure1.

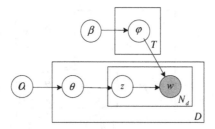

Fig. 1. LDA model

The shaded circles in Figure 1 represent the observed variable and the unshaded circles represent the latent variables. The arrows between two variables represent the conditional-dependence and boxes represent repeated sampling, the number of repetitions is located at the lower right corner of the box. There are two parameters need to be inferred in this model, one is the distribution θ of document-topic and the other is the distribution φ of topic-word. Since it is difficult to obtain precise parameters, VEM[2], Gibbs Sampling[6] and Expectation propagation[10] are often applied to estimate the parameters. Gibbs Sampling has been widely used for its simple implementation.

3.2 Topic Discovery Model ----MLDA Model

Compared with other texts, microblog texts have special symbols, such as "@"、 "# " and "retweet". @ indicates the author's relationship of a microblog post. For example, a message "@Sandy Congratulations, you get a good job." and another message "@Sandy Can you teach me some IT knowledge used in a work". When considering the author's relationship, we can set a connection between these two seemingly unrelated microblogs and consider that "job" in the first message is related to "IT" in the second message. # indicates topic tag in a microblog. For example a message "Sunyang, come on! #Olympic Games#". If considering the topic tag, "Sunyang" and "Olympic Games" are related. "retweet" indicates microblog documents' relation. For example, a message "Chinese Dream, retweet @Sandy the best popular new vocabulary …". It is difficult to analyze the specific meaning for "Chinese Dream", but compared with the original microblog, we know that "Chinese Dream" is a kind of "new vocabulary".

An MLDA model, extended from LDA, is proposed in this paper, which takes microblog document relation, topic tag and authors' relationships into consideration. The parameters of MLDA model is shown in table 1. The Bayesian network of MLDA is showed in figure 2. c indicates the author's relationship, α_r indicates microblog document relation. t indicates topic tag in mircoblog. α_c is the parameter of distribution, θ_c associate with authors' relation. θ_c is computed by Dirichlet distribution with parameter α_c . α is the parameter of distribution θ_d of document-topic. α_t is the parameter of distribution θ_d associate with special topic.

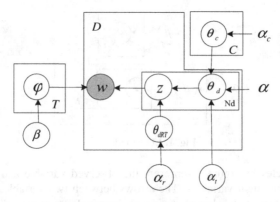

Fig. 2 Bayesian network of MLDA

α_r is the parameter that decide whether it is a retweet message. The distribution θ of topics in microblog is shown as follow.

$$P(\theta \mid \alpha, \alpha_c, \alpha_t, \pi_c, t) = [P(\theta_c \mid \alpha_c)^{\pi_c} P(\theta_d \mid \alpha)^{1-\pi_c}]^{1-t} P(\theta_d \mid \alpha_t)^t \qquad (1)$$

If there exists a symbol "@" in microblog, the value of π_c is 1, otherwise π_c is 0. If there exists a symbol "#" in microblog, boolean variable t is 1, otherwise t is 0. If there exists "retweet", the distribution of $z_{d,i}$ over topics depend on θ_{dRT} and θ_d, sampling from a Bernoulli distribution with parameter a_r to decide θ_{dRT} and θ_d, and based on them we extract the current topic $z_{d,i}$. Otherwise, we extract topics $z_{d,i}$ directly from Multinomial distribution with parameter θ_d.

For a microblog, the joint probability distribution over all words and their topics is computed as following equation (2).

$$P(w, z \mid \alpha_r, \theta, \beta) = P(\theta_{dRT} \mid \alpha_r)P(z \mid \theta)P(w \mid z, \beta)$$
$$= P(\theta_{dRT} \mid \alpha_r)P(z \mid \theta_d)^{1-\alpha_r} P(z \mid \theta_{dRT})^{\alpha_r} P(w \mid z, \beta) \qquad (2)$$

The description of generative process in MLDA is as follows:

(1) For each document $d \in D$, if there exists "#topic#", we compute $\theta_d = \theta_d \sim Dri(a_t)$. If there exists a symbol "@", we compute $\theta_d = \theta_c \sim Dir(\alpha_c)$. In other cases, we compute $\theta_d = \theta_d \sim Dir(\alpha)$.

(2) For each topic $z_{d,i}$, we judge document relation with "retweet". If there exists "retweet", then $z_{d,j} \sim Multiomial(\theta_{dRT})$. Based on $z_{d,i}$, we generate $\varphi \sim Dir(\beta)$.

(3) For each word $w_{d,i}$ in a document, firstly, we choose a topic $z_{d,j} \sim Multiomial(\theta_d)$; and then we choose a word with $w_{d,j} \sim Multiomial(\varphi)$.

Table 1. Parameters in MLDA

Parameter	Definition
α, α_t α_c, β^t	The prior parameters of $\theta_c, \theta_d, \varphi$
θ_c	Topic distribution associate with authors' relation
θ_d	Topic distribution over microblog d
φ	The distribution of topic-word
$z_{d,i}$	The i th topic in document d
$w_{d,i}$	The i th word in document d
D	The number of documents
T	The number of topics
N_d	The number of words
π_c	Bool parameter, decide whether there exists a @ message. If so, π_c is 1, else π_c is 0.
α_r	Decide whether it is a retweet message

3.3   Algorithm Implementation

We apply Gibbs sampling to estimate the parameters θ, φ. Gibbs sampling is one simple realization method of MCMC (Markov chain Monte Carlo), which is a fast and effective algorithm for estimating the parameters[11]. Gibbs sampling is based on posterior distribution of words given a topic $p(z\,|\,w)$. Repeated iterations on the probability distribution, the parameters are derived. The process is computed with equation (3).

$$P(z_i = j\,|\,z_{-i}, w_{d,i}, \alpha, \beta) \propto \frac{n^{(w)}_{-i,j} + \beta}{n^{(\cdot)}_{-i,j} + N_d\beta} * \frac{n^{(d)}_{-i,j} + \alpha}{n^{(d)}_{-i,\cdot} + N_d\alpha} \tag{3}$$

Where, $n^{(w)}_{-i,j}$ is the word counts assigned to topic j and w. $n^{(\cdot)}_{(-i,j)}$ is the counts of words assigned to topic j. $n^{(d)}_{-i,j}$ is the counts of words assigned to topic j in document d. $n^{(d)}_{-i,\cdot}$ is the counts of words assigned to topic j. All counts do not include the current iteration. After iteration finished, we can estimate θ and φ from the value z using equation (4).

$$\theta = \frac{n^{(d)}_j + \alpha}{n^{(d)}_{\cdot} + T\alpha} \qquad \varphi = \frac{n^{(w)}_j + \beta}{n^{(\cdot)}_j + N_d\beta} \tag{4}$$

Where, $n_j^{(d)}$ denotes the number of words that assigned to topic j in document d. $n_.^{(d)}$ denotes the number of words appeared in document d. $n_j^{(w)}$ denotes the number that word w is assigned to topic j. $n_j^{(.)}$ is the number of words that assigned to topic j.

The Gibbs sampling procedure of MLDA is implemented and shown in Fig.3. The Gibbs sampling algorithm consists of three parts: initialization, burn-in and sampling.

Algorithm: MLDA_Gibbs Sampling

Input: $V, T, \alpha, \alpha_c, \alpha_t, \lambda, \beta$

Output: associated topic Z, the distribution θ of document-topic, the distribution φ of topic-word.

//initialization

$n_j^{(d)} = 0, n_.^{(d)} = 0, n_j^{(w)} = 0, n_j^{(.)} = 0$;

  for : all document $d \in [1, D]$ do

    for: all word $n \in [1, N_d]$ in document d do

      $z_{d,j} \sim Multiomial(\theta_{dRT})$ //computed by Equation 1, Equation 2

      $n_j^{(d)} += 1;$

      $n_.^{(d)} += 1;$

      $n_j^{(w)} += 1;$

      $n_j^{(.)} += 1;$

//burn-in and sampling period

  While (not reach maximum iteration) do

    for : all document $d \in [1, D]$ do

      for: all word $n \in [1, N_d]$ in document d do

        //decrement counts and sums:

        $n_j^{(d)} -= 1; n_.^{(d)} -= 1; n_j^{(w)} -= 1; n_j^{(.)} -= 1;$

        $Z \sim P(z_i = j \mid z_{-i}, w_{d,j})$ //multinomial sampling according to

          equation 3

        //for the new assignment Z to the term t for word $w_{d,j}$

        $n_j^{(d)} += 1; n_.^{(d)} += 1; n_j^{(w)} += 1; n_j^{(.)} += 1;$// increment counts and sums

    If(finish sampling) then

      Output θ, φ // according to equation 4.

Fig. 3. Gibbs sampling algorithm

3.4   Topic Evolution

Topic evolution indicates that the same topic shows dynamism and difference with time going. The evolution of topic reflects in two aspects. First, the topic intensity changes over time. For example, the topic is the Olympic Games which take place every four years, which are active for Olympic years, other years is low. On the other hand, the topic content changes with the passage of time. For example, on the eve of the Olympic Games, when we pay more attention to the Olympic preparations, on the middle of Olympic, gold medal topics will be focused. In the end of Olympic Games, more attention should be paid to the summary and inventory of the Olympic Games. We use the distribution of topic j with time t to define the intensity of topic.

$$\delta_j^t = \frac{1}{D^t} \sum_{d \in D^t} \theta_{d,j} \tag{5}$$

Where δ_j^t is the intensity of a topic, D^t is the number of document, $\theta_{d,j}$ is the distribution of topic j.

The content of topic evolution is characterized as the degree associated with topics. The differences of the distribution of two topics is described with KL(Kullback-Leibler) distance[12]. The smaller the difference is, the higher degree of the two topics will be associated. Assuming that the probability distribution of topic A is $A = (A_1, A_2..., A_n)$, and B is $B = (B_1, B_2..., B_n)$, the KL distance between topic A and B   is computed as follow.

$$KL(A,B) = D(A \| B) = \sum_i^N A_i \log \frac{A_i}{B_i} \tag{6}$$

4   Experiments

The experimental data are captured from a large microblog site named Sina microblog. Through API provided by Sina, about 8 million original microblog data from July 1th 2012 to August 29th 2012 are captured. We mark five topics, i.e., "London Olympics", "heavy rains in Beijing", "Price War", "Diaoyu Island Event", "The Voice of China". In order to save space or to speed up sampling, the punctuation and stop words in the original microblog dataset must be removed before experiments. We complete this preprocessing work by using a punctuation list and a stop words dictionary.

4.1   Topic Discovery

Effectiveness experiments are conducted on dataset mainly to examine the performance of MLDA. The parameters are set up with $\alpha = 0.2$, $\alpha_c = \alpha_t = \lambda = 1$, $\beta = 0.1$, $T = 5, N_d = 716$. The key words of each topic are extracted by MLDA models. The results are shown in table 2.

Table 2. The list of representative words for each topic

topics	Most representative words
Topic1: London Olympic	Olympic Games, participation, London, champion, gold metal
Topic2: Heavy rains in Beijing	heavy rains, disaster, rescue, flood protection, emergency
Topic3: Price War	Sunning, jingdong, price war, E-commerce, fund
Topic4: Diaoyu Island Event	Diaoyu Island, Japan, China, Sovereignty, protection
Topic5: The Voice of China	Voice, China, music, pleasant music, jury

4.2   Intensity Evolution of Topics

We analyse the topics from July to August in 2012. The different change of the intensity of topics will be calculated in different times, which is divided into eight time slice by each a week. The intensity of each topic is shown in figure 3.The ordinate represents the intensity of the topic at each time slice, which is calculated with equation 5.

Figure 3 shows us that the intensity of topic 1 is higher than other topics during each time slice. This demonstrates that the influence of topic 1 is higher and is a hot topic during the two months. Besides, for each topic, the intensity of topics will change with time. This shows that each topic will go through a evolutional process that is consistency with reality.

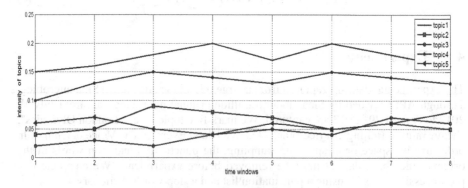

Fig. 4. The intensity evolution of topics

4.3   Content Evolution of Topics

For topic 1, the intensity is higher over time. The similarity distance of a topic during the adjacent time slice indicates the content evolution of the topic. KL distance is calculated according to the probability distribution of words in eight time intervals. Figure 4 presents the KL distance of topic 1 changes over the intervals, which indicates that the content of the topic changes over intervals. As can be seen from the figure 4 that the KL distance of topic 1 enlarge abruptly during the time slice 4 and 5, which indicates that the content of topic 1 changes quite a few. Figure 5 shows that the content of topic 1 have changed from "gold metal" into "open ceremory". During time slice 6 and7, the KL distance enlarge abruptly, which indicated that the content of topic1 have changed into "closing ceremony". The topic changing process is shown in figure 5.

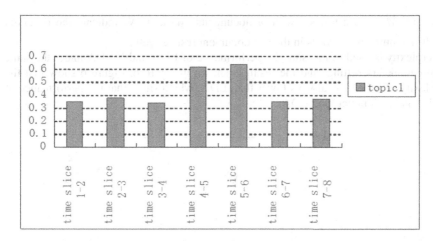

Fig. 5. The KL distance of topic 1

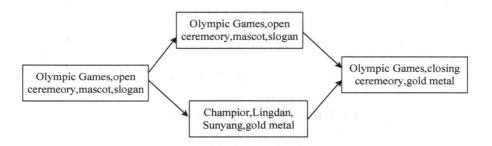

Fig. 6. The chart of topic 1 evolution

In summary, the trend of topic evolution can be captured quite accurately with MLDA model. It indicates not only the intensity change of the topic over time, but also the content change of the topic over time.

4.4   Comparisons of Algorithm Performance

We conduct a comparative experiment between MLDA and LDA, which is a baseline model in the field of topic modeling. The metric Perplexity [13] is a standard measure of performance for statistical models, which indicates the uncertainty in predicting a single word; the model with lower value is better in performance. Perplexity is defined as follows:

$$Perplexity(M_{test}) = \exp\{-\frac{\sum_{m}\ln p(w_m)}{\sum_{m}N_m}\} \qquad (7)$$

where M_{test} is a test set with m documents, w_m and N_m indicate the observed words and number of words in the test document respectively.

Perplexity is used to measure the performance of LDA and MLDA under the same hyperparameters setting, and the result is shown in figure 6. Figure 6 show that the perplexity of MLDA is always lower than LDA, which show that the performance of MLDA is much better than LDA.

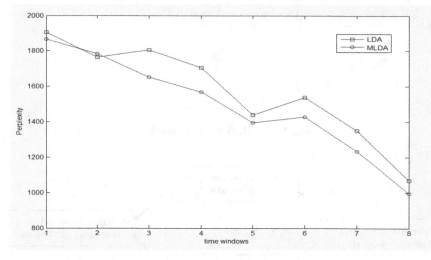

Fig. 7. The performance of MLDA and LDA

5   Conclusion

An MLDA model is proposed in this paper, which takes microblog document relation, topic tag and authors' relation into consideration. Based on MLDA, the topic evolution in content and intensity is studied. The experiments on microblog data have shown the effectiveness and efficiency of the approach to depict topic evolution. In future work, we will improve efficiency of this algorithm. Besides, the scalability of the proposed model will be further investigated.

References

[1] Allan, J., PaPka, R., Lavrenko, V.: Online new event detection and tracking. In: Proceedings of the 21st Annual International ACM SIGIR Conference on Research and Development in Information Retrieval, pp. 37–45. ACM Press, Melbou

[2] Blei, D., Ng, A., Jordan, M.: Latent Dirichlet allocation. Journal of Machine Learning Research, 993–1022 (2003)

[3] Chu, K., Li, F.: Topic Evolution Based on LDA and Topic Association. Journal of Shanghai Jiao Tong University 44(11), 1496–1500 (2010)

[4] Wang, X., McCallum, A.: Topic over time: A Non-Markov Continuous-Time Model of Topical Trends. In: ACM SIGKDD, pp. 424–433 (2006)

[5] Song, X., Lin, C.Y., Tseng, B.L., et al.: Modeling and predicting personal information dissemination behavior. In: Proceedings of the Eleventh ACM SIGKDD International Conference on Knowledge Discovery in Data Mining, pp. 479–488 (2005)

[6] Mark, S., Tom, G.: Probabilistic Topic Models. Latent Semantic Analysis: A Road to Meaning (2006)

[7] AlSumait, L., Barbará, D., Domeniconi, C.: On-line LDA: adaptive topic models for mining text streams with applications to topic detection and tracking. In: Data Mining, ICDM 2008, pp. 3–12 (2008)

[8] Griffiths, T.L., Steyvers, M.: Finding scientific topics. In: Proceedings of the National Academy of Sciences of the United States of America, pp. 5221–5228 (2004)

[9] Cui, K., Zhou, B., Jia, Y., Liang, Z.: LDA-based Model for Online Topic Evolution Mining. Computer Science 37(11), 156–193 (2010)

[10] Minka, T., Lafferty, J.: Expectation-Propagation for the Generative Aspect Model. Uncertainty in Artificial Intelligence (UAI) (2002)

[11] Blei, D., Lafferty, J.: Visualizing Topics With Multi-Word Expressions. Stat, 1050–1055 (2009)

[12] Hu, Y., Hu, L., Zhang, W.: OLDA-based Method for Online Topic Evolution in Network Public Opinion Analysis. Journal of National University of Defense Technology 34(1), 150–154 (2012)

[13] Blei, D., Lafferty, J.: Dynamic topic models. In: Proceeding of the 23rd in Conference on Machine Learning, pp. 113–120. ACM, New York (2006)

A DBN-Based Classifying Approach
to Discover the Internet Water Army

Weiqiang Sun[1,*], Weizhong Zhao[1], Wenjia Niu[2], and Liang Chang[3]

[1] College of Information Engineering,
Xiangtan University,
Xiangtan City, Hunan, 411105, China
{swq.sdh.llh,zhaoweizhong}@gmail.com
[2] Institute of Information Engineering,
Chinese Academy of Sciences,
Minzhuang Road 89, Beijing, 100093, China
niuwenjia@iie.ac.cn
[3] Guangxi Key Laboratory of Trusted Software,
Guilin University of Electronic Technology,
Guilin, 541004, China

Abstract. The Internet water army (IWA) usually refers to hidden paid posters and collusive spammers, which has already generated big threats for cyber security. Many researchers begin to study how to effectively identify the IWA. Currently, most efforts to distinguish non-IWA and IWA in data mining context focus on utilizing classification-based algorithms, including Bayesian Network, SVM, KNN and etc... However, Bayesian Network need strong conditional independence assumption, KNN has big computation costs, above approach may affect the effectiveness to some extent in real industrial applications. Hence, Neural Networks-like deep approach for IWA identification gradually becomes an emerging but possible direction and attempt. Unfortunately, there also exists one main problem, which is how to balance the deep learning and computation costs in hierarchical architecture. More specially, combine leaning-level heuristic training design and computing-level concurrent computation is a challenging issue. In this paper, we propose a collaborative hierarchical approach based on the deep belief network (DBN) for IWA identification. Firstly, a DBN-based collaborative model with hierarchical classifying mechanism is built. Then towards Hadoop platform, the Downpour Stochastic gradient descent (Downpour SGD) is exploited for DBN pre-training. Finally, the dynamical workflow will be designed for managing the whole learning-based classifying process. The experimental evaluation shows that the valid of our approach.

Keywords: Internet Water Army, DBN, Downpour SGD, Collaboration, Hadoop.

1   Introduction

With the rapid development of the Internet, cyber security issues become more and more prominent. Recently, the internet water army (IWA) draws more and more

---

* Corresponding author.

Z. Shi et al. (Eds.): IIP 2014, IFIP AICT 432, pp. 78–89, 2014.

attentions. The IWA refers to post online comments or articles with particular content on social media for some hidden purpose in order to affect public opinion. They intentionally or unintentionally spread rumors and attack others, causing extreme emotions and national antagonism, which belong to an emerging threat for cyber security. Therefore, effective regulatory on IWA is desperately needed.

To defend the IWA, the basic challenge is to identify IWA firstly. Actually, IWA identification is essentially a classification, which should analyze the users' historical behaviors and compute the inner-class and inter-class differences between IWA and normal user. Most common classification algorithms have been exploited in IWA identification, such as Bayesian Network, SVM, KNN, and Neural Networks and so on. However, for real industrial application of big data context, these approaches have some disadvantages to some extent. For instance, the Bayesian network is based on probability and statistics, which predict samples classes using Bayesian principles. However, Bayesian principles need a strong conditional independence assumption which is always invalid for IWA in real scenario. SVM requires prior calculation of the samples' space vector and the weight of each dimension, while the weight setting always relies on experience and problem analysis and directly affects the accuracy of the result. KNN is a lazy learning method, which stores samples until they are needed to learning. It may lead to a large computational overhead if the sample set is complex.

With the development of deep learning, the work based on neural network algorithm draws special attention again. As a commonly used classification algorithm, it determines its model parameters by training and objectively reflects the impact of various factors on the final result. However the basic neural network model is so complicated. When the training date set scales up, the training process will take too much time to complete, and furthermore, improper initializations of weight easily result in the local minimum problem. These disadvantages can bring poor convergence, low accuracy, time-consuming problems when using neural network-type algorithm to detect IWA.

Hence, a collaborative hierarchical approach based on DBN will be proposed in this work. Our approach will build a parallel-improved DBN model and defines the mechanism of DBN collaboration between the various parts of the model. The model mainly consists of three modules: user data pre-processing module, DBN model training module and the collaboration module. Especially for the collaboration module, we define the feedback process of the convergence and accuracy of the DBN in the form of workflow. Beside, parallel computing model are used in user data pre-processing module and DBN model training module. We hope that this approach can not only improve the convergence and accuracy of IWA detection, but also solves the problem of costing too much time in the model training process with the mass of sample data.

The rest of the paper is organized as follows. Section 2 discusses related work, followed by the presentation of a DBN-based classifying approach and its details in Section 3. Section 4 presents the experimental results that illustrate the benefits of the proposed scheme compared to other approaches. Finally, Section 5 provides the concluding remarks and future work.

2   Related Work

The related work of IWA identification can be grouped into two aspects: behavior-based recognition and content-based recognition. In addition, spam detection work in Web 2.0 also has close and significant inspiration for IWA identification.

Behavior-based work can be further divided into two categories: direct calculation type [1-2] and training-leaning type [3]. In the direct calculation type, Chen Kai et al [1] aimed at the account list which has forwarded hot topic. They determine the initial IWA sample collection S and fans of S by artificial judgments and choose part of the fans of S to join in S through a sample filter in next epochs. This method uses the social relations of the IWA to identify more IWA. However, it has some limitations that it ignores the few relations between the IWA. Zhang Guoqing et al [2] extend the feature vector to 16 dimensions and identify IWA by comparing the pre-set threshold to the result of the calculation of the weighted dimensions. However, in this method, the weight of dimensions is set relatively subjective. In the leaning-training type, Han Zhongming et al [3] transform the behavior and features of the micro-blog users to characteristic vector, constructing a probabilistic graph of features and behavior: $P(D|w)P(w)=(\Pi P(z^{(i)},y^{(i)}|x_1^{(i)},\ldots x_n^{(i)},w))P(w)$, where D is the number of samples, $x_n^{(i)}$ is the feature characteristic of dimension n of the i th user, $yn^{(i)}$ is the behavioural characteristics of dimension n of the i the user, w is the weight of each feature weight, z is the probability for user is IWA. Finally, use the Maximum Likelihood Estimation to get parameter w and so on in probabilistic graph, which can calculate the probability of that a given user is IWA. The above methods are largely dependent on the selected characteristics which will result in selection too artificial and subjective.

Content-based work mainly focus on posting content itself, through distinguishing IWA through statistics of normal and abnormal posting depending on language characteristic. There are three mainly ideas in this approach: the first one4 is to use emotional tendency for a single posting. In this idea, they consider IWA has a strong emotional tendency to beautify or demonize something; the second [4-7] is based on the tendency to deceive of a single posting. In this idea, posting of IWA is deceptive opinion spam with some statistical law. The work [8] is based on similarity of multiple posts. In this idea, they consider IWA tend not to write multiple posts in order to pursue post speed. Instead, IWA prefers to make slight changes to the same post and forward largely. However, these methods based on the content analysis have a common defect that a simple analysis of contents would not be conclusive evidence for IWA identification. In addition, in the evolution of fighting with IWA, count of the machine post decreased and manual post increased, which makes identification based on content analysis extremely difficult.

Spam is the term widely used to describe the phenomenon that spam messages spread everywhere. Spam, broadly defined, refers the behavior of sending information actively with no specific target through the electronic information system, involving micro-blog and forums. Non-target greedy spread for behavior and spam for content, there are similarities between Spam and IWA. Anti-Spam research has a history of ten years, mainly including text-based, behavior-based and image-based feature extraction. There has been a lot of typical work: a) text-based mainly uses bag-of-words

(BoW) [9], space binary polynomial hashing (SBPH) [10], orthogonal space bigrams (OSB) [11], and biological immune system (BIS) [12]; b) behavior-based spam detection is to filter spam by extracting the behavioral characteristics of the email. These commonly used methods are based on system log and message header information [13], attachment [14], and the network [15]; c) image-based feature extraction mainly focuses on extracting the key characteristics of the picture. There are a lot of the typical approaches to extract e-mail feature based on machine learning in intelligent spam detection, involving Bayesian [16,17], K-nearest neighbor[18,19], Boosting Trees [20], SVM [21,22], Rcchio [23] and Artificial Neural Networks [24], and so on.

It is worth noting that the Neural Network has been a very special research in social media Spam, due to the surprising improved recognition effect. However, it is difficult to be widely used because of the non-ignorable locally optimal, slow convergence [25] and other problem. In the works of J.Clark et al [24], the email classification based on the artificial neural network received 99.13% recognition accuracy on Ling-Spam corporace10 dataset. Guangchen Ruan et al [26] use BP Neural Network to identify Spam. First, they extract feature from email using KLGA algorithm and reduce the dimension of the vector space model of message. Then they classify the message by three layers BP Neural network. This approach has a 97%-99% recognition rate on PU and Ling-Spam corpora10 dataset. But its main shortcomings include the need for more time to compute and the great impact on network convergence the initial settings have.

With the breakthrough that Hinton et al did on Neural Network in 2006, deep learning technology, especially DBN [27], gained success in image recognition, voice recognition, and natural language content comprehension. So this has inspired us to take advantage of the new generation of neural networks to extract behavior and feature, exchanging time for the calculation of neurons for accuracy and adaptive performance in IWA recognition.

3   Overview of the Classifying Approach

The classifying approach is designed to consist of three modules: user data pre-processing module, DBN model training modules and collaboration module. The user data pre-processing module convert the original user descriptive information to vector, and divide the user data set into two parts: DBN model training data and test data; DBN model training module trains DBN model with training data set which is get in user data pre-processing module, includes two processes: model pre-training process and model fine tuning process; Collaboration module defines the feedback process of the convergence and accuracy of the first two modules in the form of workflow. Schematic overview of the approach is shown in Figure 1.

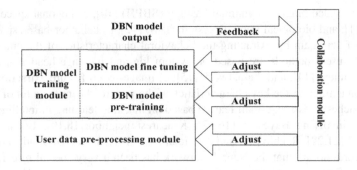

Fig. 1. Schematic Approach Overview

3.1   Data Pre-processing

User information needs to be expressed in the form of a mathematical transformation before the classification. Objectively, the user relevant information contains user name, user registration time, previous login time, login IP, browsing history records, post history records, replies history records, friends' records, followers' records, followings' records. We select the representative information as the reference to classify users and accordingly propose a multi-attribute description user information framework.

Using the multi-attribute descriptive user information framework, user descriptive information can be transformed into a vector which is mathematical representations of the user information. In addition, in order to facilitate setting initial weights in DBN model training, value of each dimension in user information vector need to be in [-1,1]. We present a normalization to deal with each dimension of the vector. That is to say we extract the value range of each dimension of the vector, find and normalized the dimensions which value range is beyond [-1, 1].

As shown in Figure.2, the vector generation process and normalization process can use parallel computing model. In user descriptive vector generation phase, all user information is randomly divided into m group to parallel process. Each group will be responsible for transforming user descriptive information to descriptive vector and each vector will be assigned an ID number then. Finally, we obtain a collection which contains pairs of user ID and descriptive vector. In the user descriptive vector normalization process, we first use MapReduce parallel frame to find the dimensions which value range is beyond [-1, 1] of the vector as well as the biggest absolute values of these dimensions. Then we use these values to normalize these dimensions. Normalization process can also divide user descriptive vector into m group to normalize in parallel.

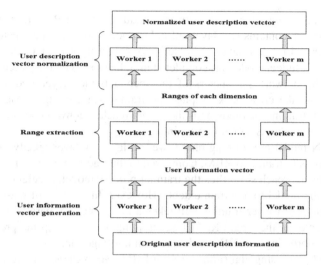

Fig. 2. Parallel Pre-processing Diagram

Through the above process, the normalized descriptive user vector collection can be obtained. A part of the collection data is classified, that is, this part of users has been marked as IWA or normal. This type of date is called "classified data sets". The "classified data sets" is divided into two parts for the subsequent DBN model training. One part called "training data set" is for the training while the other part called "test data set" is for accuracy tests for the DBN model obtained after training. As the DBN model needs enough samples to learn so that hiding law in these samples can be simulate, the "training data set" generally is given more samples. However, too many samples will increase the amount of computation and bring an over calculate disadvantage.

3.2   DBN Training

DBN (Deep Belief Network) model is a kind of deep neural network model and is a probability generation model that consists of multilayer random variables. Basic DBN model consists of two layers of RBM (Restricted Boltzmann Machines) and a layer of BP neural network (Back Propagation Neural Network). The training process of the DBN model is divided into two phases: pre-training phase and fine-tuning phase.

Pre-training phase use layer by layer unsupervised greedy learning methods to train the two RBM layers in the model: First, use the input data and the first layer if hidden layers as a RBM to train to get the parameters, then fix the parameter of the RBM and take the first hidden layer as the visible layer and the second as the hidden layer to train the new RBM. Finally, we obtain the parameters of the second RBM and determine the parameters of the two RBM at the moment and complete the pre-training process of the DBN model. In this process, the training process of each RBM is independent, which greatly simplifies the process of training the model.

After the pre-training, the entire network is equivalent to BP neural network. This BP neural network contains two layers of hidden nodes, network parameters between the input layer and the first layer of hidden nodes, as well as the parameters between the two layers of hidden node, has completed initialization. You only need to randomly initialize the network parameters of the second hidden layer nodes and output nodes, you can take error back propagation trainings according to the normal BP neural network training methods until the model reaches convergence or termination conditions, this process is called fine-tuning.

In the DBN model pre-training phase, we use layer by layer greedy unsupervised learning methods to train two RBM. Compared to the traditional multi-feedback training model, this approach simplifies the training process model, accelerating the training speed of the model to a certain extent. However, In the face of massive training data set, single RBM layer training still takes a long time. So, parallel processing is applied to accelerate the single RBM layer training and speed up the pre-training of DBN model, shortening the DBN model pre-training stage time.

For parallel processing algorithms used in RBM, we examined the following three kinds of schemes: MapReduce model, the traditional SGD mode and Downpour SGD model. MapReduce model is good at data parallel processing in the usual sense, but not suitable for depth iterative calculation in deep network training; Traditional SGD (Stochastic gradient descent) model is the most commonly used optimized method for training the Deep Neural Networks. However, this model is essentially serial, which means data movement between machines would be very time-consuming; Downpour SGD is a parallel optimization to the traditional model and is an asynchronous stochastic gradient descent variant which use single DistBelief model and has many distributed copies. It is a good choice for large-scale data parallel computing. Accordingly, we choose DownPour SGD for parallel processing to train RBM training process.

As shown in Figure 4, parallel RBM training based on Downpour SGD has a basic idea: the training data is divided into several subsets and distributed on multiple Worker servers; each Worker server runs a copy of RBM model and just simply does communication with parameter server. The parameter server stores the current state of model parameters. Parameters of the model can be updated by updating the parameters storing in the parameter server. At training phase, each Worker server obtains the parameters of current state of the model from the parameter server and performs minbatch according to these parameters, calculate the gradient and push the results back to parameter server. In a simple implementation of Downpour SGD, you can set the Work server to obtain the updated parameters from parameter server after every "nfetch" times of mini-batch and push the new calculated gradient back to parameter server to update.

The gradient update process is performed asynchronously in DownPour SGD, in this way, even a Worker server is down, it will not affect the work of other Worker servers. Although asynchronous update process will lead to the parameters of each Worker server slight difference, the algorithm has a good stability in current implementation.

After the training of the two RBM, the pre-training process of the DBN model is complete. The model is equivalent to a four-layer BP neural network now. The parameters between the three bottom layers have been initialized. Initialize the parameters between the top two layers and train the BP neural network with the training data set, which is fine-tuning process of the DBN model.

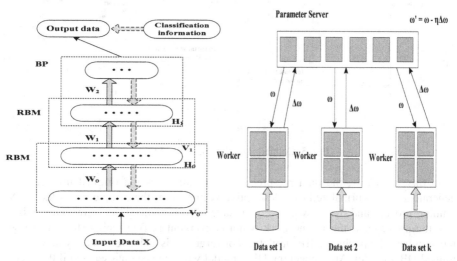

Fig. 3. Basic DBN Model   **Fig. 4.** Downpour SGD

3.3 Workflow-Based Collaborative Learning

In the model training process, different settings of parameters in each module may affect the output of subsequent module so that it will affect the accuracy of the final DBN model. For example, if the proportion of the training data set is too low, it will prevent the IWA feature extraction, resulting in a low accuracy; if the maximum number of epochs is too low, it will make the network immature. In the collaborative module, we define the feedback process of the convergence and accuracy of the first two modules in the form of workflow. According to the convergence and accuracy of DBN model, we determine whether to make a reverse adjustment to the parameters of the user data pre-processing module and DBN training module or not, finally, improving the performance of the DBN model.

Workflow is a class of business processes that can be completely or partially performed automatically. Documents, information, and tasks can be passed and executed between different actors based on a series of procedural rules. The workflow we defined is shown as Figure 5.

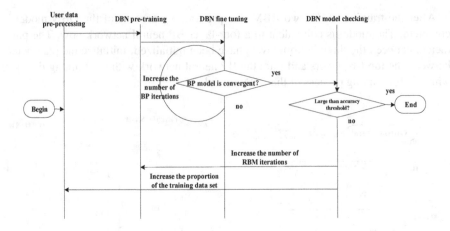

Fig. 5. Collaborative Learning Workflow

In this workflow, the data pre-processing module uses the original user data to generate a set of normalized user descriptive vector. After the completion of the DBN training process and the network weights have been initialized, the flow enters DBN fine tuning stage. If the BP network is not convergent in fine tuning stage, increase the number of epochs of BP until the convergence is achieved, then you get the trained DBN model. After checking DBN model with the test data set, and if the accuracy of DBN model does not achieve the desired threshold, increase the proportion of the training data set in classified data set in the user data pre-processing stage and increase the number of epochs of RBM in DBN pre-training stage. Retrain DBN model until the DBN model has a desired accuracy.

4   Experiment

4.1   Setting Up

We get 3019 user information from Sina WeiBo by using web crawlers. The information contains 26 behaviour related dimensions which including duration of registration, total number of micro-blog, the number of forwards, the number of collections, member level, the number of Tab, location information, the number of self-describing words, total number of links contained in all blogs, the number of followers, the number of followings, the ratio of mutual followers, rate of interactions, the time of most blogs, and so on. Then we try to mark these users manually which includes 588 IWA and 2431 normal users in three months, this is very challenging. All DBNs, which containing 4 hidden layers each with 2048 units, were trained using SGD with a mini-batch size of 128 training cases in pre-training. For simplicity, we use a default learning rate of 0.005 in Gaussian-binary RBMs while using a learning rate of 0.08 in binary-binary RBMs. For fine-tuning, we used SGD with the same mini-batch size in pre-training. The learning rate started at 0.1. If the accuracy fells at the end of each epoch, the learning rate is halved. This process continues until the leaning rate fell below 0.001.

To evaluate the performance, we designed two experiments. We hope to check: 1) whether our improved DBN has acceptable training cost;2)whether our improved DBN has an averagely better identification accuracy than normal DBN.

4.2   Training Cost

We used 60% of the 3019 user information as the training data while the rest 40% as test data both in our improved DBN and normal DBN. In this experiment, we ran 100 epochs for Gaussian-binary RBMs and 50 epochs for binary-binary RBMs in pre-training. We determined that our improved DBN should take an accuracy of 85% to 90% as a threshold in DBN model checking part of the workflow. Figure 6 shows the time cost of DBN models in pre-training and fine-tuning when our improved DBN complete training and achieve the accuracy goal. We can observe that the time under the accuracy of 90% is just double of the time under 85%. It is acceptable to get more five percent accuracy to archive an accuracy of 90%.

4.3   Identification Accuracy

The second experiment was designed to show identification Accuracy with different number of training data and test data. The proportion of the training data in this experiment increased from 10% to 90% and the rest were used as test data. The figure 7 shows that our improved DBN always has an averagely better accuracy than normal DBN, especially when the proportion of training data is small than 40%. We can see that the accuracy will begin falling when the proportion is larger than 75%. When the proportion is 60%, both our improved DBN and normal DBN nearly have the same accuracy.

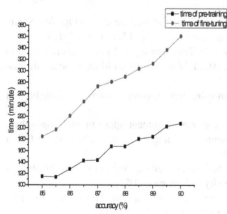

Fig. 6. Training Cost

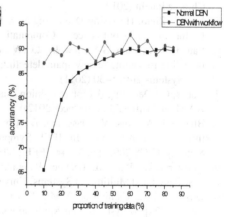

Fig. 7. Identification Accuracy

5   Conclusions

This paper proposes a collaborative hierarchical approach based on the DBN for IWA identification. We found that parallel computing can accelerate the user data pre-process and the training process when the simple set is complex and large. We believe that this research still runs its beginning stage, and in the future, we will put more study and discussion on how to control the balance point between computing cost and identification accuracy.

Acknowledgements. This research is supported by the National Natural Science Foundation of China (No. 61105052, 61103158, 61202398, 61272295), the Strategic Priority Research Program of the Chinese Academy of Sciences Grant (XDA06030200), Guangxi Key Laboratory of Trusted Software.

References

1. Chen, K., Zhou, Q., Zhou, Y., Lin, C.: Method for capturing water armies on microblog platforms, Univ. Shanghai Jiaotong, CN103095499A, Chinese Patent (2013)
2. Zhang, G., Bian, J., Fu, C., Li, Y.: Microblog ghostwriter identifying method and device, INST Computing Tech CN Academy, CN103198161A, Chinese Patent (2013)
3. Han, Z., Wan, Y., Microblog, F.X.: Microblog water army identifying method based on probabilistic graphical model, Univ. Beijing Tech & Business, CN103077240A, Chinese Patent (2013)
4. Zhang, W., Zheng, Z., Gao, W., Shuai, Z., Zhou, Y.: Detection and determination method of network navy, Anhui Boryou Information Technology Co. Ltd., Cn102629904A, Chinese Patent (2012)
5. Xu, Q., Zhao, H.: Using Deep Linguistic Features for Finding Deceptive Opinion Spam. In: International Conference on Computational Linguistics, pp. 1341–1350 (2012)
6. Lau, R.Y.K., Liao, S.Y., Kwok, R.C.W., et al.: Text mining and probabilistic language modeling for online review spam detecting. ACM Transactions on Management Information Systems 2(4), 1–30 (2011)
7. Harris, C.: Detecting deceptive opinion spam using human computation. In: Workshops at AAAI on Artificial Intelligence (2012)
8. Bhattarai, A., Rus, V., Dasgupta, D.: Characterizing comment spam in the blogosphere through content analysis. In: IEEE Symposium on Computational Intelligence in Cyber Security, CICS 2009, pp. 37–44. IEEE (2009)
9. Gansterer, W., Ilger, M., Lechner, P., et al.: Anti-spam methods-state of the art. Institute of Distributed and Multimedia Systems, University of Vienna (2005)
10. Guzella, T.S., Caminhas, W.M.: A review of machine learning approaches to spam filtering. Expert Systems with Applications 36(7), 10206–10222 (2009)
11. Siefkes, C., Assis, F., Chhabra, S., Yerazunis, W.S.: Combining winnow and orthogonal sparse bigrams for incremental spam filtering. In: Boulicaut, J.-F., Esposito, F., Giannotti, F., Pedreschi, D. (eds.) PKDD 2004. LNCS (LNAI), vol. 3202, pp. 410–421. Springer, Heidelberg (2004)
12. Ruan, G., Tan, Y.: A three-layer back-propagation neural network for spam detection using artificial immune concentration. Soft Computing 14(2), 139–150 (2010)

13. Yeh, C.Y., Wu, C.H., Doong, S.H.: Effective spam classification based on meta-heuristics. In: 2005 IEEE International Conference on Systems, Man and Cybernetics, pp. 3872–3877. IEEE (2005)
14. Hershkop, S.: Behavior-based email analysis with application to spam detection. Columbia University (2006)
15. Ramachandran, A., Feamster, N.: Understanding the network-level behavior of spammers. In: ACM SIGCOMM Computer Communication Review, pp. 291–302. ACM (2006)
16. Ming, L., Yunchun, L., Wei, L.: Spam filtering by stages. In: International Conference on Convergence Information Technology, pp. 2209–2213. IEEE (2007)
17. Wang, M., Li, Z., Wu, H.: An improved Bayes algorithm for filtering spam email. Journal of Huazhong University of Science and Technology (Nature Science Edition) (8), 27–30 (2009)
18. Li, X., Tian, Y., Duan, H.: Implementation and evaluation of Chinese spam filtering system. Journal of Dalian University of Technology (z1), 189–195 (2008)
19. Sakkis, G., Androutsopoulos, I., Paliouras, G., et al.: A memory-based approach to anti-spam filtering for mailing lists. Information Retrieval 6(1), 49–73 (2003)
20. Schapire, R.E., Singer, Y.: BoosTexter. A boosting-based system for text categorization. Machine Learning 39(2-3), 135–168 (2000)
21. Drueker, H., Donghui, W.W., Vapnik, V.N.: Support vector machines for spam categorization. IEEE Transactions Neural Networks 10(5), 1048–1054 (1999)
22. Tang, Z.-H., Fu, J.-M., Du, N.-S.: Design and Analysis of Spam-Filtering System Based on Words Segmentation. Journal of Wuhan University (Natural Science Edition) 51(S2), 191–194 (2005)
23. Schapire, R.E., Singer, Y., Singhal, A.: Boosting and Rocchio applied to text filtering. In: Proceedings of the 21st Annual International ACM SIGIR Conference on Research and Development in Information Retrieval, pp. 215–223. ACM (1998)
24. Clark, J., Koprinska, I., Poon, J.: A Neural Network Based Approach to Automated E-Mail Classification. In: Web Intelligence, pp. 702–705 (2003)
25. Xu, Z.B., Zhang, R., Jing, W.F.: When does online BP training converge? IEEE Transactions on Neural Networks 20(10), 1529–1539 (2009)
26. Guangchen, R., Ying, T.: A three-layer back-propagation neural network for spam detection using artificial immune concentration. Soft Computing 14(2), 139–150 (2010)
27. Hinton, G., Osindero, S., The, A.: A fast learning algorithm for deep belief nets. Neural Computation 18(7), 1527–1554 (2006)

An Efficient Microblog Hot Topic Detection Algorithm Based on Two Stage Clustering

Yuexin Sun[1], Huifang Ma[1], Meihuizi Jia[1], and Wang Peiqing[2]

[1] College of Computer Science and Engineering,
Northwest Normal University Lanzhou Gansu, China
{sunyuexintc,jmhuizi0424}@163.com, mahuifang@yeah.net
[2] Anding District People's Armed Forces Department, Dingxi Gansu, China
wangwsnwnu@126.com

Abstract. Microblog has the characteristic of short length, complex structure and words deformation. In this paper, a two stage clustering algorithm based on probabilistic latent semantic analysis (pLSA) and K-means clustering (K-means) is proposed. Besides, this paper also presents the definition of popularity and mechanism of sorting the topics. Experiments show that our method can effectively cluster topics and be applied to microblog hot topic detection.

Keywords: Probabilistic Latent Semantic Analysis, Topic detection, Microblog, K-means.

1   Introduction

Nowadays, Micro-blogging is increasingly extending its role from a daily chatting tool into a critical platform of spreading real-time information during emergencies. It provides a short and convenient way for users to express and share their attitudes instantly. Since it enables users to acquire near real-time information, it has become a major source for producing and spreading hot events on the internet. However, acquisition of information regarding real-time hot events and news poses significant challenges due to the extensive irrelevant personal messages and unstructured short texts. Recent research of hot topic detection started to focused on the micro-blog [1-3]. Vector space model is always adopted as the way to represent microblog [4]. A great amount of efforts have been made for microblog hot topic detection in a semantic analyzed way. Some people extend short texts information by Knowledge Base, such as WordNet or Wikipedia [5-8]. In addition, topic model is also an effective way for microblog hot topic detection [9]. However, all of the above methods take advantage of external knowledge and they fail to consider some prior knowledge for microblogs.

In this paper, we formulate the task of microblog hot topic detection as a two-stage clustering algorithm based on probabilistic latent semantic analysis (pLSA)[10] and K-means[11] clustering (K-means). Furthermore, the definition of popularity(hotness) and mechanism of sorting of the topics are presented.

The basic outline of this paper is as follows: Section 2 presents details of our approach. The experiments and results are given in Section 3. Lastly, we conclude our paper in Section 4.

Z. Shi et al. (Eds.): IIP 2014, IFIP AICT 432, pp. 90–95, 2014.
© IFIP International Federation for Information Processing 2014

2   The Proposed Method

In this section, we will present the details of our algorithm for building topic model and hot microblog acquisition.

2.1   Basic Idea

The conservative pre-process approach is taken to be the first step, which means that the microblog will be filtered and segmented first. Then the stopwords are eliminated and the traditional Chinese are translated into simplified Chinese. All terms occurring less than 20 times are removed.

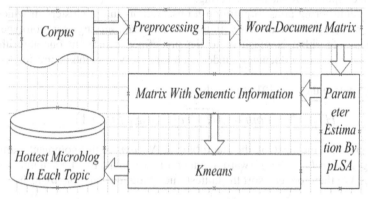

Fig. 1. Algorithm process

Specifically, pLSA model is first applied on the traditional term-document matrix to estimate two matrix denoting the distribution of topic under certain document and distribution of term under given topic, respectively. Then we multiply these two matrixes to obtain the word-document matrix with latent semantics, which becomes the input for the next stage clustering algorithm. We sort the probability of a document under each topic, the biggest one will be regarded as the centre for K-means clustering. Figure 1 shows the entire algorithm follow:

2.2   Our Approach

2.2.1   Pre-process of Microblogs

a). For microblog which is directed at another user, as denoted by the use of the @username token, we believe that the content of this field has little to do with real content. Therefore we just simply delete this field.

b). For microblog that contains "RT @" field, these microblogs are produced by someone else but the user copies, or forwards, them in order to spread it in his network. This is not treated as the real content. So we just keep the real content of this microblog.

c). For microblog that contains "# certain topic (user) #" microblog, the field can be considered as the noise, we just delete it.

d). We take advantage of stemmer for segmentation and remove the stop words at the same time. The stop list is based on 1028 stop words provided by sina.

2.2.2   pLSA Parameter Estimation and K-means Clustering

We first use TF*IDF[4] to represent the importance of each word, and the word-document matrix is constructed. pLSA is then used to estimate the parameters: distribution of topic under document d_i $P(z_k/d_i)$ and distribution of term under topic z_k $P(w_j/z_k)$.

pLSA uses a generative latent model to perform a probabilistic mixture decomposition. This results in a more principled approach with a solid foundation in statistical inference. More precisely, pLSA makes use of a temperature controlled version of the Expectation Maximization algorithm for model fitting, which has shown excellent performance in practice. The word-document matrix with semantic information is then obtained by multiplication of these two matrixes. This matrix also becomes the input for K-means clustering.

K-means clustering aims to partition n observations into k clusters in which each observation belongs to the cluster with the nearest mean, serving as a prototype of the cluster. This algorithm is sensitive to the predefined centre. In our algorithm the biggest probability of a document under each topic serves as the centre for K-means clustering.

We also define the heat[12] of microblog:

$$HT_i = N\_com_i + \sqrt{N\_rel_i} + \log(fan_i + 1) \tag{1}$$

HT_i represents the heat of the microblog i, N_com_i is the comment number of microblog i. N_rel_i is the forward number of i. fan_i is the fans number of the microblog. We therefore can sort the hottest microblog in each topic by its heat.

3   Experiments

3.1   Data Corpus

Two thousand two hundred and twenty one microblog entries were collected in two phases. Both the training and test data of microblog entries are manually collected from sina microblog sites during 2011-11-01 to 2011-11-03 and 2013-03-10, respectively. The training dataset is composed of 37 topics.

Two baseline methods K-means and LSA+Kmeans are adopted to verify effectiveness of our model. As for the parameters of our approach, to give these algorithms some advantage, we set the number of clusters equal to the real number of microblog clusters.

3.2   The Experimental Results and Analysis

The experiments include three parts: 1) Overall evaluation of our model by comparing with that of 2 other algorithm; 2) Clustering results of our method; 3) Hot topic Visualization.

3.2.1   Overall Evaluation

We run our algorithm on the training data. We have run 5 times for each evaluation, and the average performance scores are reported. Four clusters with largest number of documents are selected. To evaluate the performance of our algorithm, we use precision[13] as our measures. Table 1 and Table 2 show the results.

From the experimental results, we can see that our method is superior to the former two algorithms in terms of precision. This is because the traditional VSM method simply considers the term occurrence information while ignores the semantic information, therefore it has difficulty in represent high dimensional and sparse vector. LSA performs decomposition of matrix with some negative numbers. All these characters are not in conformity with the microblog' feature.

Table 1. Performance of Precision on training datasets of our method

Topic	Correct cluster	Incorrect cluster	Precision
1	25	12	0.68
2	22	13	0.63
3	33	6	0.85
4	26	10	0.72

Table 2. Performance of Precision on training datasets of 3 methods

Methods	Precision
Kmeans	0.55
LSA+Kmeans	0.65
Our method	0.72

3.2.2   Clustering Results

Figure 2 demonstrates the clustering results on the test data. In the figure, each colour stands for one topic. Since the coordinate system is two-dimensional, we select each microblogs' two largest vector dimensions to show the cluster results. We can see that the clustering result is basically in conformity with clustering standard.

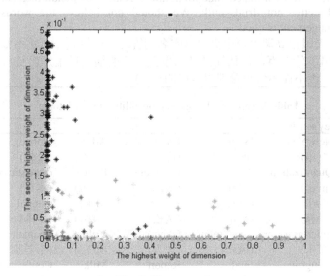

Fig. 2. Cluster result

3.2.3   Hot Topic Visualization

We choose the hot topic with highest HT value from five clusters with largest number of microblogs. Table 3 lists the five hottest topics on the test data, while table 4 lists terms with highest probabilities for each topic. It is clear that our method is able to discover meaningful topics with much better readability and the experimental results are consistent to our intuition.

Table 3. Hot topic with highest HT value from five clusters

Topic	Microblog
1	Happy New Year 2013!
	(祝大家2013年新年快乐! 欢度元旦节!)
2	Since many countries' policies are difficult to change, China's Internet should find reasons from itself.
	(中国互联网应从自身找原因，很多国家政策是难改变的，只能适应。)
3	In order to implement the value of its currency reduction goals, zero interest rate policy guides the domestic capital to high interest rates countries by the means of carry trade.
	(零利率政策主要借助套利交易，引导境内资本流向高利率国家，实现降低本币汇率的目标。)
4	Luo Yufeng is recently named the most shameless person in China by CNN, who has defeated Yang Erche Namu and sister Furong, following Xiao Shenyang who has achieved the China's most vulgar person. It seems that Americans appear to be fair on the issue of shame.
	(继小沈阳被美国国家有线广播电视台CNN评为中国最低俗的人之后，最近罗玉凤凤姐一路打败杨二车娜姆和芙蓉姐姐,被CNN评为中国最不要脸的人，看来美国人在判断不要脸这个问题上非常之公正。)
5	Yesterday, China's government would manage melamine by the real-name system. It's estimated that Mengniu is related to melamine. Anyway, we had better not drink milk. We should not believe the words that our management is good.
	(昨天国内高调报道中央要以实名制管理三聚氰胺。虽然不知道蒙牛到底出了什么事，但估计又是三聚氰胺惹的祸。反正牛奶是再也不能吃了。绝不能相信说管理得很好的话。)

Table 4. Terms with highest probabilities for each topic

Topic	Keywords				
1	New Year	Wish	Promised	Adjust	Notice
	(新年)	(愿望)	(许下)	(调整)	(公告)
2	Quantization	Lead	Carry out	Policy	domestic currency
	(量化)	(引导)	(实施)	(政策)	(本币)
3	World	Currency	Loose	Risk	Capital
	(世界)	(货币)	(宽松)	(风险)	(资本)
4	Entertainment	Luo Yufeng	Choose	Vulgar	Shameless
	(娱乐)	(凤姐)	(评为)	(低俗)	(不要脸)
5	Mengniu	Find	Report	Milk	Confiscate
	(蒙牛)	(发现)	(报导)	(牛奶)	(查收)

4   Conclusion

In this paper, we have described a new algorithm for hot topic detection on microblogs. While it is based on two stages clustering, the second stage Kmeans can fully take advantage of the prior information provided from the first stage pLSA. We also define the concept of topic hotness. In the future, we will take users' role into consideration.

Acknowledgement. Supported by the National Natural Science Foundation of China (61363058), the Scientific Research Fund for Colleges and Universities of Gansu Province (2013B-007,2013A-016), Youth Science and technology support program of Gansu Province (145RJZA232,145RJYA259), and Promotion Funds for Young Teachers in Northwest Normal University (NWNU-LKQN-12-14).

References

1. Sakaki, T., Okazaki, M., Matsuo, Y.: Earthquake shakes Twitter users: real-time event detection by social sensors. In: Proceedings of the 19th International Conference on World Wide Web, New York, USA, pp. 851–860 (April 2010)
2. Chen, J.F., Yu, J.J., Shen, Y.: Towards Topic Trend Prediction on a Topic Evolution Model with Social Connection. In: Proceedings of the 2012 IEEE/WIC/ACM International Joint Conferences on Web Intelligence and Intelligent Agent Technology, vol. 1, pp. 153–157. IEEE Computer Society, Washington (2012)
3. Chen, Y., Xu, B., Hao, H., Zhou, S., Cao, J.: User-defined hot topic detection in microblogging. In: Proceedings of the Fifth International Conference on Internet Multimedia Computing and Service, New York, USA, pp. 183–186 (August 2013)
4. Salton, G.: The SMART retrieval system—experiments in automatic document processing, Upper Saddle River, NJ, USA (1971)
5. Sahami, M., Heilman, T.D.: A web-based kernel function for measuring the similarity of short text snippets. In: Proceedings of the 15th International Conference on World Wide Web, New York, USA, pp. 377–386 (May 2006)
6. Yih, W.T., Meek, C.: Improving similarity measures for short segments of text. AAAI 7(7), 1489–1494 (2007)
7. Zhai, Y.D., Wang, K.P., Zhang, D.N., Hunag, L., Zhou, C.G.: An algorithm for semantic similarity of short text based on WordNet. Acta Electronica Sinica 40(3), 617–620 (2012)
8. Banerjee, S., Ramanathan, K.: Clustering short texts using wikipedia. In: Proceedings of the 30th Annual International ACM SIGIR Conference on Research and Development in Information Retrieval, New York, USA, pp. 787–788 (July 2007)
9. Ma, H.F., Wang, B.: Microblog Online Event Analysis Based on Incremental Topic Model. Computer Engineering 39(3), 191–196 (2013)
10. Hofmann, T.: Unsupervised learning by probabilistic latent semantic analysis. Machine Learning 42(1-2), 177–196 (2001)
11. MacQueen, J.: Some methods for classification and analysis of multivariate observations. In: Proceedings of the Fifth Berkeley Symposium on Mathematical Statistics and Probability, vol. 1(281-297), p. 14 (1967)
12. Sun, S.P.: Chinese microblog hot topic detection and tracking technology. Beijing Jiaotong University, Beijing (2011)
13. Sebastiani, F.: Machine learning in automated text categorization. ACM Computing Surveys 34(1), 1–47 (2002)

Collecting Valuable Information from Fast Text Streams

Baoyuan Qi[1,2], Gang Ma[1,2], Zhongzhi Shi[1], and Wei Wang[3]

[1] Key Lab of Intelligent Information Processing, Institute of Computing Technology,
CAS, Beijing 100190, China
[2] University of Chinese Academy of Sciences, Beijing 100190, China
[3] Beijing Lexo Technologies Co., Ltd. Beijing 100080, China
{qiby,mag,shizz}@ics.ict.ac.cn, waywang@hotmail.com

Abstract. It has become a challenging work to collect valuable information from fast text streams. In this work, we propose a method which gains useful information effectively and efficiently. Firstly, we maintain an analyzer based on the Trie structure and the dynamic N-Gram tokenizer; secondly, unlike the traditional search engine principle, we consider the documents as a query by building the indexes for the whole query base. The experimental results show that it has the strong adaption ability, low latency and high quality support for the complex query combination compared with the conventional methods.

Keywords: Fast Text Stream, Information Collection, Trie, N-Gram.

1 Introduction

Text streams is a kind of data stream, which is composed of texts, such as news feed, blog, weibo, etc.. Knowledge discovery from data stream has attracted more research recent years [1-4]. Data stream mining in the time and space constraints challenges traditional methods and algorithms[5, 6]. There is always concept drift in the data streams, which make the prediction accuracy decrease along with the time [7, 8].

It is always easy for a user to retrieve relative records from a static document set. The inverted document index methods are always fast for querying. However, under the streaming, many problems always arise in the evolving environments. The first one comes from the analyzer that has to be used during indexing and search process. There are always new words emerging and combining in the coming documents and querying, the system has to reindex all the documents periodically to help improve search accuracy. Another problem among the conventional practice is the response time for the query with the increasing size of documents. The time for a query will expand excessively when the index base receives endless incoming documents.

User may change his query at any time and he only concerns the results since the newly query is settled. The case always holds in the monitor systems like network traffic and public safety. Under these systems, the speed of incoming data is very fast and the queries waiting for processed are plenty. To tackle this problem, we propose a

Z. Shi et al. (Eds.): IIP 2014, IFIP AICT 432, pp. 96–105, 2014.

method to effectively collect valuable information from text streams which cost less time and space and dynamically update the query base in time. In this paper we make contributions on various aspects:

- We design a new flexible analyzer to automatically expand and quickly adapted to user defined dictionary. We use Trie tree data structure to insert new items and at the same time the max length are recorded for further N-Gram tokenizer. The dynamically maintained length is used to implement a maximum segmentation analyzer and the tree is used as a fast dictionary loo-kup table. The advantage is new entry can be handled as soon as new data ar-rives, so no data will be lost. Still this can reduce the index size as we only use the user defined term to index the incoming documents.

- A new angle to collect valuable information from the endless text streams. Since there is always the time consuming and space demanding to index all the documents when user indexes the query base set. While constructing the query index, the analyzer changes its parameters accordingly. Newly coming document are tokenized by the analyzer and then searched in the index repo-sitory. The advantage is that each query is often short and simple; it is often easy to rebuild query index.

2   Related Work

Information retrieval (IR) obtains information relevant to a need from many docu-ments. Many models have been proposed like Vector Space Model(VSM)[9], Okapi BM2.5, etc.. The web search engines like Google, Baidu are the most significant ap-plications. The indexed documents are always static and need to be reindexed once new terms are used for indexing. To return the result in time, there is always query length limit in the input box as long as the operation between the query terms, like OR, AND, etc.. Batch-incremental index mechanism like Zoie[1] is always used in dynamic environment. The newly coming documents come first into the small index, it will merge when the small index becomes big enough. This strategy is helpful when the process time of the merge operation doesn't take too much time and it will hang the system when the index size goes larger.

The terminology prospective search[10]   is widely used in text stream. There are a lot of applications based on this method, like Google News Alerts, Google Search. Rather than targeting full text search of infrequent queries over huge persistent data archives (historic search), this class targets full text search of huge numbers of query-ing over comparatively small transient real-time data. Our method extends this idea and we propose dynamical N-Gram tokenizer to reduce the index size and reduce the response time.

Trie tree, as a data structure makes its good performance in unknown words and variant length conditions[11, 12]. In this paper we use Trie tree to maintain the added

---

[1] http://javasoze.github.io/zoie/

terms extracted from the query phrases and make a fast lookup for prefix check. N-Gram, as another useful technology, is always used in computational language and translation[13, 14]. We integrate these two technologies to construct a dynamic analyzer that can be further used for indexing and searching.

3   The Proposed Method

In this paper we propose a method to effectively collect valuable information from text streams and it is composed of two key elements supporting the idea: the first one is maintaining an analyzer that adjust dynamically to the query and the maximum length used for N-Gram tokenizer; secondly, based on the tailored analyzer, the method collecting valuable data proposed by us is different from the conventional ones.

3.1   Dynamic Analyzer

The analyzer is essential for tokenizing the document in inverted document search system. The conventional analyzer is always static and doesn't keep pace with the input documents. Under the text stream environment, the data flow is always endless and the processing server should never stop. This causes serious problems including new words that will hurt the performance of the system. The common solution is to periodically rebuild all the documents or to use batch-incremental indexing method. These solutions will incur the high latency of the accurate results, the memory and space cost may be heavily grown along with the time goes by. The size of the data index is always expanding excessively and the valuable data always hides from normal search strategy.

To cope with the problems, we build and maintain a dynamic analyzer. We integrate the commonly used Trie data structure and N-Gram terminology to gain an effective and efficient analyzer.

Trie data structure coms from the word 'Retrieval' and it is a string tree. Its philosophy is to reduce time complexity by improving space complexity. The operations like INSERT and QUERY can be processed instantly and so it is commonly used as a lookup tables especially in PREFIX/SUFFIX computation settings. As an example to build a Trie, suppose the input terms are $T=[b, abc, abd, bcd, abcd, efg, hii]$, the constructed Trie is illustrated in Figure 1. The red circle indicates a complete term through the root, the white circle indicates an intermediate node.

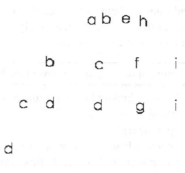

Fig. 1. An Illustrative Trie Example

From Figure 1 we can see it is easy to retrieve all term that begin with a specified prefix, for example *b* and *bcd* are all begin with letter *b*. Trie can be used in suggestion complete to improve the search experience. In this paper we use them as a lookup table to store and search the query base.

N-Gram partitions the input into N continuous segments used for predicting in the form of (N-1)-Markov. N-Gram is very simple to understand and easy to scale up. We can simply increase the parameter *N*, then a model can be used to store more context with a space–time tradeoff, enabling experiments to scale up very efficiently. As an example to generate an N-Gram sequence, suppose the sample *S=ABCA*, then the 1-gram sequences (also known as *unigram*) are [*A, B, C, A*], 2-gram sequences (also known as *bigram*) are [*AB, BC, CA*]. In this paper the base parameter *N* is set during the Trie construction process, so, it does not need predefined by an expert.

We implement the dynamic trie update algorithm in details in Algorithm 1. We sort the query to the leaf of the trie and update the maximum length of the query along with the construction process.

Algorithm 1. Dynamically update the Trie

Input: Given an empty or already existed Trie *trie* and query set *qset*.
Output: An updated *trie* and maximum length *N* in the query base.
    1. Extract all meaning and single words from the *qset*. We do this because there may be operators and quoted keywords. For simplicity, all the quoted letters are extracted to the set *trimedQSet*.
    2. **if** *trie* is empty **then**
    3.   Initialize *trie*;
    4.   *N*=0;
    5. **for** each *trimedQ* in *trimedQSet* **do**
    6.   Use *trie* to *trimedQ* to the *trie* leaf;
    7.   **if** the length of *trimedQ* is larger than *N* **then**
    8.   *N:=length(trimedQ)*;
    9. Output *trie* and *N*.

We segment the upcoming document according to the trie obtained in Algorithm 1, then extract the hit term and transfer them to the indexing or searching process. Algorithm 2 details how to segment the document accordingly. We segment the document content and return all the tokens existing in the trie.

Algorithm 2. Segment a document using analyzer based on N-Gram

---

Input: Given a document *doc*. *trie*, lookup tables for token search. N, the current maximum length among all the queries.
Output: The tokens existing in the trie.
1. Extract the content *content* from *doc*. For indexing there may be other meta information, like *timestamp, author* etc.. In this step we just use content part like *title* and *text*.
2. *tokens*=[];
3. **for** i:=0 to length(*content*) **do**
4.   **for** j:=i+1 to i+N **do**
5.   token:=*content*.substring(i,j);
6.   **if** *trie*.contains(*token*) **then**
7.   *tokens*.add(token);
8. Output *tokens*.

In this part, we give two algorithms to show how to update trie and segment the document using N-Gram. The dynamic analyzer can insert the query fast and segment the input document, which makes the index size smaller than conventional method and collects valuable information according to the condition user specified.

3.2   Information Collecting Method

In the dawn of big data, the speed of the content generated grows fast. In some specific domains, people always want to focus the valuable information and filter out the information not relating to them. The traditional IR base methods require operating on the full data don't adapt to the data flow format. Another problem is the complex query conditions that will cost too much time to compute and load all the relative record from the index.

We propose an information-collecting method from another angle based on the previous dynamic analyzer. In the old-fashioned roadmap, documents are collected and indexed by the analyzer, then, the query are sent to the index repository, the possible BOOLEAN operation goes and the different RANK algorithms are used to sort the relative documents. Our work, however, index all the user query beforehand, we consider the incoming document as a query, if the score is larger than zero, we will collect it and store it, otherwise drop it. The score is a similarity between the query and the document, we could use VSM model to help calculating it easily. We illustrate the logic in Algorithm 3.

Algorithm 3. The method to collect valuable information from text streams

---

Input: Given endless text stream *ts. analyzers*, generated by each query.
Output: Add the valuable document to its own list.

   1. **while true do**
   2.   Pick up a document *document* from the text stream;
   3.   **for** each *query* in the *querySet* **do**
   4.   Segment *document* using the corresponding *analyzer*, we get *SEG;*
   5.   *Similarity_Score:=sim(query,SEG);*
   6.   **if** *Similarity_Score* > 0 **then**
   7.   Add document to the document list of this query.

We assume there are plenty of query sets and they are complex, which is very common in the platform of monitoring system, like traffic control of network, sensitive information detecting for a corporation. We obtain the valuable information through checking the each document to the query base from which the score is calculated as a metric for similarity distance.

4   Experimental Results

We conduct experiments on real data and make comparative results according to various settings.

4.1   Data Sets

We use Chinese patent data to evaluate the method proposed in this paper. The date range of the data is between 1985 and 2011 years. We extract all the keywords from the raw data and remove duplicated ones, and we get 350,000 distinct keywords. To simulate the text stream, we sort the document by the date and then iterate them in order.

4.2   Measurements of Trie Implementations

Since Trie is essential for our method, we evaluate the state-of-the-art implementations. The patricia-trie(for short PAT)[2] and concurrent-trees (for short CT)[3] are used, and we compare the time cost and memory cost in each insertion with the number of keywords.

---

[2] https://code.google.com/p/patricia-trie/
[3] https://code.google.com/p/concurrent-trees/

All the keywords are inserted under the same condition; the results are illustrated in Figure 2. The time and memory cost change with the growth of the dictionary number. It is clear from the figure that PAT is always better than CT on both time and memory factors.

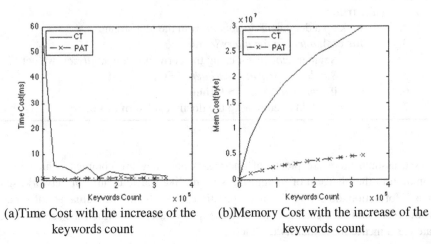

| (a)Time Cost with the increase of the keywords count | (b)Memory Cost with the increase of the keywords count |

Fig. 2. Comparison of Time Cost and Memory Cost between CT and PAT with the Number of Keywords

4.3   Performances of Analyzers

We evaluate our proposed dynamic analyzer(for short DA) with ICTCLAS[15] analyzer. ICTCLAS is based on HHMM and is considered as a baseline for many analyzers. In this paper we focus on not its performance but the index growth ratio and time cost about indexing speed with documents increase.

We use Lucene[4] as an index engine. The document set is constrained to the subset of the total data. We used all the CLAIM and KEYWORD field in the patent during 2012s. We got totally 219,144 patents and 105744 keywords for PAT and the final maximum length is 7. Additionally, the maximum number can be adjusted according to the current dictionary and is able to cover the entire possible window for indexing.

Figure 3 illustrates the results, from the figures we can see our proposed analyzer DA could reduce the index size drastically and still it could reduce the index latency in each update. We should notice that the keyword is added dynamically when the patent document arrives; it proves the applicability of the dynamic mechanism proposed in this paper.

---

[4] http://lucene.apache.org

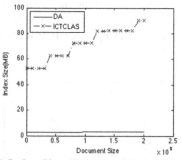

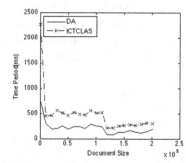

(a) Index Size changes with the increase of document size   (b) Time Period changes with the increase of document size in each update

Fig. 3. Comparison of Index Size and Time Period between DA and ICTCLAS when Index Changes with the Document Size

4.4   Performances of Collecting Methods

We compare our proposed method (for short *Stream Search*) and the traditional batch-incremental method (for short *BI Search*). We test totally 40,000 patent documents. For text stream, the documents are sorted chronologically and are sent in order. For batch-incremental method, the size of documents is 1000 for each batches and the small index is merged to the big index. The Figure 4 gives an illustration of the two methods in time cost, what we can see is our method reduces the time cost to half of the batch-incremental method.

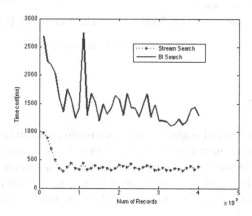

Fig. 4. Time cost between Stream Search and BI Search with the document Size

To evaluate the soundness when the query is very complex and the combination of different joint operations, we increase the number of keywords and use OR and AND operation to evaluate the time cost for the batches with different document size, the results are listed in Figure 5. From the figures, we can see that the time increases linearly with the number of keywords, obviously, this phenomenon is reasonable in a practical production environment. Additionally, the AND operator is always cost much time than the OR operator.

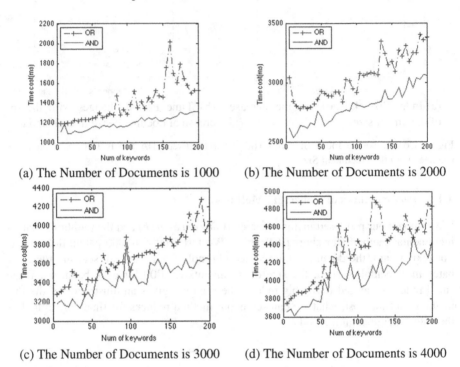

(a) The Number of Documents is 1000   (b) The Number of Documents is 2000

(c) The Number of Documents is 3000   (d) The Number of Documents is 4000

Fig. 5. Comparison of Time Cost for OR and AND Operators with New Documents Added to the Text Stream

5   Conclusions

In this paper, we propose and evaluate a new information-collecting method from fast text streams. Our experimental results indicate that dynamic analyzer based on N-Gram performs well in the speed for tokenizing, smaller index size and its automatic adaption character. The method considering the query set as index base and the incoming document as query can significantly reduce the time cost in fast text stream environment.

Acknowledgments. The work was financed by the National Program on Key Basic Research Project (973) (No. 2013CB329502), National Natural Science Foundation of China (No. 61035003, 61202212, 61072085), National High-tech R&D Program of China (863 Program) (No.2012AA011003), National Science and Technology Support Program (2012BA107B02).

References

1. Gama, J., et al.: Knowledge discovery from data streams, Citeseer (2010)
2. Aggarwal, C.C.: Data streams: models and algorithms. Springer (2006)
3. Graham, C., et al.: Synopses for Massive Data: Samples, Histograms, Wavelets, Sketches. Found. Trends Databases 4(1-3), 1–294 (2012)
4. Muthukrishnan, S.: Data streams: Algorithms and applications. Now Publishers Inc. (2005)
5. Li, M., et al.: Time and space efficient spectral clustering via column sampling. In: 2011 IEEE Conference on Computer Vision and Pattern Recognition (CVPR). IEEE (2011)
6. Zhang, Y., et al.: Space-efficient relative error order sketch over data streams. In: Proceedings of the 22nd International Conference on Data Engineering, ICDE 2006. IEEE (2006)
7. Xioufis, E.S., et al.: Dealing with concept drift and class imbalance in multi-label stream classification. In: Proceedings of the Twenty-Second International Joint Conference on Artificial Intelligence, vol. 2. AAAI Press (2011)
8. Gama, J., Medas, P., Castillo, G., Rodrigues, P.: Learning with Drift Detection. In: Bazzan, A.L.C., Labidi, S. (eds.) SBIA 2004. LNCS (LNAI), vol. 3171, pp. 286–295. Springer, Heidelberg (2004)
9. Salton, G., Wong, A., Yang, C.-S.: A vector space model for automatic indexing. Communications of the ACM 18(11), 613–620 (1975)
10. Irmak, U., et al.: Efficient query subscription processing for prospective search engines. In: Proceedings of the 15th International Conference on World Wide Web. ACM (2006)
11. Kanlayanawat, W., Prasitjutrakul, S.: Automatic indexing for Thai text with unknown words using trie structure. In: Proceedings of the Natural Language Processing Pacific Rim Symposium (NLPRS 1997) (1997)
12. Kijkanjanarat, T., Chao, H.: Fast IP lookups using a two-trie data structure. In: Global Telecommunications Conference, GLOBECOM 1999. IEEE (1999)
13. Doddington, G.: Automatic evaluation of machine translation quality using n-gram co-occurrence statistics. In: Proceedings of the Second International Conference on Human Language Technology Research. Morgan Kaufmann Publishers Inc. (2002)
14. Brown, P.F., et al.: Class-based n-gram models of natural language. Computational Linguistics 18(4), 467–479 (1992)
15. Zhang, H.-P., et al.: HHMM-based Chinese lexical analyzer ICTCLAS. In: Proceedings of the Second SIGHAN Workshop on Chinese Language Processing, vol. 17, p. 2003. Association for Computational Linguistics (2003)

An AUML State Machine Based Method
for Multi-agent Systems Model Checking

Dapeng Zhang[1,2], Xiang Ji[1], and Xinsheng Wang[1]

[1] Institute of Information Science and Engineering, Yanshan University,
Qinhuangdao, 066004, China
[2] Guangxi Key Laboratory of Trusted Software, Guilin University of Electronic Technology,
Guilin, 541004, China
daniao@ysu.edu.cn, 907887343@qq.com, wxs@ysu.edu

Abstract. This paper firstly proposes a Multi-agent System Model Checking
Framework, which is based on AUML (Agent Unified Modeling Language)
state machine model and temporal logics of knowledge and provides a method
using AUML state machine for Multi-Agent Systems formal modeling. Then a
method for the conversion from AUML state machine formal description
to ISPL language is proposed. Finally a simulation is accomplished with the
conversion tool AUML2ISPL.

Keywords: Model Checking, MCMAS, AUML state machine, ISPL, Multi-
agent System.

1   Introduction

Currently, agent and Multi-agent has become the focus of the research in artificial
intelligence. Multi-agent System has been used more widely in computer and related
fields than before. So it has become a very urgent issue to ensure the correctness and
reliability for Multi-agent System. As a method of automatic verification for finite-
state system, model checking [1] has drawn wide attention. The basic idea for model
checking is to establish finite state models for the system that is to be verified,
using some logical formula to describe the specification properties of the system
which are expected, then put them into model checking tools to verify whether the
model meets the specification properties by automatic checking. Model checking has
a high degree of automation implementation and it can provide a counter-example
path when the system does not meet the specification properties. So it can decrease
the workload for researchers and it is also convenient for researchers to trace the error
path for the system.

2   Related Work and the Issue

Since the idea of model checking has been proposed, many researchers have paid a lot
of attention for the consistency of model checking for Multi-agent System. Dong [2]

Z. Shi et al. (Eds.): IIP 2014, IFIP AICT 432, pp. 106–112, 2014.

has promoted a method about formal description of state machine use UML to defining the extension hierarchical automata and the label transfer system, then verified the correctness of the UML state machine model based on this method. Luo [3] has provided a method for automatic detection of the interaction Web service, and modeled the MTELG by 5-tuple, then proposed a rule to make the 5-tuple formal description automatic convertion to the input language for the model checking tool MCTK. Du [4] has promoted a method for UML consistency and tested case generation based on model checking. Abdel and Gherb [5] have provided a consistency test framework for UML/SPT model. This framework gives a method to verify the model state chart for UML/SPT, Which includes the consistency of syntax and semantics.

For Multi-Agent System, although UML is a popular modeling language, it can't describe the agents' knowledge and their collaborations well. Therefore scientists have extended the UML language and proposed a language for modeling Multi-agent System, which is called AUML (Agent Unified Modeling Language). Although scientists have developed a number of model checking tools, such as MCK [6], MCMAS [7] and MCTK [8], the input language for those tools is incompatible with UML, especially AUML.

3   A Model Checking Framework for Multi-agent Systems

In order to solve the problem proposed in section 2, we proposed a Multi-agent Systems Model Checking Framework that is based on AUML state machine model and temporal logics of knowledge [9]. Fig.1 shows this framework.

In this framework, we choose MCMAS as the Model Checking tool. It is because MCMAS is more efficiency, which is based on the logic of CTLKD-A^{DC}, It can verify the knowledge for Multi-agent System, and it has the ability to verify the correctness behaviors of agents and the collaboration among agents for Multi-agent System. MCMAS uses ISPL (Interpreted Systems Programming Language) [10] as the input language.

This Model Checking framework can be achieved in the following steps:

(1)Symbolic the Multi-agent System model to AUML state chart.

(2)Formalize the AUML state chart into AUML state machine formal description.

(3)Making a syntax conversion between AUML state machine formal description and the ISPL language.

(4)Formalized system specifications into CTLKD-A^{DC} formulas, respectively put the CTLK-A^{DC} formulas and the ISPL description that the third step obtained into MCMAS to check whether this model satisfied the formulas.

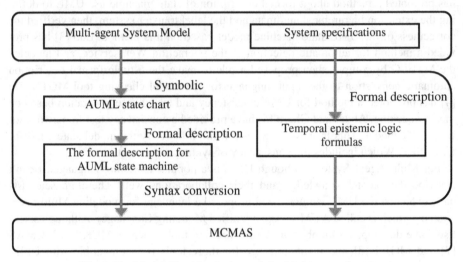

Fig. 1. The Model Checking framework for Multi-agent System

4   The Conversion from AUML to ISPL Language

4.1   The Formal Description of AUML State Machine

AUML state machine model is very important. It can reflect specific circumstances from the interaction of agents. We extend the AUML state machine description syntax in reference [11]. Firstly we add the observed variable into AUML state machine description. Secondly we redefine the transition rules. The syntax for AUML state machine description is as follows.

$StateChartModel \triangleq agentname: String; states: StateSet;$
$\quad\quad actions: Actionset; goals: GoalSet; transitions: TransitionSet$
$StateSet \triangleq \{State\}$
$ActionSet \triangleq \{Action\}$
$GoalSet \triangleq \{Goal\}$
$TransitionSet \triangleq \{Transition\}$
$State \triangleq statename: String; statetype: StateType;$
$\quad\quad boolformula: BoolFormulaSet$
$BoolFormulaSet \triangleq \{BoolFormula\}$
$BoolFormula \triangleq var: Var; relation: Relation; value: Value$
$Var \triangleq \{Variable\}$
$Relation \triangleq \{'\leq'; '\geq'; '\neq'; '>'; '<'\}$
$Value \triangleq \{Int; String; Float; Bool\}$
$StateType \triangleq \{Initial; Final; Common; FaultState\}$

Transition description is the instance of the connection of two states. The rules consist of the triggered state transitions preconditions and the results after transitions. The preconditions include variable assignments for the state before the state transition and the other agent actions that triggered state transition (represented by the message receiving events), while the results include variable assignment after state transition and the agent collection that other agents receive messages from. Their syntax is described below.

$Rule \triangleq condition: Cdt; agentaction: AtentAction; result: Rst$
$Cdt \triangleq \{BoolFormula\}$
$AgentAction \triangleq agent: Agent; action: Action$
$Rst \triangleq \{varassign: VarAssignSet \}; \{agent: Agent \}$
$VarAssignSet \triangleq \{VarAssign\}$
$VarAssign \triangleq var: Var; value: Value$

4.2   The Conversion from AUML State Machine Formal Description to ISPL Language

Here we decompose the description of state transition for AUML state machine, and give protocol description and evolution description for ISPL respectively. Finally we propose a conversion algorithm between AUML state machine and ISPL language.

The description of protocol is represent by protocol functions: protdef::= Protocol : protdeflist. Here protdef is composed of a condition Protocol and a list of executable actions protdeist.

The evolution is represent by evolution functions: evline::= boolresult if gboolcond. Here each evolution function is composed of a group of boolresultes and the preconditions for evolution gboolcond.

The conversion algorithm is given in Algorithm 1.

5   Simulation

5.1   The State Machine Formal Description for Transmission System

The bit transmission system [12] contains two agents (Receiver, Sender), which is shown in Figure 2. One protocol to achieve communication is as follows: S immediately starts sending a bit to R, and continues to do this until it receives an acknowledgement from R. R does nothing until it receives the bit. Then it sends acknowledging to S. S stops sending the bit to R when it receives the first acknowledgement from R, and the protocol terminates. We encode each agent for the bit transmission system in AUML state machine formal description respectively. Here we give part of each agent encoding result in AUML state machine formal description.

Algorithm 1. Conversion from AUML state machine formal description to ISPL language

---

1. Input: StateChartModel
2. Output: Protocol, Evolution
3. For each transition ∈ StateChartModel.transitions
4.   if transitions.source = state, then protdeflist = protdeflist + ' ' +transition.rule.act.name + ',';
5. For each boolformula ∈ state.boolformula, order protocol = protocol + ' ' + boolformula.vas.name + ' ' + relation + ' ' + String (boolformula. vas.value) + ' and ';
6. Add protdef into Protocol;
7. For each transition ∈ StateChartModel.transitions
8.   Output protocol;
9. For each vasassign ∈ transition.rule.result. vasassign, order boolresult = boolresult + '(' + vasassign.vas.name + ' = ' + String (vasassign.value) + ') ' + ' and ';
10. gboolcond = ' if ', for every boolformla ∈ transition.rule. condition.boolformula, order gboolcond = gboolcond + '(' + boolformula.vas.name + ' ' + relation + ' ' + String(boolformula.value) + ' and ';
11. gboolcond = gboolcond + transition.rule.condition.action + ')';
12. evline = boolresult + gboolcond;
13. Add evline into Evolution;
14. Output: Evolution

For the Sender, its state is expressed by two variables: bit and ack. An emulation type bit encodes the value of the bit the Sender wants to send, which consists b0 and b1. Ack is a boolean variable encoding whether or not an ack has been received from Receiver. If the Sender receives the ack, ack equals true. The actions collection for Sender is expressed by Actions which consists sb0, sb1 and noting, respectively indicating the value of the bit the Sender sends. There are two initial state situations for the Sender: transmit b0 or transmit b1. Here we encode one state transition for Sender as follows.

　　　　Transition = S00, S1; rule : Rule;

　　　　Cdt = { boolformula: b0 }

　　　　Agent : Environment, Receiver;

　　　　Action : (Receiver.Action = sendack and Environment.Action = R)
or (Receiver.Action = sendack and Environment.Action = SR);

　　　　Rst = {(b0, ack = false) }; { agent: Receiver, Environment }

　　　　For Receiver, one enumeration variable is declared for the agent, representing whether the bit has been received or not and the value it's received, the collection of enumeration variable is {r0, r1, empty}. The collection of actions is {sendack, noting}. Sendack indicates Receiver send an acknowledgement to Sender.noting indicate Receiver don't send anything. Here we encode one state transition for Receiver as follows.

Transition = R00, R0 ; rule : Rule;

Cdt = { boolformular : empty };

Agent : Sender , Environment;

Action : (Sender.Action = sb0 and Receive.state = R00 and Environment.Action = S) or (Sender.Action = sb0 and Receiver.action = noting and Environment.Action = SR)

Rst = { boolformula: r0 }; { agent: Environment, Sender }

5.2   Specification Verification

We use the CTLK-A^{D,C} formula to formalize the specifications of bit transmission System. It can describe the correctness of actions and the knowledge for the Agents. The specification is usually based on system requirement. Here we manually input the following specifications.

1、E((bit0 or bit1) U recack);

2、EF(EG(recbit and !recack));

3、AF(recack);

4、AF(recbit -> AF (recack));

5、AF(recbit -> K(Sender,K(Receiver,bit0) or K(Receiver,bit1)));

Formulas 1-4 are CTL formulas and they can describe the correctness actions for agents. Formula 5 is CTLK-A^{D,C} formula and it can describe the reasoning ability for agents. For example, formula 1 means if agent Sender sends message, it can get the acknowledgement. Formula 5 means it is always true. If an acknowledgement is received, then the sender knows that the receiver knows the value of the bit.

5.3   Simulation

We realize the conversion tool AUML2ISPL according to algorithm 1. Then we put the whole AUML state machine formal description for bit transmission System into this tool, and get the conversion result as Fig.2 shows.

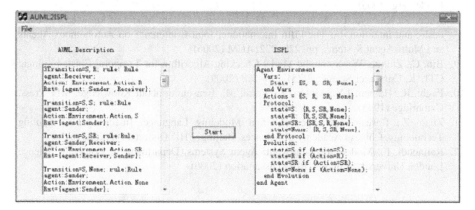

Fig. 2. The conversion result from AUML2ISPL

We put the conversion result and the system specifications into MCMAS and run a simulation. The model checking result shows that out modeling is correct.

6   Conclusion

This paper uses AUML language for the modeling language of Multi-agent System, proposes a method for AUML state machine formal description and the conversion algorithm from the AUML formal description to ISPL, and the conversion tool is realized. Simulation shows that out method is feasible.

Acknowledgements. This work is supported by the Natural Science Foundation of China (No. 61303129) and Open Foundation of Guangxi Key Laboratory of Trusted Software (No. 201211).

References

1. Huimin, L., Wenhui, Z.: Model Checking: Theories, techniques, and applications. Acta Electronica Sinica 12(30), 1906–1912 (2002)
2. Dong, W., Wang, J., Qi, X., Qi, Z.C.: Model checking UML statecharts. In: Proc. of 8th Asia-Pacific Software Engineering Conference (2001)
3. Luo, X.Y., Tan, Z., Dong, S.R.: Automatic Detection Method for Web Services Feature Interaction. Computer Science 37(12), 106–110 (2010)
4. Du. UML consistency and test case generation based on model checking. College of Computer Science and Technology, NanJing, Master Dissertation (2011)
5. Gherbi, A., Khendek, F.: Consistency of UML/SPT Models. In: Gaudin, E., Najm, E., Reed, R. (eds.) SDL 2007. LNCS, vol. 4745, pp. 203–224. Springer, Heidelberg (2007)
6. Gammie, P., van der Meyden, R.: Mck: Model Checking the logic of Knowledge. In: Alur, R., Peled, D.A. (eds.) CAV 2004. LNCS, vol. 3114, pp. 479–483. Springer, Heidelberg (2004)
7. Lomuscio, A., Raimondi, F.: MCMAS: A model checker for multi-agent systems. In: Hermanns, H., Palsberg, J. (eds.) TACAS 2006. LNCS, vol. 3920, pp. 450–454. Springer, Heidelberg (2006)
8. Kaile, S., Xiangyu, L., Scatter, A.: The Interpreted System Model of knowledge, belief, desire and intention. In: The Fifth International Joint Conference on Autonomous Agents and Multi-Agent Systems, pp. 220–222. ACM (2006)
9. Bin, C., Zhixue, W.: Symbolic Model Checking Algorithm for Temporal Epistemic logic CTL*K. Computer Science 36(5), 214–219 (2009)
10. Fagin, R., Halpen, J.Y., Moses, Y., Vardi, M.: Reasoning about Knowledge. MIT Press, Cambridge (1995)
11. Zhao, Z.: Research on Agent Unified Modeling Language. Institute of Computing Technology, Chinese Academy of Sciences, Beijing, PHD Dissertation (2003)
12. Raimoudi, F.: Model Checking Multi-Agent Systems, Department of Computer Science London University, London, PHD Dissertation (2006)

Adaptive Mechanism Based on Shared Learning in Multi-agent System

Qingshan Li, Hua Chu, Liang Diao, and Lu Wang

Software Engineering Institute, Xidian University, Xi'an 710071, China
{qshli,hchu}@mail.xidian.edu.cn,
{diaoliang,Lu_Wang}@stu.xidian.edu.cn

Abstract. In view of the deployment environment of the adaptive system is complex, dynamic, unpredictable, focusing on the construction of dynamic, uncertain environment adaptive system, and this paper combines the reinforcement learning technology and software agent technology to propose an adaptive mechanism based on shared learning in multiple agent system. Based on this, framework for constructing adaptive systems and shared learning algorithm of agent are given. Finally, by conducting a comparative experiment and result analysis to verify the feasibility of the theory put forward by this article.

Keywords: adaptive mechanism, shared learning, Multi-agent System, system framework.

1 Introduction

In the key application areas of software system, system environment is very complex, and the specific performance features are dynamic, changeable and difficult to control [1]. Software systems with self-adaptive capacities can adapt to the changing environment better [2]. In dynamic and uncertain environment, traditional techniques have been difficult to meet the development needs of self-adaptive systems [3]. In view of the above application requirements and technical challenges, this article uses the following two techniques to support the development of self-adaptive systems: Agent-based technology and reinforcement learning technology.

The second chapter of this paper introduces the self-adaptive system development framework; the third chapter focuses on the description of self-adaptive mechanisms based on shared learning, including self-adaptive processes and shared learning algorithm; the fourth chapter verifies the share learning-based self-adaptive mechanism proposed in this paper, combining case analysis; the fifth chapter summarizes the full text and discusses the further research.

2 Supporting Framework for Self-adaptive System Development and Running

By referencing FIPA specifications [4], we put forward the support framework of self-adaptive system for development and running as shown in figure 1. The entire frame consists of two core parts: supporting platform and development tools.

Z. Shi et al. (Eds.): IIP 2014, IFIP AICT 432, pp. 113–121, 2014.

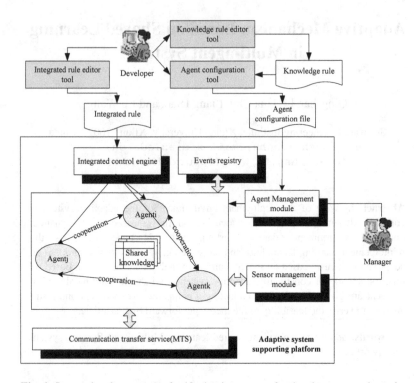

Fig. 1. Supporting framework of self-adaptive system for development and running

The supporting platform provides communication services, event services, and capabilities of the agent lifecycle management, the interaction of distributed multi-agent system and the perception of agent environmental.

Integrated control engine loads the user-specified integrated rules and distributes them to agents, so that the agents can collaborate independently under the guidance of integrated rules. Event registry maintains a list of all events in the system, which agent is interested in, and notifies the agent when the event occurs. Agent library management tool is a special agent in the platform. It is primarily responsible for the static management of local agent and the instantiation of agent.Sensor management module manages the different sensors, therefore the agent can dynamically load different sensors to active perception information of non-active in environmental.

Developers can develop needed configuration files for the system by using self-adaptive system development tools. User also can define collaborative relationships between agents by using a scripting language or graphical way. And user can abstract knowledge rules inside the appropriate agent, using scripting language describe knowledge rules to control the agent to process information automatically, so that the system can complete the decision-making process fast and accurately. Agent configuration tool mainly is used for defining simulation Agent basic information which is required.

3   Self-adaptive Mechanism Based on Shared-Learning

Reinforcement learning is considered to be one of the core technologies to design the intelligent agent [5]. To cope with unpredictable, dynamic environment in the process, self-adaptive system based on reinforcement learning will interact with the environment continually at runtime, and judge self-adaptive operations performed in a different state to formed knowledge.

3.1   Self-adaptive Process Based on Shared-Learning

In the self-adaptive system, agent is constantly monitoring the environment and sensing the current state of the environment, thus constantly triggering "perception-select-do-learning" process. At the same time, knowledge table is also constantly updated and maintained, which enables the agent to choose actions more responsive to environmental change according to knowledge table, improving effects that agent adapt to environment the gradually.

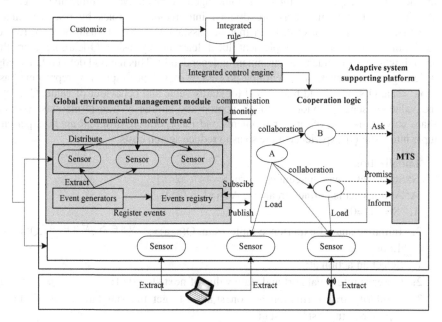

Fig. 2. Self-adaptive process based on shared-learning

Self-adaptive process based on shared-learning is shown in figure 3, the whole process can be divided into the following four steps:

Step1: (sensing stage) Agent interacts with the environment in real time to access the environmental information.

Step2: (decision-making stage) Agent sends the state of the environment (assume that the state is s) to behaviour selecting unit after perceiving environmental change; behaviour selecting unit selects a most appropriate behaviour (assume the action is b) from multiple behaviours which might be performed according to selecting policy and submitted to the execution units.

Step3: (implementation stage) Execution unit executes behaviour directly or executes by calling a number of external interfaces (such as communication interfaces, and so on). Implementation of behaviours may affect the state of the environment;

Step4: (learning stage) Learner is learning unit of self-adaptive agent, Learner has learning algorithm internal. Using state of environment and environment return value of action as entered, learner uses learning algorithm to learn optimal acts policy of self-adaptive agent in environment states, updates Q value that implement acts b at states. When the environment changes, self-adaptive agent calls actions or behaviours that are defined in the agent to adapt to the changes of environment according to knowledge exist in learner.

3.2   Algorithm of Shared-Learning

In the shared experience tuple of multi-agent cooperative reinforcement learning algorithms, all team members are divided into reciprocal roles by role assignment procedure. The task that each team member assumed is isomorphism and not rely on each other. Each member implement local learning by using Q-learning algorithm according to return value of environment independently. This method defined knowledge which agent accumulate in Q-learning process as experience tuple, it is represented as a triple <s, a, Q(s, a)>, s and n are stand for status value and behaviour of value, Q<s, a>is the value of state-behaviour. Though appropriate similarity transformations, you can share solving method of similar problems and incorporating different agent experience tuple into a shared-experience tuple so as to reduce search costs.

Initialization:
S: States set is which agent concern, S= {s1, s2 ...};
A: Actions set is defined in agent, A= {a1, a2 ...};
1. S= s0,t=0
2. Initialize shared-experience tuple <s, a, Q(s, a)>: $\forall s \in S$, $\forall a \in A$,Q(s, a)=0
    Loop:
1. Select an action at;
2. Observe return value rt and state value of next step st+1;
3. similarity transformation act on st,at,st+1, get the standard form of shared-experience tuple st*,at*,st+1*;
4. update shared-experience tuple to<s t*, a t*, Qt '(s t*,a t*)>:
    Qt '(s t*,a t*)=(1-αt) Qt-1 '(s t*,a t*)+αt(rt+γQt-1 '(s t+1*,πQ(s t+1*)))
    πQ(s t+1*) is optimization policy based on experience under state value s t+1*;γ is discount factor;
    Learning rate is visits of state-behaviour;
5. t:=t+l, st:= st+1 repeat step 1-4 until reach final state to complete a learning process;
6. Reset initial state of agent, repeat step 1-4 until all the values in the table of sharing-experience tuple become convergence.

In the initialization part of the algorithm, we build the Q table and initialize the Q value of each state-behaviour to 0. Of course, Q can also be set into a more meaningful value for experienced designers to speed up the convergence of learning. When algorithm get into the loop part, it will execute the process "choice behaviour-implement behaviour-receive return value-watch the new state-update Q table-set up the next loop starting state-choice behaviour ...-update Q table ..." repetitively. When all the values in the table of sharing-experience tuple become convergence, the loop is end.

4   Self-adaptive Mechanism Based on Shared-Learning

In this chapter, we implement an air defence command and control system to intercept intrusion target. As shown in Figure 3, our three UVAs are flying in the area near the border and decide flight behaviour according to the current location of the intrusion target. Learning trail of shared learning algorithm is the return value from the simulated environment of simulation system and three aircraft shares experience tuple what they have learned.

Fig. 3. Situation maps of UAV formation intercept invasion target

4.1   System Running

When agent initialization is completed, each agent receives information from the outside and works in terms of internal knowledge rule file. The generated messages is sent out though collaborative relationship. Agents finish the blocking process through collaborative interaction. Agent communication processes is shown in Figure 4.

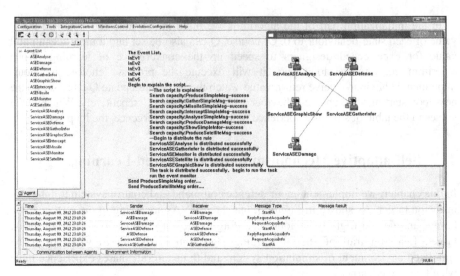

Fig. 4. Interface of explain and distribute integrated script

4.2   Results and Analysis

Set the appropriate parameters for the problem of interception of UAVs to make shared learning algorithm meet the required conditions. Experimental basic conditions are as follows, a series of successor experiments have been conducted on the basis of change different basic conditions.

(1)The UAV and intrusion target selecting one walking behaviour randomly, behaviour collections are represented as {eastbound, westbound, northbound, southbound, stationary}; (2)The UAV has the ability to learn and make independent decisions respectively, by receiving feedback information of environment to improve its policy; (3)Do not have shared experiences tuple; (4)Assume that initial Q value of each state-behaviour to 0;(5)Instantaneous return functions which environment feedbacks to the UAV are as follows:

Table 1. Return value which environment feedbacks to the UAV

	arrive at the target	the distance to target reduce one step	the distance to target is decreasing	the distance to target keep unchanged	the distance to target increase one step	the distance to target is increased
r	300	100	100	0	-1	-10

This section verifies the effectiveness of shared learning and different return values impact of the system learning though two contrast experiment.

Comparative experiment 1 UAVs execute task of blocked, comparison curve of moving times between UAVs which does not use shared-learning and UAVs which use shared-learning.

The result is shown in Figure 5. In the figure, horizontal represents the number of experimental group, each group represents 10 flight simulation, and ordinate represents the moving times of unmanned plane in each set of experiment.

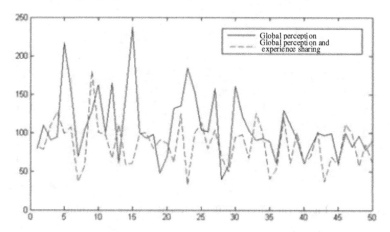

Fig. 5. Verification for effectiveness of shared-learning

In the case of all area perceptions, UAVs can reach agreement on how to intercept invasion targets as soon as possible; learning between the UAVs is the synchronization. After a running process of learning, agent is able to achieve the effect of convergence of expectations.

This experiment will change the basic conditions (3) form not shared experiences tuple to share experiences tuple, each UVA uses a shared learning algorithm for learning. Seen from the experimental results, compared with does not use shared experiences, agent uses shared experiences can convergence to stable values faster, this indicates that the shared learning algorithm to significantly improve the efficiency of collaboration.

Comparative experiment 2 UAVs execute task of blocked, comparison curve of moving times between before change return values and after change return values.

On basis of experiment 1, we change return values set to try to observe the influence of return value on learning. We change return value set of basic condition (5) to:

Table 2. Amendatory return value which environment feedbacks to the UAV

	arrive at the target	the distance to target reduce one step	the distance to target is decrescendo	the distance to target increase one step
r	300	100	100	-50

The result is shown in Figure 6. In the figure, horizontal represents the number of experimental group, each group represents 10 flight simulation, and ordinate represents the moving times of unmanned plane in each set of experiment.

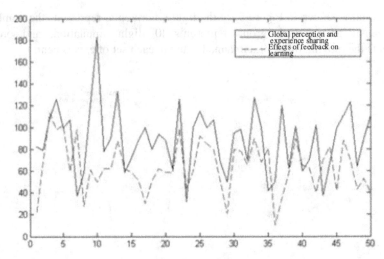

Fig. 6. The effect of different feedback value to learning

Setting return value of increase distance between ion task and target location with a step to-50 make the algorithm get a faster convergence effect.

Through these two experiments, the following conclusions can be obtained: shared-learning algorithm is effective for multi-agent team and come to convergence under certain conditions; Collaborative learning between agents can improve the learning efficiency of multi-agent team. The return value which acts as the only clue of learning, different set of return values can bring to different effect of the system's learning.

5   Conclusion

This paper proposes an operation mechanism and learning algorithm for self-adaptive multi-agent system, which makes the self-adaptive agent have the learning capacity and can use the learning algorithm to learn related knowledge of environment online, can develop out effective self-adaptive policy according to the change of environment. All of these enhances the self-adaptive capacity of the system in dynamic and uncertain environment greatly.

This major work of this paper can be summarized in several areas:

- This paper introduces the reinforcement learning technology and agent technology in the area of self-adaptive systems research, proposes self-adaptive mechanism of self-adaptive system under uncertainty environment, and the corresponding learning algorithm for self-adaptive agent to support the learning process.
- This paper develops a specific application, presented the development process of the case, showing the actual performance of the self-adaptive agent with learning ability, verify the legitimacy and effectiveness of the methods described in this paper.

Acknowledgment. This work is supported by the Projects (61173026, 61373045, 61202039) supported by the National Natural Science Foundation of China; Projects (BDY221411, K5051223008, K5051223002) supported by the Fundamental Research Funds for the Central Universities of China; Project (513***103E) supported by the Pre-Research Project of the "Twelfth Five-Year-Plan" of China; Project (2012AA02A603) supported by the National High Technology Research and Development Program of China.

References

[1] Szepesvári, C.: Algorithms for reinforcement learning. Synthesis Lectures on Artificial Intelligence and Machine Learning 4(1), 1–103 (2010)

[2] Meignan, D., Koukam, A., Créput, J.C.: Coalition-based metaheuristic: a self-adaptive metaheuristic using reinforcement learning and mimetism. Journal of Heuristics 16(6), 859–879 (2010)

[3] Salehie, M., Tahvildari, L.: Self-Adaptive Software: Landscape and Research Challenges. ACM Transactions on Autonomous and Adaptive Systems 4(2), 1–42 (2009)

[4] Kabysh, A., Golovko, V., Lipnickas, A.: Influence Learning for Multi-Agent Systems Based on Reinforcement Learning. International Journal of Computing 11(1), 39–44 (2012)

[5] Smarsly, K., Law, K.H., Hartmann, D.: Multiagent-based collaborative framework for a self-managing structural health monitoring system. Journal of Computing in Civil Engineering 26(1), 76–89 (2011)

An Agent-Based Autonomous Management Approach to Dynamic Services

Fu Hou, Xinjun Mao, Junwen Yin, and Wei Wu

College of Computer, National University of Defense Technology,
NUDT Changsha, Hunan, China
{fu.houmail,mao.xinjun,wuei08}@gmail.com, jwyin_cn@163.com

Abstract. The services over Internet are often dynamic, evolving and growing due to the changes of the requirements and operating contexts. Therefore, it is necessary to effectively manage the dynamics of services in order to support flexible, efficient, and transparent services access for applications at run-time. This paper proposes an agent-based technical framework of managing dynamics of services, in which agent as the bridge between the applications and the target services is responsible for encapsulating the services as its behaviours, monitoring the dynamic of services, and providing proper services for applications. The situation model of services and implementation architecture of service manager are designed to specify the expected dynamic aspects of services and support the monitoring and access on services. Based on Jade and Tomcat, we have developed a prototype platform called *ServiceAutoManager* to implement the above technologies and studied a case to show the effectiveness of our proposed approach.

Keywords: dynamic services, software agent, autonomous management, service situation.

1 Introduction

Service-Oriented Computing (SOC) paradigm has gained popularity today for supporting the development of rapid, low-cost, interoperable, evolvable, and massively distributed applications. It is promised to easily assemble application components into a loosely coupled network of services that span organizations and computing platforms etc. [1]. Web services are currently the most promising SOC based technology. Several technologies and standards have been proposed to manage the web services, e.g. WSDL to describe the services, SOAP to access service etc., for service-based applications development and invocation. Based on the technologies and standards, service provider registers web service in registry with some basic information e.g. operations, binding port, etc., if application intends to invoke some service, it will look up the registry to get the required service information, and then access the service in term of SOAP.

However, the services over Internet are dynamic, evolving and growing due to the changes of the requirements and operating contexts. One typical case is the

Z. Shi et al. (Eds.): IIP 2014, IFIP AICT 432, pp. 122–132, 2014.

evolving changes of web service's QoS. Doubtlessly, to determine the service to be accessed at design-time or only by considering the functionality information of service seems unfeasible, because the expected service may be unavailable or more suitable services for applications may arise due to the dynamic of services at runtime. This leads to the requirements of managing the dynamics of services in an autonomous and transparent way. Obviously, the autonomous management of services requires to perceive, organize, acquire the dynamic of services, and to decide which service should be accessed and invoked to satisfy the application requirements according to the dynamic information. There are several ways to tackle the issues: one is to re-construct web services and provide them with the capabilities of managing dynamics and providing autonomous decision. The other is to enrich the applications with the capabilities of monitoring the dynamics of services and of autonomy on the access of the expected services. However, both of these methods are complex because they need to modify the existing services or applications.

Software agent technology gives us inspirations about how to manage and organize the dynamic services. First, agent is context-aware, which means it can sense the situated environment. If we provide the monitoring capabilities for agents, they can perceive the dynamic information of services. Second, agent is autonomous on behaviours, which means agent can decide which behaviours should be performed according to its internal state and external requests. Therefore, we can access services by agent in an autonomous and flexible way. Third, agent is expected to be social, which means multiple agent can interact and cooperate with each other in order to achieve global objectives. Therefore, the complex management or organization can be achieved in term of the interaction of agents. Moreover, software agent technology enable legacy systems to be incorporated with new functions and capabilities.

The remainder of this paper is structured as follows. Section 2 discusses the related works about service management methods and the utilization of agent technology into service-oriented computing. Section 3 overviews the technical framework to manage dynamics of services. Section 4 details the management model and situation model of dynamic services, and presents the implementation architecture and the autonomous management process of dynamic services. Section 5 introduces a prototype platform - *ServiceAutoManager* that implements the above technologies and supports the autonomous management of dynamic services, and a web service application of travel searching is discussed to show the effectiveness of our proposed approach. Conclusions and future works are made in section 6.

2 Related Work

As more and more evolving and dynamic services are deployed on Internet, how to manage the dynamic of services has obtained great attentions from both industry and academic fields. Several attempts have been made in the past years, most of them focus on the service operation management. A typical work is to establish the registry management pattern based on SOA [2], e.g. ESB, and there are also some researchers propose web service distributed management which essentially defines a protocol for interoperability of management information and capabilities in a distributed environment via Web services [3]. Recently seeding autonomic capabilities for service level management is an evolutionary service level management approach where autonomic computing capabilities anticipate IT system requirements and resolve problems with minimal human intervention [1]. E.g. Yu Cheng et al. propose a prototype architecture and discuss related implementation issues for an autonomic service management framework [4]. Bhakti et al. present the idea of adapting the autonomic computing paradigm into SOA to dynamically (re)organizing its topologies of interactions between the services with little human intervention [5].

Software agent techniques have been extensively exploited to solve issues occurring in several areas. They use software agent to implement the adaptive process, service composition, service selection and even a broker. E.g. Yu Fei Ma et al. present a lightweight autonomous agent fabric for Web service [6]. Lopes and Botelho present a framework to enable the execution of semantic Web services using a context-aware broker agent [7]. Sreenath and Singh proposes a new agentbased approach in which agents cooperate to evaluate service providers [8]. Chainbi et al present an agent-based framework for autonomic web services which is based on two agent-based systems collaborating to enrich web services and registries with self-* capabilities [9].

3   Agent-Based Technical Framework of Managing Dynamics of Services

In order to manage the dynamic of services and support applications to flexibly access services, we propose an agent-base technical framework (see Fig. 1). In this framework, services are to be managed by a number of managers that are actually software agents. Each service manager can manage one or more services, and each service corresponds to at most one manager. Developer should explicitly describe the management relationship between the service and service manager at deploy-time. Such relationship can be dynamically adjusted at runtime in term of self-organizing of multiple service managers in order to achieve the re-organization and optimization of the management relationships. When the relationship is established, the infrastructure of services is responsible for monitoring the dynamics of the managed services and sending the dynamic information to the service managers. The managed services together with their dynamic information consist of the internal resources and states of service manager. The service managers should register at the service manager center. The registered information includes the agent ID and address of manager, the managed services, etc. Therefore, service manager center knows what the functions services provide, and who are responsible for managing these services. Applications that intend to access some service should first look up service manager center to obtain the information about which agent is responsible for managing

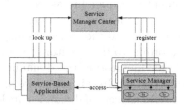

Fig. 1. Agent-based Technical Framework of Managing Dynamics of Services

the service that satisfies the expected functions. Then applications send request that describes the expected functions, constraints and corresponding parameters to the service manager. According to the requests and the dynamics of the managed services, service manager decides which service is more suitable for providing the function, and then access the service and return the service results.

4   Approaches to Autonomous Management of Services

4.1   Model of Services Management

Fig. 2 depicts the abstract model of service management respecting the above technical framework, in which service managers are designed to manage the dynamics of services, decide and invoke the services to satisfy the applications requests. The whole management consists of three aspects: service enrolment, service monitoring and service invocation.

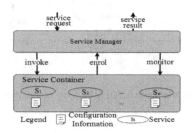

Fig. 2. Model of Services Management

a) *Service Enrolment.* The purpose of service enrolment is to establish the management relationship between service manager and the managed services. Before service manager begins to manage services, developers should deploy the service into service container and explicitly define service's configuration file. There are two types information in configuration file: service access information and the monitored dynamic aspects of the service. When the services are deployed in the container, container is responsible for enrolling the configuration information into the service manager.

b) *Service Monitoring.* According to the monitored aspects of dynamics, service manager is responsible for monitoring the dynamics of services and maintaining the dynamic information. In this paper, we provide a situation model of service that covers several aspects of dynamics, including satiated context, quality status, and business capability. When service manager gets the configuration information of service, it will monitor the services and obtain the dynamic information of services based on the infrastructure.

c) *Services Invocation.* Once the enrolment of management relationship and service monitoring are established, service manager can autonomously manage and invoke the services. When receiving service requests from applications, service manager will find and access the proper services that satisfy the request.

4.2   Implementation Architecture of Service Manager

In the above abstract model, service manager is the core to achieve the objective of service's autonomous management. Based on the reactive agent architecture we propose the implementation architecture of service manager (see Fig. 3). In Fig. 3 we can see that service manager is comprised of six basic modules:

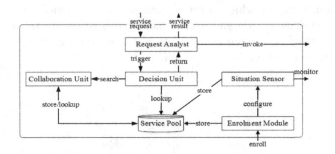

Fig. 3. Implementation Architecture of Service manager

a) *Enrolment Module.* Once the enrolment of management relationship and service monitoring are established, service manager can autonomously manage and invoke the services. When receiving service requests from applications, service manager will find and access the proper services that satisfy the request.

b) *Service Pool.* Service pool is used to store the managed services and their enrolment and dynamic information. The monitored information of services are sent and processed by the service pool. Therefore it saves up the latest information about the situation of the managed services. The decision center and collaboration unit can look up the service pool to get the dynamic information of the services.

c) *Collaboration Unit.* Collaboration unit is to perform the interactions among service managers in order to achieve the collaborations. The purpose of collaboration may be querying services managed by other managers or self-organizing among managers to reorganize the managed services in different managers.

d) *Decision Unit.* Decision unit is the component to decide which service managed by the manager is more suitable for the service request. The decision process is based on the composite considerations on the service request and the service dynamics and triggered by the service request.

e) *Request Analyst.* Request analyst is responsible for analysing the service request from application, interacting with the decision unit to obtain the expected service, and returning the service results to the applications.

f) *Situation Sensor.* Situation sensor is in charge of monitoring the managed services and obtain their dynamic information. It is also responsible for interacting with the service pool to store the perceived information.

4.3   Autonomous Management Process of Services

The autonomous management process for services consists of several phases in which different components and agents are involved and cooperated (see Fig. 4).

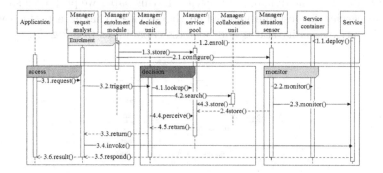

Fig. 4. Autonomous Management Process Sequence Diagram of Services

a) *Service Enrolment Phase.* First services are to be deployed into service container that supports the running of services and maintains the meta information of services. Then service container sends enrolment request to manager in order to establish the management relationship between the service and the manager.

b) *Service Monitoring Phase.* First services are to be deployed into service container that supports the running of services and maintains the meta information of services. Then service container sends enrolment request to manager in order to establish the management relationship between the service and the manager.

c) *Service Access Phase.* Request from application will trigger the process of accessing service. Typically, such request will be pre-processed by the request analyst, and request analyst will interact with the decision unit to find the proper service. Then service analyst will invoke the service and return result to application.

d) *Autonomous Decision Phase.* Once an application request trigger the decision unit, the autonomous decision process will begin. First decision unit will look up the service pool and perceive the service situation if there exist the suitable service that satisfies the request, decision unit will return the service. If there is no proper service, the decision unit will collaborate with other managers to obtain the proper service.

4.4   Service Situation Model

With the proliferation of services as a business solution to enterprise application integration, the QoS is becoming increasingly important to service providers [10]. However, various applications with different QoS requirements will compete for network and system resources e.g. bandwidth and processing time, so QoS is not enough to describe the dynamics of service. In this paper we propose a situation model of service to describe and define the service dynamic aspects covering situated context, quality status and business capabilities (see Fig. 5).

Fig. 5. Situation Model of Dynamic Service

Situated context describes the utilization status of network and system resource (e.g. bandwidth etc.). Quality status describes the service qualities aspects status information, including performance (e.g. throughput, response time etc.), capacity (e.g. the limit of the number of simultaneous requests), accuracy (e.g. the number of errors), and availability. Business capability describes business capability status information of service. For example, the current positioning precision of GPS's position service.

5   ServiceAutoManager Platform and Case Study

In order to support the autonomous management of services and corresponding development tasks, we have developed a prototype platform called *ServiceAutoManager*. *ServiceAutoManager* integrates Jade and Tomcat together, and provides the autonomous management functions for services. The implementation framework of *ServiceAutoManager* consists of four layers (see Fig. 6).

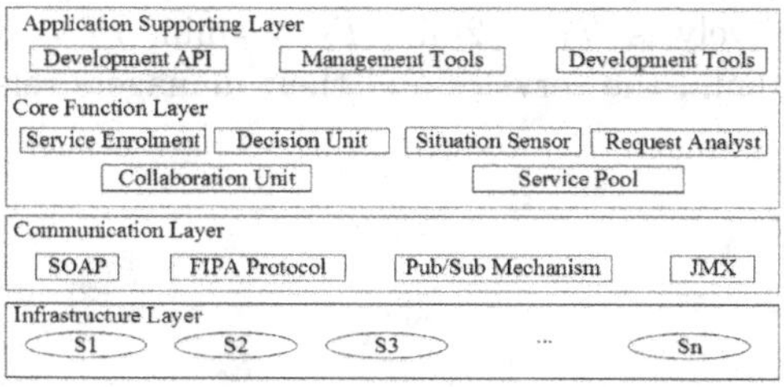

Fig. 6. Implementation Framework of SAutManager Platform

a) *Infrastructure Layer*. Infrastructure layer implements service container. It is in charge of service deployment and running. We adopt Apache Tomcat as the service running environment and JMX technique as the service monitoring engine.

b) *Communication Layer*. This layer is the basis of *ServiceAutoManager* and provides the FIPA protocol supporting service managers' collaboration. We implement this layer by means of expending Jade platform [11]. This layer also provides the basic mechanisms like JMX, SOAP and Sub/Pub to implement service enrolment and monitoring.

c) *Core Function Layer*. This layer is the core function layer of *ServiceAutoManager*. It is the concrete function implementation of service manager, including components like service enrolment, decision unit, request analyst, collaboration unit, service pool and service sensor.

d) *Application Supporting Layer*. In this layer we provide several supports to aid the development and management of applications, such as development API, management tools and development tools. Using the API and tools either the developers or managers can develop and manage the services quickly and efficiently, and can implement the transparent access.

In the following we give a case study to illustrate the above technologies and platform. The case is about travel searching problem, namely from **A** position to **B** position to find a path which satisfies user requirement with cellphone map searching application. In this paper we assume that the searching process needs three types of services: position service, path information service and taxi service. To reach **B** with the shortest time we need a position service to locate the position of **B**, and based on the path information service we will get some paths, and depending on the path status we need select one path and take taxi to **B**. This process is an artificial process, we hope this process can be implemented on the cellphone application, we assume that at the time there are three position services, four path information services and six taxi services, where the positioning precision P_1 of position service PS$_1$ satisfies $P_1 \geq$ 30m, position service PS$_2$ satisfies $P_2 \geq$ 10m and position service PS$_3$ satisfies $P_3 \geq$ 25m; the current ratio of traffic jam respectively is: $R_{PIS1} = 6\%$, $R_{PIS2} = 3\%$, $R_{PIS3} = 8\%$, $R_{PIS4} = 7\%$, and the ratio of traffic jam is dynamic ; the response time of taxi respectively is: $T_1 = 2$min, $T_2 = 4$min, $T_3 = 6$min, $T_4 = 5$min, $T_5 = 3$min, $T_6 = 10$min, the services and their managers relationship diagram sees Fig. 7.

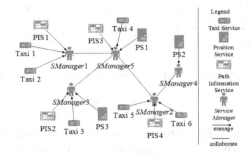

Fig. 7. Diagram of Travel Searching Case Services and Service Managers Relationship

We design the travel searching application under above hypothesis, the application accesses services process is: **Firstly**, find a Service Manager namely *SManager5* from service manager centre. **Secondly**, define the service requirements of application, the requirements just are the service situation constraints: 1).The requirement of position service: the positioning precision P of position service satisfies $P \leq$ 20m; 2).The requirement of path information service: the current ratio R of traffic jam satisfies $R \leq 5\%$; 3).The requirement of taxi service: the response time T satisfies $T <$ 3min. **Thirdly**, access the services based on requirements, in this process service manager will autonomously and transparently return the service results, in this case the results are **PositionService2**, **PathInformationService4** and **TaxiService1**.

6   Conclusions and Future Work

This paper researches the autonomous management technologies of services due to the dynamic of services and the requirements of flexible and transparent service access. Different from related works, we adopt agent as technical solution and design service managers to perceive, organize and maintain the dynamics of services, and to further decide which service should be invoked according to the service request and situation. Service managers are actually agents that encapsulate services, manage their dynamic and provide autonomous service access. Developer only needs to explicitly claim the management relationship between service managers and the managed services. Therefore, our approach enables us to manage services promote the reuse and re-organization of legacy service systems. In order to support service's autonomous management based on the agent-based technical framework, we have presented situation model of service and implementation architecture of service manager that answer what the dynamic information should be managed and how to manage the dynamic service in an autonomous way respectively. We have developed a prototype platform called *ServiceAutoManager* based on Jade and Tomcat, and studied a travel searching case to illustrate the proposed approach.

The future works include: (1) the self-organization and collaboration among service managers to re-organize the managed services; (2) self-adaptation technologies for accessing web services in term of the changes of both requests, situated contexts and the dynamic of service itself; (3) more experiments to analyse the richness and weakness of our proposed approach.

Acknowledgments. The research leading to these results has received funding from National Nature and Science Foundation of China under Granted Nos.61379051 and 61133001, Program for New Century Excellent Talents in University under Granted No. NCET-10-0898 and Open Fund State Key Laboratory of Software Development Environment under Granted No. SKLSDE-2012KF-0X.

References

1. Papazoglou, M.P., Traverso, P., Dustdar, S., Leymann, F.: Service-oriented computing: a research roadmap. International Journal of Cooperative Information Systems 17, 223–255 (2008)
2. Papazoglou, M.P., van den Heuvel, W.J.: Web Services Management: A Survey. IEEE Internet Computing 9, 58–64 (2005)
3. Kreger, H., et al.: Management Using Web Services: A Proposed Architecture and Roadmap, IBM, HP and Computer Associates (2005), www-128.ibm.com/developerworks/library/specification/ws-mroadmap
4. Cheng, Y., Leon-Garcia, A., Foster, I.: Toward an autonomic service management framework: A holistic vision of SOA, AON, and autonomic computing. IEEE Communications Magazine 46, 138–146 (2008)
5. Bhakti, M.A.C., Abdullah, A.B., Jung, L.T.: Autonomic, self-organizing service-Oriented Architecture in service ecosystem. In: The 4th IEEE International Conference on Digital Ecosystems and Technologies (DEST), pp. 153–158 (2010)
6. Ma, Y.F., Li, H.X., Sun, P.: A lightweight agent fabric for service autonomy. In: Proc. of the AAMAS 2007 Workshop Service-Oriented Computing: Agents, Semantics and Engineering, pp. 63–77 (2007)
7. Lopes, A.L., Botelho, L.M.: Executing semantic Web services with a context-aware service execution agent. In: Proc. of the AAMAS 2007 Workshop Service-Oriented Computing: Agents, Semantics and Engineering, pp. 1–15 (2007)

8. Oyenan, W.H., DeLoach, S.A.: Design and evaluation of a multi-agent autonomic information system. In: International Conference on Intelligent Agent Technology, pp. 182–188 (2007)
9. Chainbi, W., Mezni, H., Ghedira, K.: AFAWS: An Agent based Framework for Autonomic Web Services. Multi Agent and Grid Systems 8, 45–68 (2012)
10. Lee, K.C., Jeon, J.H., Lee, W.S., Jeoong, S.-H., Park, S.-W.: QoS for Web Services: Requirements and Possible Approaches (2004), http://www.w3c.or.kr/kr-office/TR/2003/ws-qos/
11. Bellifemine, F.L., Caire, G., Greenwood, D.: Developing multi-agent systems with JADE, vol. 7. Wiley. com (2007)

Research and Application Analysis of Feature Binding Mechanism

Youzhen Han[1] and Shifei Ding [1,2]

[1] School of Computer Science and Technology,
China University of Mining and Technology,
221116, Xuzhou, Jiangsu, China
15152114529@126.com
[2] Key Laboratory of Intelligence Information Processing,
Institute of Computer Technology, Academy of Sciences,
100190, Beijing City, Beijing, China
dingsf@cumt.edu.cn

Abstract. In order to recognize the object of the external world, brain need to integrate the information of different cortical areas, then form a complete world. And binding problem study on a process to percept a complete object by integrating information, which scattering on different cortical areas. As a central problem of cognitive science and neuroscience, the concept of feature binding is becoming to the focus of consciousness argument. At the beginning of this paper, we introduced the concept, characteristics and theory source of feature binding. And according to the main theoretical research methods of this mechanism, combining the latest advance at home and abroad, we made a systematical review of the research situation of feature binding problem applying in perceptual learning. At last, we pointed out the key point of the further study, which may be refined research of bundled brain mechanism on different cognition process and systematical study of general bundled brain mechanism.

Keywords: feature binding, vision perception, perceptual learning, synchronous neural oscillations, γ-band oscillations, cognitive neuroscience.

1   Introduction

The so-called binding problem is to study a way, integrating the information scattering on different cortical areas to percept a complete object. This theory supports rich theoretical and experimental bases for the research of image understanding based on visual hierarchical perceptual mechanism. Thereby, the research on the cognitive process and practical application of feature binding mechanism has an important theoretical breakthrough and application value. This paper attempts to outline the principle theoretical bases, and make a more systematic review about the research on the binding problem and the functional relations of cognition. We hope to raise the domestic scholars' concern in this field.

Z. Shi et al. (Eds.): IIP 2014, IFIP AICT 432, pp. 133–140, 2014.

The organizational structure of this paper is as follows: after outlined the basic theory of feature binding; we introduced the origin and progress of binding mechanism, and proposed the further research direction, in section 2; in section 3, we introduced the current main theory, including the feature-integration theory, neural synchronous activation theory and bundled neural network model; section 4 mainly introduced the crucial effect of binding mechanism in the cognition process, concretely include distinguish perception, the selection of perceptual information, the memory and consciousness wakeup etc.; at last, in conjunction with our own experience in the research, we made a prospect of the progress of binding mechanism.

2   Research Situation of Binding Mechanism

There is a long history of the exploiting of cognition; our ancestors already have thinking about the origin and nature of humans' recognition, long time ago. Limited by the level of science, the knowledge of brain is mainly stay on the observation and guesses, lacking the study of the brain's internal theory. Until 20th century, with the breakthrough of neuron anatomy, neurophysiology and electrophysiological processes of neuron, people have a deeper understand of the structure and the basic unit of brain.In 1999, Riesenhuber and Poggi(Riesenhuber and Poggio 1999) proposed a standard quatitative model based on visual perception mechanism and visual feature computing, their theory has the error-tolerant about the location, size and shape, this give people a quantitative mathematical understanding of this bio-inspired model. Shi Zhongzhi et al., based on PCNN, proposed a computering model of feature binding-Bayesian Linking Field model, research on the feature binding in visual perception. Since the PCNN extremely close to the process method which used by biological visual cortex, many scholars improved PCNN itself and have imposed a lot of modified PCNN model (Zhan et al. 2009, Huang and Jing 2007) through analyzing this model, and made it performing excellent properties.

3   Main Approaches of Feature Binding

There are many theories about binding mechanism. Inside, the feature-integration theory of cognitive psychology, synchronous neuronal activation theory which influential in neuroscience and also called temporal synchronization theory, and recent years proposed bundled neural network model are some most typical theories.

3.1   Feature-Integration Theory

Treisman's Feature-integration theory tries to use selective attention mechanism of visual space - the attention window to explain the feature binding problem(Bouvier and Treisman 2010, Di Lollo 2012). The theory considers the features from the parallel processing of early stage to the late integration are realized through the spatial attention model. The model includes a position main-map and a set of independent feature-maps. The position main-map is used for registering the location of the object, but not understands the features of that location. The feature maps mainly include two types of information: one is the "flag", which is used to mark whether the feature is existed in somewhere of the scope; the other is about some implicit information in the feature space arrangement. Every feature detector in feature map connected to a unit of main map. After established the unified object representation, the stored template can be matched and recognized, and the related behavior can also be completed.With further research, people realized that there are three deficiencies in the traditional feature integration theory : (1) a traditional feature binding is measured through conscious report, there is no use of unconscious implicit indicators; (2) all are inspected according to the results of immediate report, there is no inspection based on the results of the memory report; (3) the information involved only biting characteristics and location, did not involve the bundle between behavior and perception. These studies are more significance to the general bundling mechanisms.

3.2   Synchronous Neuronal Activation Theory

Synchronous neural activation theory (temporal synchronization) is currently considered the most influential binding theories. This theory is that, feature binding based on the synchronization activates neurons, realization through the synchronization activation of neural activity(Singer and Gray 1995, Von Der Malsburg 1994). The neurons which are characterization of the same object or trajectory can activate the corresponding behavior in time synchronization to the millisecond. And this synchronous activation is not generated in charge of different cells or between cells. This instant synchronous activation is the key of binding mechanism. Synchronous activation can make selective mark to the neuronal that are responsible for encoding, and distinguish from other objects' activated neuronal responses. This highly selective temporal structure that allows different cell populations in the same neural networks activates synchronously and differentiates from each other. In addition, time-based synchronization binding also serves as a mechanism of selecting next processing object. Because of accurate synchronous activation, the specific content become a desired activation result, and make other brain neurons perceive such a result. These selectively activated neurons will complete binding operations in different regions , and make this activation model in the neural network get the further process.

This theory is that the binding is accomplished by the synchronization of neural activity(Singer and Gray 1995, Von Der Malsburg 1994).The neural basis of time synchronization is the connection of the cells which react to the features stimulating in senses at the same time. The features may be bound quickly by the connection through the accurate millisecond time synchronization and spatially parallel manner. This mechanism has the advantages of low-performance requirements to the functional structure, high flexibility and adapt to a wide range.

3.3   The Neural Network Model of Binding

The neural network model of binding was developed by Watanabe, Nakanishi and Aihara(Watanabe et al. 2001). The model is composed of a primary map (primary map) and two advanced modules. Two modules are responsible for two different characteristics in the primary map, and each module is divided into three different levels, between all levels are pair-wise two-way connected. The external representation of the object, it is represented by the "overall dynamic cell collection" which are distributed in the primary map and two advanced modules. The details of the information are encoded in the main map, and rough, complex feature information is characterized in the two advanced modules. Feature binding is completed in this 'overall dynamic cell clusters'.

The bases of the model are "functional links" and "two-way link" of neural networks. Functional link is dynamic link between specific neurons which use coupled sensor encoding neural network's time pulse (temporal spike). This link is different from traditional synaptic connections for its value can be changed with the excited level of neural network dynamics change, its intensity is decided by the time consistency level of occasional action potentials and each neuron, directly affected by the spatial and temporal dynamics. These cells are called the 'dynamic cell clusters' that are linked by the functional link. The model uses two-way link, it can explain time synchronization nerve release phenomena without introducing neural oscillation unit of the time synchronization theory.

4   The Functions of Binding Mechanism in the Perceptual Processes

As for the specific function of binding mechanism of perceptual processes, different researchers come to different conclusions. As Friedman-Hill found that right parietal is related with visual features binding in the perception report(Friedman-Hill et al. 2003); while Ashbridge found that right parietal and binding have no direct relationship in the visual search task, related with the selection of spatial position only, the ventral temporal cortex is responsible for the feature binding(Ashbridge et al. 1999); And numerous studies show that the hippocampus is related with features binding in memory(Mitchell et al. 2000); Prabhakaran, etc. observed in working memory storage consolidation feature, prefrontal got activation(Prabhakaran et al. 2000); Llinas etc. found that nonspecific thalamic circuits is responsible for perceiving a wake-up binding(Llinás 1988). In addition, in the specific binding task , the left middle temporal lobe , cerebellum, left frontal lobe, the parietal lobe , left medial frontal gyrus, right anterior cingulate gyrus, right frontal gyrus , right prefrontal , anterior cingulate cortex , left center back and other brain regions can be activated. It can be seen in different cognitive tasks clearly, brain mechanism of binding is extremely complex.

4.1   Binding and Perception Distinction

In the neurophysiologic level, it was found binding is related with the situational perception distinction height; external sensory input formed structural characterization under the function of binding. There is evidence that nerve released simultaneously depends on the presentation of stimulation. Spatially separate cells shown strong synchronization only in response to the same object . If the two cell populations made reaction to the two independent objects which were moving in the different directions, then these two cell populations' activation almost had no time relation(Castelo-Branco et al. 2000). These results indicate that each object corresponds to its own synchronous activation mode,different objects have different neural activation patterns. Therefore, it can achieve different objects' perception distinction based on synchronous neural release binding mechanism. Tallon-Baudry through EEG studies has found that the perception of the object's consistency is associated with a specific γ-band activity(Tallon-Baudry et al. 1996). This relationship between high-frequency activity and the perception of the object's consistency also confirmed by a lot of other paradigms studies. Since γ-band activities reflect the information of time synchronization, it also reflect binding mechanisms' role situation.

4.2   Binding and the Selection of Perception Information

Fries record the cat's visual cortex neural responses under the conditions of binocular rivalry(Fries et al. 1997). They give two stimulations that moved in different directions to the wakefulness cats' left and right eye at the same time. According to the direction of eye movement to speculate perception advantage, making the grating deviation as identification of the eye movement direction perceives the selection result. Studies found that neuronal activity stimulated a high degree of synchronization, these neuronal activity are used to characterize the perception; while nerve cells that used to process the ignored stimulation exhibit only weak time relationship. Meanwhile, the simultaneously released of the recording point is accompanied with the advantaged γ oscillations, characterized the selected stimulated neurons, and the intensity of the γ-band increased, but the neurons which make a reaction to the neglect stimulation, its strength decreased. The results show that, at least in the early stages of visual perception, sensory information's dynamic selection and inhibition are associated with the synchronization regulation of neural activity. Thus, this also proved that binding and the selection of perception information are closely related.

4.3   Binding and Memory

4.3.1   Binding and Working Memory

Research shows that the function of working memory depends on the nerve cells' temporal coordination on the time (temporal coordination). Tallon-Baudry etc. studied the relationship between synchronization of γ-band and working memory(Tallon-Baudry et al. 1998) They found that in a task of visual delayed matching, the accurate synchronization of ventral occipital and frontal area was strengthened. Sarnthein's etc. study found that in a visual-spatial working memory task, time consistency of forehead and posterior parietal parts' electrodes has also been strengthened. And this increase occurred in γ-band, but in the lower frequency bands (for example θ frequency) does not occur. It shows that the time binding mechanisms also plays an important role in working memory.

Recent studies of cognitive aging found that compared with the young, on the memory of binding characteristics the elderly has the obvious shortages, while on the memory of single feature there was no difference. This shows that the binding mechanism has a relationship with multi-feature combination's information coding, save or extract. This conclusion has also been supported by brain imaging studies. This shows that for normal subjects' binding capacity, the working memory tasks that need for feature binding can lead to greater hippocampus activation. The study not only proved the presence in working memory binding problems, but also made important contributions to reveal the neural basis of the binding.

4.3.2   Binding and Episodic Memory

Lukeu etc. pointed out that episodic memory depends on the ability of individual tied up the central message of an event with contextual information(Lekeu et al. 2002). Center information refers to the theme of the event (such as the main content of the conversation). Contextual information includes spatial and temporal characteristics of events, encoding mode (such as visual or auditory, etc.), physical characteristics (such as pitch or color) and emotional status. Contextual information can act as the clues to extract center information. In order to form memory traces which is complete and relevant a particular event, it is not enough to code two kinds of information independently. You must also establish a binding between central feature and contextual feature. The possibility of an event extracted depends on the number and strength of these two kinds of information bundled. It was also found that there were two forms of binding to central feature and contextual feature: Intentional and unintentional binding. These two forms of binding between different characteristics have different effects. They found that under the conditions of the unintentional binding, the left middle temporal lobe and cerebellum activated in the word recognition; the left frontal and parietal areas activated in combination recognition. Under the conditions of the intentional binding, in the word recognition, the left middle frontal gyrus, right anterior cingulate and right frontal gyrus activated; in combination recognition, the right prefrontal activated.

4.4   Binding and Consciousness Awakening

4.4.1   Binding and Perception Awakening

Some studies have found that in the awakening state and rapid eye movement sleep (REM) state, in the EEG (electroencephalogram) or MEG (magneto encephalography) appears frequencies of γ-band. In the deep sleep state such frequencies disappears. The similarity of REM state and wakefulness on the high-frequency activity indicates that in these two states, the synchronization of the neural activity frequency band is associated with the generating process of the arousal state. Another study found that, under the height anesthesia, the high frequency components of the perception awakening also disappeared. Since γ-band activity is associated with synchronous activation of nerve cells. From the above results we can speculate that the binding mechanisms involved in the process of the sense change from the depth anesthesia or deep sleep state to the awake state.

4.4.2   Binding and Consciousness

At present, a main point in the study of consciousness holds that sense should be understood as a number of interacting systems' functions. These systems include sensing systems, memory structure, the executive control center and emotional and motivational systems(Engel and Singer 2001). The specific functions of consciousness mainly for two aspects(Huisong 2012): (1) cross-system coordination function, namely how a variety of different nature process integrated, and how they are in harmony and consistently organized in the distributed neural activity model; (2) dynamic selection function. Since in a given time, only a part of information can enter into the consciousness. So how consciousnesses achieve the dynamical selection to the specific content? Obviously, from the meaning of binding, these two functions of consciousness are closely related with binding mechanisms. In fact, the previously mentioned perception distinction, information selection, working memory, awakening and other mental processes are the main content of consciousness(Engel and Singer 2001). The relationship between these processes and binding mechanisms also confirmed the function of binding in the consciousness.

5   Conclusion and Prospect

Basing on the mechanism of feature binding, this paper systematically introduced the present study about feature binding used in perceptual learning, combining the latest research advances at home and abroad. According to our paper, the binding problem has got more and more academic attention, also is an important field in cognitive science, neuroscience and brain science. But the current research is in its infancy, there is no theory which can perfectly solve the binding problem. In addition, the brain areas which are responsible for different forms of binding may have different working mechanism in the different cognitive tasks. Therefore, with the development of neuroscience, and the increasingly deep of the research on binding problem and relationship of cognitive function. Hence force, on one hand we need to do more in-depth and meticulous research to the brain regions which are responsible for different forms of binding, on the other hand, it must strengthen systematic study of the overall brain function which is responsible for general binding problem, making connections to the variety of fragmented brain function location research, and unifying the systematic framework.

Acknowledgements. This work is supported by the National Natural Science Foundation of China (No.61379101), the National Key Basic Research Program of China (No.2013CB329502), the Basic Research Program (Natural Science Foundation) of Jiangsu Province of China (No.BK2013093), and the Students Innovation and Entrepreneurship Foundation of China University of Mining and Technology (No.201446).

References

1. Ashbridge, E., Cowey, A., Wade, D.: Does parietal cortex contribute to feature binding? Neuropsychologia 37(9), 999–1004 (1999)
2. Bouvier, S., Treisman, A.: Feature binding signals in visual cortex. Journal of Vision 10(7), 96–96 (2010)
3. Castelo-Branco, M., Goebel, R., Neuenschwander, S., et al.: Neural synchrony correlates with surface segregation rules. Nature 405(6787), 685–689 (2000)
4. Di Lollo, V.: The feature-binding problem is an ill-posed problem. Trends in Cognitive Sciences 16(6), 317–321 (2012)
5. Engel, A.K., Singer, W.: Temporal binding and the neural correlates of sensory awareness. Trends in Cognitive Sciences 5(1), 16–25 (2001)
6. Friedman-Hill, S.R., Robertson, L.C., Desimone, R., et al.: Posterior parietal cortex and the filtering of distractors. Proceedings of the National Academy of Sciences 100(7), 4263–4268 (2003)
7. Fries, P., Roelfsema, P.R., Engel, A.K., et al.: Synchronization of oscillatory responses in visual cortex correlates with perception in interocular rivalry. Proceedings of the National Academy of Sciences 94(23), 12699–12704 (1997)
8. Huang, W., Jing, Z.: Multi-focus image fusion using pulse coupled neural network. Pattern Recognition Letters 28(9), 1123–1132 (2007)
9. Huisong, W.: affinity propagation algorithm to build context-aware learning system analysis. Fuqing Branch of Fujian Normal University (5), 46–51 (2012)
10. Lekeu, F., Van Der Linden, M., Collette, F., et al.: Effects of incidental and intentional feature binding on recognition: a behavioural and PET activation study. Neuropsychologia 40(2) (2002)
11. Llinás, R.R.: The intrinsic electrophysiological properties of mammalian neurons: insights into central nervous system function. Science 242(4886), 1654–1664 (1988)
12. Mitchell, K.J., Johnson, M.K., Raye, C.L., et al.: fMRI evidence of age-related hippocampal dysfunction in feature binding in working memory. Cognitive Brain Research 10(1), 197–206 (2000)
13. Prabhakaran, V., Narayanan, K., Zhao, Z., et al.: Integration of diverse information in working memory within the frontal lobe. Nature Neuroscience 3(1), 85–90 (2000)
14. Riesenhuber, M., Poggio, T.: Hierarchical models of object recognition in cortex. Nature Neuroscience 2(11), 1019–1025 (1999)
15. Singer, W., Gray, C.M.: Visual feature integration and the temporal correlation hypothesis. Annual Review of Neuroscience 18(1), 555–586 (1995)
16. Tallon-Baudry, C., Bertrand, O., Delpuech, C., et al.: Stimulus specificity of phase-locked and non-phase-locked 40 Hz visual responses in human. The Journal of Neuroscience 16(13), 4240–4249 (1996)
17. Tallon-Baudry, C., Bertrand, O., Peronnet, F., et al.: Induced γ-band activity during the delay of a visual short-term memory task in humans. The Journal of Neuroscience 18(11), 4244–4254 (1998)
18. Von Der Malsburg, C.: The correlation theory of brain function. Springer (1994)
19. Watanabe, M., Nakanishi, K., Aihara, K.: Solving the binding problem of the brain with bi-directional functional connectivity. Neural Networks 14(4), 395–406 (2001)
20. Zhan, K., Zhang, H., Ma, Y.: New spiking cortical model for invariant texture retrieval and image processing. IEEE Transactions on Neural Networks 20(12), 1980–1986 (2009)

The Correspondence between Propositional Modal Logic with Axiom $\Box\varphi \leftrightarrow \Diamond\varphi$ and the Propositional Logic

Meiying Sun[1,2], Shaobo Deng[1,2], and Yuefei Sui[1]

[1] Key Laboratory of Intelligent Information Processing,
Institute of Computing Technology, Chinese Academy of Sciences,
Beijing China, 100190
[2] University of Chinese Academy of Sciences,
Beijing China, 100049
sunmeiying07@mails.ucas.ac.cn

Abstract. The propositional modal logic is obtained by adding the necessity operator $\Box$ to the propositional logic. Each formula in the propositional logic is equivalent to a formula in the disjunctive normal form. In order to obtain the correspondence between the propositional modal logic and the propositional logic, we add the axiom $\Box\varphi \leftrightarrow \Diamond\varphi$ to K and get a new system K^+. Each formula in such a logic is equivalent to a formula in the disjunctive normal form, where $\Box^k (k \geq 0)$ only occurs before an atomic formula p, and $\neg$ only occurs before a pseudo-atomic formula of form $\Box^k p$. Maximally consistent sets of K^+ have a property holding in the propositional logic: a set of pseudo-atom-complete formulas uniquely determines a maximally consistent set. When a pseudo-atomic formula $\Box^k p_i (k, i \geq 0)$ is corresponding to a propositional variable q_{ki}, each formula in K^+ then can be corresponding to a formula in the propositional logic P^+. We can also get the correspondence of models between K^+ and P^+. Then we get correspondences of theorems and valid formulas between them. So, the soundness theorem and the completeness theorem of K^+ follow directly from those of P^+.

Keywords: propositional modal logic, propositional logic, disjunctive normal form, pseudo-atomic formula, pseudo-atom-complete, completeness.

1 Introduction

Van Benthem gave the standard translation[1, 2] from the propositional modal logic into the first-order logic. The necessity operator $\Box$ corresponds to the universial quantifier $\forall$ via the standard translation and every modal formula corresponds to a first-order formula. Benthem discussed connections between modal axioms and first-order properties of the accessibility relation among worlds in the paper[3]. For example, the formula $\Box p \rightarrow p$ is valid on all reflexive frames. In

Z. Shi et al. (Eds.): IIP 2014, IFIP AICT 432, pp. 141–151, 2014.

order to solve the classical problems arising from the interplay between propositional attitudes, quantifiers and the notion of identity, Aloni proposed the conceptual covers semantics of the first-order modal logic[4, 5].

There are a number of systems of propositional modal logic, like K, D, T, S5, Triv and Ver. Trivially, K is a subsystem of all the others, and every normal modal system[1] is contained either in Triv or in Ver. We also have that every consistent system which contains D is contained in Triv[7]. Each formula in S5 can be equivalent to a formula in the modal conjunctive normal form[7], and each conjunct is the disjunction of φ or $\Box\varphi$ or $\Diamond\varphi$, where φ is a formula of the propositional logic. This makes the proof of the completeness of S5 much easier and provides an effective procedure for determining whether or not a formula is a theorem of S5 and S5-valid[7]. Similarly, Triv can collapse into the propositional logic and Ver provides a different form of collapsing into the propositional logic[7].

In the propositional logic, any formula is equivalent to a formula in the disjunctive normal form and each disjunct is the conjunction of atomic formula or its negation[8]. If there is a modal system S such that each formula can also be equivalent to a formula in the disjunctive normal form, then we can easily get a correspondence between S and the propositional logic. In K necessity can only be distributed over conjunction and possibility can only be distributed over disjunction. In order to distribute necessity over disjunction, we add the axiom $\Box\varphi \leftrightarrow \Diamond\varphi$ to K and construct a new normal modal system K^+.

In K^+ necessity can be distributed over conjunction and disjunction. And it is easy to prove that the formula $\Box\neg\varphi \leftrightarrow \neg\Box\varphi$ is a theorem of K^+. Therefore the system K^+ has a remarkable property that any formula can be equivalent to a formula in the disjunctive normal form and each disjunct is the conjunction of pseudo-atomic formulas $\Box^k p(k \geq 0)$ or the negation of a pseudo-atomic formula. Then discussions of logical properties of K^+ reduce to discussions of logical properties of the propositional logic.

We have the concept of atom-completeness in the propositional logic, and in K^+ we correspondingly give the definition of pseudo-atom-completeness. In K^+ for any pseudo-atom-complete set A of formulas, if we delete one $\Box$ from each φ in A then the new set $\Box^- A$ is also pseudo-atom-complete. Each atom-complete set of formulas in the propositional logic determines a maximally consistent set in a unique way. Similarly, each pseudo-atom-complete set of formulas in K^+ uniquely determines a maximally consistent set. So the accessibility relation in the canonical model[7] of K^+ can be simply given by eliminating one $\Box$ from each pseudo-atomic formula in a possible world.

Assume that $P^+ = \{q_{ki} : k, i \geq 0\}$ is a language of the propositional logic. When a pseudo-atomic formula $\Box^k p_i(k, i \geq 0)$ is corresponding to a propositional variable q_{ki}, each formula in K^+ then can be corresponding to a formula in P^+. Then we can prove that any theorem of K^+ can be corresponding to a theorem of P^+, and vice versa. The semantics of K^+ is the possible world semantics, which

---

[1] A normal modal system is a consistent extension of the system K which retains the rules MP and N. And the definition can be found in [6, 7].

is a set of models of P^+. For any model M of K^+ and any possible world w in M we can construct an assignement v_w of P^+ such that for any formula φ of K^+, $M, w \models \varphi$ if and only if $v_w \models \varphi'$, where φ' is the corresponding formula of φ. Then we can prove that any valid formula of K^+ can be corresponding to a valid formula of P^+, and vice versa. So, the soundness theorem and the completeness theorem of K^+ follow directly from those of P^+.

This paper is organized as follows: the next section gives the basic definition of the propositional modal logic K; the third section defines the propositional modal logic K^+ and gives the disjunctive normal form of formulas in K^+; the fourth section defines pseudo-atom-complete, and proves that each maximally consistent set of formulas is uniquely generated by a pseudo-atom-complete set of formulas; the fifth section gives correspondences between K^+ and the propositional logic P^+, and get the soundness and completeness of K^+; and the last section concludes the whole paper.

2 The Propositional Modal Logic K

This section gives the basic definitions of the propositional modal logic K, including the language, the syntax and the semantics, where the syntax gives the definitions of formulas, axioms and inference rules; and the semantics defines the models, satisfaction and validity. K is sound and complete, and proofs can be found in [7, 9].

Let the logical language L of the propositional modal logic contain the following symbols:

- propositional variables: $p_0, p_1, ...$;
- logical connectives: $\neg, \rightarrow$;
- the necessary modality: $\square$;
- auxiliary symbols: $(,)$.

A formula of the propositional modal logic is defined as follows[2]:

$$\varphi ::= p\,|\,\neg\varphi_1\,|\,\varphi_1 \rightarrow \varphi_2\,|\,\square\varphi_1.$$

The possible modality $\lozenge$ and the necessary modality $\square$ are dual. For any formula φ,

$$\lozenge\varphi =_{df} \neg\square\neg\varphi.$$

Definition 1. A *frame* F of the propositional modal logic is an ordered pair $\langle W, R \rangle$, where W is a non-empty set of possible worlds, and $R \subseteq W^2$ is a dyadic relation defined over the members of W.

For any possible worlds $w, w' \in W$, $(w, w') \in R$ is usually denoted by wRw'. Frames can be classified by the relation R, and details of the classification can be found in [7, 9].

Definition 2. A *model* M is an ordered triple $\langle W, R, I \rangle$, where $\langle W, R \rangle$ is a frame; and I is an interpretation such that for any propositional variable p, $I(p) \subseteq W$.

---

[2] The logical connectives $\wedge, \vee$ and $\leftrightarrow$ can be defined by $\neg$ and $\rightarrow$. For example, $\varphi \wedge \psi =_{df} \neg(\varphi \rightarrow \neg\psi)$, where $=_{df}$ is read as "be defined as".

A model $\langle W, R, I \rangle$ is said to be *based on* the frame $\langle W, R \rangle$.

A model we define here is called a standard model, which is different from a Kripke model [10]. You can see the difference between them in [6] and any standard model can be transformed into a Kripke model[6].

Definition 3. Given a model M, for any possible world w,

$$M, w \models \varphi \text{ if and only if} \begin{cases} w \in I(p) & \text{if } \varphi = p, \\ M, w \not\models \varphi_1 & \text{if } \varphi = \neg\varphi_1, \\ M, w \models \varphi_1 \Rightarrow M, w \models \varphi_2 & \text{if } \varphi = \varphi_1 \to \varphi_2, \\ \mathbf{A}w'((w, w') \in R \Rightarrow M, w' \models \varphi_1) & \text{if } \varphi = \Box\varphi_1, \end{cases}$$

where $\mathbf{A}w'$ means "for any possible world w'". We will give definitions of validity of formulas in a model, a frame and a class of frames.

Definition 4. A formula φ is *valid in a model* M, denoted by $M \models \varphi$, if φ is true at every world in M. A formula φ is *valid on a frame* F, denoted by $F \models \varphi$, if φ is valid in every model based on F. Given a class of frames C, a formula φ is valid with respect to C, if φ is valid on every frame in C.

Given a set Γ of formulas and a formula φ, and C is a class of frames, φ is a *logical consequence* of Γ in C, denoted by

$$\Gamma \models_C \varphi,$$

if for any frame $F \in C$ and model M based on F, $M \models \Gamma$ implies $M \models \varphi$. In this section we mainly talk about the system K, which is characterized by all frames. So we will use $\Gamma \models \varphi$ to denote $\Gamma \models_C \varphi$.

In the following we will give the axiom system of the propositional modal logic K.

Axioms:

$$\varphi_1 \to (\varphi_2 \to \varphi_1),$$
$$(\varphi_1 \to (\varphi_2 \to \varphi_3)) \to ((\varphi_1 \to \varphi_2) \to (\varphi_1 \to \varphi_3)),$$
$$(\varphi_1 \to \varphi_2) \to (\neg\varphi_2 \to \neg\varphi_1),$$
$$\Box(\varphi_1 \to \varphi_2) \to (\Box\varphi_1 \to \Box\varphi_2).$$

Inference rules:

$$(MP) \frac{\varphi_1, \varphi_1 \to \varphi_2}{\varphi_2},$$

$$(N) \frac{\varphi_1}{\Box \varphi_1}.$$

The last axiom is called the axiom **K**. The system we given above is called the system K of the propositional modal logic. It is sound and complete. We will only give the soundness theorem and the completeness theorem here, and the details of their proofs can be found in [7] and [9].

Theorem 1. (Soundness theorem of K) For any formula φ, if $\vdash \varphi$ then $\models \varphi$.

Theorem 2. (Completeness theorem of K) For any formula φ, if $\models \varphi$ then $\vdash \varphi$.

3   The New System K^+ and the Disjunctive Normal Form

Let K^+ be the propositional modal system obtained by adding to K the following axiom:

$$\mathbf{K^+} \ \Box \varphi \leftrightarrow \Diamond \varphi.$$

Then the necessary modality $\Box$ is equivalent to the possible modality $\Diamond$. In K we have the theorems about the distributivity of $\Box$ over $\wedge$ and the distributivity of $\Diamond$ over $\vee$[7]. So in K^+ we can easily get the following theorems about the distributivity of $\Box$ over $\wedge$ and $\vee$.

Proposition 1.
$$\text{(i)} \vdash \Box(\varphi \wedge \psi) \leftrightarrow (\Box \varphi \wedge \Box \psi);$$
$$\text{(ii)} \vdash \Box(\varphi \vee \psi) \leftrightarrow (\Box \varphi \vee \Box \psi).$$

$\Box$

Each formula in the propositional logic is equivalent to a formula in the disjunctive normal form[8]. Similarly, each formula in the system K^+ can also be equivalent to a formula in the disjunctive normal form, where $\Box^i (i \geq 0)$ only occurs before an atomic formula, and $\neg$ only occurs before a pseudo-atomic formula of form $\Box^i p$, where p is an atomic formula. In the following we will first give the definitions of pseudo-atomic formula and the disjunctive normal form, and then prove that each formula in K^+ is equivalent to a formula in the disjunctive normal form.

Definition 5. A formula φ is *pseudo-atomic* if $\varphi = \Box^k p$, where $k \geq 0$ and p is a propositional variable.

Definition 6. A formula φ is in the *disjunctive normal form* if

$$\varphi = \varphi_1 \vee \varphi_2 \vee \cdots \vee \varphi_n,$$

where for each $1 \leq i \leq n$,

$$\varphi_i = \psi_{i1} \wedge \cdots \wedge \psi_{im_i},$$

where for each $1 \leq j \leq m_i$,

$$\psi_{ij} = \sim^{l_{ij}} \Box^{k_{ij}} p_{ij},$$

where $\Box^{k_{ij}} = \overbrace{\Box \cdots \Box}^{k_{ij}\text{-many } \Box\text{s}}$;

$$\sim^{l_{ij}} = \begin{cases} \neg \text{ if } l_{ij} = 1, \\ \lambda \text{ if } l_{ij} = 0, \end{cases}$$

and p_{ij} is a propositional variable, where λ is the empty string.

Similarly, we can define the conjunctive normal form. In this paper we will use the disjunctive normal form only. For simplicity, we call it the normal form. Since the propositional modal logic is obtained by adding $\Box$ to the propositional logic, and $\Box$ can be distributed over $\wedge$ and $\vee$ in K$^+$, we can get the conclusion that for any formula ψ of K$^+$, we have a formula φ in the normal form which is equivalent to ψ.

Proposition 2. For any formula ψ, there is a formula φ in the normal form such that

$$\vdash \psi \leftrightarrow \varphi.$$

Proof. By induction on the construction of the formula ψ. $\Box$

4   Pseudo-Atom-Complete Set of Formulas

In this section we give the definition of pseudo-atom-complete. In the propositional logic an atom-complete set determines a maximally consistent set in a unique way. Similarly, in K$^+$ a pseudo-atom-complete set determines a maximally consistent set in a unique way. In K$^+$ for any pseudo-atom-complete set A of formulas, if we delete one $\Box$ from each φ in A then the new set $\Box^- A$ is also pseudo-atom-complete. And the accessibility relation in the canonical model of K$^+$ can be simply given by eliminating one $\Box$ from each pseudo-atomic formula in a possible world.

Let A be a maximally consistent set of formulas in the propositional logic. For any propositional variable p, exactly one member of $\{p, \neg p\}$ is in A.

Definition 7. A set A of atomic formulas or the negation of atomic formulas is *atom-complete* if for any atomic formula p, exactly one member of $\{p, \neg p\}$ is in A.

Proposition 3. An atom-complete set A determines a maximally consistent set in a unique way.

Let B be a maximally consistent set of formulas of K$^+$. For any pseudo-atomic formula $\Box^k p$, exactly one member of $\{\Box^k p, \neg \Box^k p\}$ is in B.

Definition 8. A set A of pseudo-atomic formulas or the negation of pseudo-atomic formulas is *pseudo-atom-complete* if for any pseudo-atomic formula $\Box^k p$, exactly one member of $\{\Box^k p, \neg \Box^k p\}$ is in A.

For any maximally consistent set B of formulas, let

$$\sigma(B) = \{\sim^l \Box^k p :\sim^l \Box^k p \in B\}.$$

Then, $\sigma(B)$ is pseudo-atom-complete.

Proposition 4. For any pseudo-atom-complete set B of formulas, there is a unique maximally consistent set C of formulas such that

$$\sigma(C) = B.$$

Proof. It is easy to see that B is consistent. So by the theorem of Lindenbaum we can extend B to a maximally consistent set C of formulas and it satisfies the condition that $\sigma(C) = B$. Then we can prove that C is unique by contraposition. $\Box$

Theorem 3. For any pseudo-atom-complete set $A, \Box^- A = \{\sim^l \Box^{n-1}p :\sim^l \Box^n p \in A$ and $n \geq 1\}$ is pseudo-atom-complete.

Proof. By the definition of pseudo-atom-complete. $\Box$

For any set B of formulas, let

$$\Box^-(B) = \{\varphi : \Box\varphi \in B\}.$$

Let W be the set of all maximally consistent sets, and $R_\Box$ be the binary relation on W. For any possible worlds $w, w' \in W$, $(w, w') \in R_\Box$ if and only if $\Box^-(w) \subseteq w'$.

Corollary 1. Given a pseudo-atom-complete set A, let $\tau(A)$ be the unique maximally consistent set such that $\sigma(\tau(A)) = A$. Then,

$$\{\tau(A), \tau(\Box^- A), ..., \tau(\Box^{-n} A), ...\}$$

is a set of possible worlds such that
  (i) $(\tau(A), \tau(\Box^- A)) \in R_\Box$, and
  (ii) for any $n \geq 1, (\tau(\Box^{-n+1} A), \tau(\Box^{-n} A)) \in R_\Box$.
  Proof. We can prove (i) by contraposition. From the theorem 3 and the proposition 4 we can also prove (ii) with the same method of (i). $\Box$

5 Correspondences between K$^+$ and the Propositional Logic

In this section we build correspondences between K$^+$ and the propositional logic P^+. When a pseudo-atomic formula $\Box^k p_i(k, i \geq 0)$ is corresponding to a propositional variable q_{ki}, each formula in K$^+$ then can be corresponding to a formula in the propositional logic P^+. We also get the correspondence of models between K$^+$ and P^+. Then we get correspondences of theorems and valid formulas between them. So, the soundness theorem and the completeness theorem of K$^+$ follow directly from those of P^+.

Suppose that $P^+ = \{q_{ki} : k, i \geq 0\}$ is a language of the propositional logic. In order to give the correspondence between K^+ and P^+, we firstly redefine a formula φ in the normal form of K^+.

$$\theta ::= \Box^k p_i | \neg \Box^k p_i;$$
$$\psi ::= \theta | \psi_1 \wedge \psi_2;$$
$$\varphi ::= \psi | \varphi_1 \vee \varphi_2.$$

where $k, i \geq 0$ and p_i is a propositional variable of K^+.

Then we define a mapping σ of formulas from K^+ to P^+. And σ is defined as follows:

$$\sigma(\theta) = \begin{cases} q_{ki} & \text{if } \theta = \Box^k p_i \\ \neg q_{ki} & \text{if } \theta = \neg\Box^k p_i \end{cases}$$
$$\sigma(\psi) = \begin{cases} \sigma(\theta) & \text{if } \psi = \theta \\ \sigma(\psi_1) \wedge \sigma(\psi_2) & \text{if } \psi = \psi_1 \wedge \psi_2 \end{cases}$$
$$\sigma(\varphi) = \begin{cases} \sigma(\psi) & \text{if } \varphi = \psi \\ \sigma(\varphi_1) \vee \sigma(\varphi_2) & \text{if } \varphi = \varphi_1 \vee \varphi_2 \end{cases}$$

It is easy to see that each formula φ in the normal form in K^+ can be corresponding to a formula $\sigma(\varphi)$ in the normal form in P^+. Since each formula φ in K^+ can be equivalent to a formula in the normal form in K^+, each formula φ in K^+ can be corresponding to a formula in the normal form in P^+. For simplicity, we will only consider formulas in the normal form in K^+ and formulas in the normal form in P^+. Then it is easy to prove that σ is a bijective function. And we can also get that for any formula φ of K^+, $\sigma(\neg\varphi) = \neg\sigma(\varphi)$.

By induction on the number of formulas in the sequence forming the deduction of φ from K^+, we can get that if φ is a theorem of K^+ then $\sigma(\varphi)$ is a theorem of P^+. For example, if φ is a theorem of K^+ and we get it by an application of the N rule from the theorem $\varphi_1(p_0, p_1, ..., p_n)$, then $\sigma(\varphi)$ is also a theorem of the propositional logic. Since we can distribute $\Box$ over $\wedge$ and $\vee$, we get that $\varphi = \varphi_1(\Box p_0, \Box p_1, ..., \Box p_n)$. By induction hypothesis we know that $\sigma(\varphi_1) = \varphi_1'(q_{k_0 0}, q_{k_1 1}, ..., q_{k_n n})(0 \leq i \leq n, k_i \geq 0)$ is a theorem of P^+, when we use $q_{(k_i+1)i}$ to substitute for $q_{k_i i}$ we get a formula $\varphi_1'(q_{(k_0+1)0}, q_{(k_1+1)1}, ..., q_{(k_n+1)n})(0 \leq i \leq n, k_i \geq 0)$ of P^+, which is $\sigma(\varphi)$ and also a theorem of P^+. We know that each axiom of the propositional logic is an axiom of K^+, by using the rule MP each theorem of P^+ we get is also a theorem of K^+. So we get the following proposition.

Proposition 5. For any formula φ of K^+, φ is a theorem of K^+ if and only if $\sigma(\varphi)$ is a theorem of P^+. $\qquad\square$

Suppose that A is a set of formulas in K^+, and $A' = \{\sigma(\varphi) : \varphi \in A\}$ is a set of formulas of P^+. Then we have the following propositions.

Proposition 6. A is consistent if and only if A' is consistent. $\qquad\square$

Proposition 7. A is maximally consistent if and only if A' is maximally consistent. $\qquad\square$

For any model $M = \langle W, R, I \rangle$ of K^+ and any possible world $w \in W$, there is an assignment v_w of P^+ such that for any pseudo-atomic formula $\square^k p_i$,

$$M, w \models \square^k p_i \text{ if and only if } v_w \models q_{ki}.$$

Then we have the following proposition.

Proposition 8. For any formula φ of K^+, $M, w \models \varphi$ if and only if $v_w \models \sigma(\varphi)$.

Proof. Suppose that φ is in the normal form, and

$$\begin{aligned}
\varphi &= \varphi_1 \vee \cdots \vee \varphi_n \\
&= (\psi_{11} \wedge \cdots \wedge \psi_{1m_1}) \vee \cdots \vee (\psi_{n1} \wedge \cdots \wedge \psi_{nm_n}),
\end{aligned}$$

where $\psi_{ij} = \sim^{l_{ij}} \square^{k_{ij}} p_{ij}$, $1 \leq i \leq n$, $1 \leq j \leq m_i$ and $k_{ij} \geq 0$. Then

$$\begin{aligned}
\sigma(\varphi) &= \sigma(\varphi_1) \vee \cdots \vee \sigma(\varphi_n) \\
&= (\sigma(\psi_{11}) \wedge \cdots \wedge \sigma(\psi_{1m_1})) \vee \cdots \vee (\sigma(\psi_{n1}) \wedge \cdots \wedge \sigma(\psi_{nm_n})).
\end{aligned}$$

Obviously, $M, w \models \psi_{ij}$ if and only if $v_w \models \sigma(\psi_{ij})$.

$v_w \models \sigma(\varphi)$ if and only if there is an i such that $v_w \models \sigma(\varphi_i)$; if and only if there is an i and for any j, $v_w \models \sigma(\psi_{ij})$; if and only if there is an i and for any j, $M, w \models \psi_{ij}$; if and only if there is an i such that $M, w \models \varphi_i$; if and only if $M, w \models \varphi$. $\qquad\square$

For any assignment v of P^+ we can construct a model of K^+.

Let $W = \{v_0, v_1, v_2, ...\}$ be a set of assignments such that for any propositional variable $p_i (i \geq 0)$ of K^+,

$$\begin{aligned}
&v_0(p_i) = 1 \text{ if and only if } v(q_{0i}) = 1; \\
&v_1(p_i) = 1 \text{ if and only if } v(q_{1i}) = 1; \\
&\cdots \\
&v_k(p_i) = 1 \text{ if and only if } v(q_{ki}) = 1; \\
&\cdots
\end{aligned}$$

Let $M = \langle W, R, I \rangle$, where $R = \{(v_n, v_{n+1}) : n \geq 0\}$; for any propositional variable $p_i (i \geq 0)$ and any $k \geq 0$, $v_k \in I(p_i)$ if and only if $v_k(p_i) = 1$. Then we have the following proposition.

Proposition 9. For any formula φ of K^+, $M, v_0 \models \varphi$ if and only if $v \models \sigma(\varphi)$. $\qquad\square$

It is easy to get the following proposition.

Proposition 10. For any formula φ of K^+, $\models \varphi$ if and only if $\models \sigma(\varphi)$.

Proof. The proof is by contraposition. $\qquad\square$

Let C be the class of frames in which each world can only see one world, itself or another. We should prove that K^+ is sound and complete with respect to C. And theorems of K^+ are also theorems of P^+, and vice versa; valid formulas of K^+ are also valid formulas of P^+, and vice versa. So we can easily get the following theorems from the soundness and completeness of P^+.

Theorem 4.(Soundness of K^+) For any formula φ of K^+, if $\vdash \varphi$ then $\models_C \varphi$.

Theorem 5.(Completeness of K^+) For any formula φ of K^+, if $\models_C \varphi$ then $\vdash \varphi$.

6   Conclusions

In this paper we construct a new propositional modal system K^+, in which maximally consistent sets have a property holding in the propositional logic: a maximally consistent set is uniquely determined by a set of pseudo-atom-complete formulas. Each formula in such a logic is equivalent to a formula in the disjunctive normal form, where each disjunct is the conjunction of pseudo-atomic formula or the negation of pseudo-atomic formula. The accessibility relation in the canonical model of K^+ can be simply given by eliminating one $\square$ from each pseudo-atomic formula or its negation in a possible world. We also build correspondences of formulas, theorems and valid formulas between K^+ and the propositional logic P^+. So, the soundness theorem and the completeness theorem of K^+ follow directly from those of P^+.

Acknowledgements. This work is supported by National Natural Science Foundation of China under grant No. 60573064, 91224006, 61203284, and 61173063.

References

1. Van Benthem, J.: Modal correspondence theory [Ph.D. Thesis]. University of Amsterdam, Netherlands (1976)
2. Blackburn, P., Van Benthem, J., Wolter, F.: Handbook of Modal Logic. Elsevier Science Ltd (2006)
3. Van Benthem, J.: Correspondence Theory. In: Gabbay, D., Guenthner, F. (eds.) Handbook of Philosophical Logic, pp. 325–408. Kluwer Academic Publishers (2001)
4. Aloni, M.: Quantification under Conceptual Covers [Ph.D. Thesis]. University of Amsterdam, Amsterdam (2001)
5. Aloni, M.: Individual Concepts in Modal Predicate Logic. Journal of Philosophical Logic 34(1), 1–64 (2005)
6. Hazen, A., Rin, B., Wehmeier, K.: Actuality in Propositional Modal Logic. Studia Logica 101(3), 487–503 (2013)
7. Hughes, G.E., Cresswell, M.J.: A New introduction to Modal Logic. Routledge, Lodon (1996)
8. Hamilton, A.G.: Logic for Mathematicians. Cambridge University Press (1988)
9. Fitting, M., Mendelsohn, R.: First-order Modal Logic. Kluwer Academic Publishers, The Netherlands (1998)

10. Kripke, S.: Semantical analysis of modal logic I: Normal modal propositional calculi. Zeitschrift für Mathematische Logik und Grundlagen der Mathematik 9, 67–96 (1963)

A Sound and Complete Axiomatic System for Modality $\Box\varphi \equiv \Box_1\varphi \wedge \Box_2\varphi$

Shaobo Deng[1,2], Meiying Sun[1,2], Cungen Cao[1], and Yuefei Sui[1]

[1] Key Laboratory of Intelligent Information Processing,
Institute of Computing Technology, Chinese Academy of Science,
Beijing, 100190, China
houjiyuan2002@163.com
[2] University of Chinese Academy of Sciences,
Beijing, 100049, China

Abstract. An axiomatic system is presented in this paper, which has a modal operator $\Box$ such that $\Box\varphi \equiv \Box_1\varphi \wedge \Box_2\varphi$, where $\Box_1$ and $\Box_2$ are the modal operators of the language for the axiom system $S5$. The axiomatic system for $\Box$ is proved to be sound and complete.

Keywords: Modal logic, Axiomatic system $S5$, Soundness, Completeness, Canonical model.

1 Introduction

The modal logic has many axiomatic systems, such as $K, T, D, B, S4$ and $S5$ ([1]). The axiom system $S5$ is characterized by all equivalence frames([1]). The approximation spaces for Rough sets can be used as the possible-world semantics for $S5$. Let an approximate space (U, R) be an equivalence frame $\langle W, R \rangle$ for $S5$, i.e., $U = W$. Then for any formula φ, if the interpretation of φ corresponds to a subset X of U, then the lower and upper approximations of X correspond to the interpretations of $\Box\varphi$ and $\Diamond\varphi$, respectively, and the equivalence relation R corresponds to the accessibility relation for $\Box([2])$.

Given two approximation spaces (U, R_1) and (U, R_2), $R_1 \cup R_2$ may not be an equivalence relation. Given two modal operators $\Box_1$ and $\Box_2$, let R_1 and R_2 be the accessibility relations for $\Box_1$ and $\Box_2$, respectively. Let $\Box$ be a modal operator such that $R_1 \cup R_2$ is the accessibility relation for $\Box$, that is, $M, w \vDash \Box\varphi$ iff for any $w' \in W$ if $(w, w') \in R_1 \cup R_2$ then $M, w' \vDash \varphi$, which implies and is implied by that for any $w' \in W$ if $(w, w') \in R_1$ then $M, w' \vDash \varphi$ and for any $w' \in W$ if $(w, w') \in R_2$ then $M, w' \vDash \varphi$, i.e., $M, w \vDash \Box_1\varphi$ and $M, w \vDash \Box_2\varphi$ if and only if $M, w \vDash \Box\varphi$.

Let $\Box$ be a modal operator such that for any possible world $w, M, w \vDash \Box\varphi$ iff $M, w \vDash \Box_1\varphi$ and $M, w \vDash \Box_2\varphi$, i.e., for any formula φ, $\Box\varphi \equiv \Box_1\varphi \wedge \Box_2\varphi$. In this paper, we consider the modal operator $\Box\varphi \equiv \Box_1\varphi \wedge \Box_2\varphi$. We shall give the language, the syntax and the semantics for the modal logic with modal operator $\Box\varphi \equiv \Box_1\varphi \wedge \Box_2\varphi$. The axiomatic system for $\Box$ will be given and proved to be sound and complete.

Z. Shi et al. (Eds.): IIP 2014, IFIP AICT 432, pp. 152–160, 2014.

The main contribution of this paper is that a propositional modal logic with a modal operator $\Box\varphi \equiv \Box_1\varphi \wedge \Box_2\varphi$. The axiomatic system for $\Box$ is sound, and complete with respect to the class of all reflective and symmetric frames, where the accessibility relation R for $\Box$ is equivalent to $R_1 \cup R_2$, where R_i is the equivalence relation for $\Box_i$ and $i = 1, 2$.

If $R_1 = R_2$, then the axiomatic system for $\Box$ turns out to be $S5$.

This paper is organized as follows: the propositional modal logic with modal operator $\Box\varphi \equiv \Box_1\varphi \wedge \Box_2\varphi$ is described in section 2, including the language, the syntax and the semantics for the logic. Then we shall give the axiomatic system for $\Box$ and prove the soundness theorem and the completeness theorem. Section 3 summaries results of the paper and discusses some possible extension of the logic.

2   The Propositional Modal Logic with Modal Operator $\Box\varphi \equiv \Box_1\varphi \wedge \Box_2\varphi$

In this section, we shall give the language, the syntax and the semantics for the propositional modal logic with the modality $\Box\varphi \equiv \Box_1\varphi \wedge \Box_2\varphi$. The axiomatic system for $\Box$ is denoted by $^{S5_1}\wedge^{S5_2}$, then we prove that $^{S5_1}\wedge^{S5_2}$ is sound and complete.

2.1   The Language, Syntax and Semantics for the Logic

The language for $^{S5_1}\wedge^{S5_2}$ contains the following symbols:
- propositional variables: $p_0, p_1, ...$;
- logical connectives: $\neg, \rightarrow$;
- modalities: $\Box, \Box_1, \Box_2$;
- auxiliary symbols: $(,)$.

Formulas:

$$\varphi := p | \varphi_1 \rightarrow \varphi_2 | \neg\varphi_1 | \Box\varphi_1;$$
$$\Box\varphi := \Box_1\varphi_1 \wedge \Box_2\varphi_1.$$

Other operators:

$$(\alpha \vee \beta) =_{\text{def}} (\neg\alpha \rightarrow \beta);$$
$$(\alpha \wedge \beta) =_{\text{def}} \neg(\alpha \rightarrow \neg\beta);$$
$$(\alpha \leftrightarrow \beta) =_{\text{def}} ((\alpha \rightarrow \beta) \wedge (\beta \rightarrow \alpha));$$
$$(\Diamond\alpha) =_{\text{def}} (\neg\Box\neg\alpha).$$

Definition 2.1. A frame F is a triple $\langle W, R_1, R_2 \rangle$, where W is a non-empty set of possible worlds, and $R_1 \subseteq W^2$ and $R_2 \subseteq W^2$ are the equivalence relations defined over the members of W and the accessibility relations for $\Box_1$ and $\Box_2$, respectively.

Definition 2.2. A model M is a quadruple $\langle W, R_1, R_2, I \rangle$, where $\langle W, R_1, R_2 \rangle$ is a frame and I is an interpretation such that for any propositional variable p $I(p) \subseteq W$ and for any $w \in I(p)$ p is true in w.

A satisfaction relation $\vDash$, between any formula φ and any possible world w, is defined as follows:

Definition 2.3. Given any model M, any possible world $w \in W$ and any formula φ,

$$M, w \vDash \varphi \text{ iff } \begin{cases} w \in I(p) & \text{if } \varphi = p \\ M, w \nvDash \varphi_1 & \text{if } \varphi = \neg\varphi_1 \\ M, w \vDash \varphi_1 \Rightarrow M, w \vDash \varphi_2 & \text{if } \varphi = \varphi_1 \to \varphi_2 \\ \text{for all } w' \in W \text{ if } wR_1w' \text{ then } M, w' \vDash \varphi_1, \text{ and} \\ \quad \text{for all } w' \in W \text{ if } wR_2w' \text{ then } M, w' \vDash \varphi_1 & \text{if } \varphi = \Box\varphi_1 \end{cases}$$

By the definition of the satisfaction relation, we can give the following definition:

Definition 2.4. A formula φ is valid in a model M, dented by $M \vDash \varphi$, iff for any $w \in W$ $M, w \vDash \varphi$; a formula φ is valid in a frame F, denoted by $F \vDash \varphi$, iff for any model M based on F $M \vDash \varphi$; let C be a class of frames. A formula φ is valid in C iff for any $F \in C$ $F \vDash \varphi$; $\Sigma \vDash_C \varphi$ iff for any frame $F \in C$ if $F \vDash \Sigma$ then $F \vDash \varphi$. If $\Sigma = \emptyset$ then $\vDash_C \varphi$.

Now we give the following axiom schemas and inference rules for $^{S5_1} \wedge ^{S5_2}$:

- **Axiom schemes:**

$$L1 \quad \varphi \to (\psi \to \varphi)$$
$$L2 \quad (\varphi \to (\psi \to \mu)) \to ((\varphi \to \psi) \to (\varphi \to \mu))$$
$$L3 \quad (\neg\psi \to \neg\varphi) \to (\varphi \to \psi)$$
$$L4 \quad \Box(\varphi \to \psi) \to (\Box\varphi \to \Box\psi)$$
$$L5 \quad \Box\varphi \to \varphi$$
$$L6 \quad \varphi \to \Box\Diamond\varphi$$
$$L7_1 \quad \Box_1\varphi \to \Box_1\Box_1\varphi$$
$$L7_2 \quad \Box_2\varphi \to \Box_2\Box_2\varphi$$

- **Inference rules:**

$$(MP) \quad \frac{\varphi, \varphi \to \psi}{\psi}$$

$$(N) \quad \frac{\varphi}{\Box\varphi}$$

Definition 2.5. A formula φ is provable from Γ, denoted by $\Gamma \vdash \varphi$, if there is a sequence of formulas $\varphi_1, ..., \varphi_n$ such that $\varphi = \varphi_n$, and for each $1 \leq i \leq n$, either φ_i is an axiom or a formula in Γ, or is deduced from the previous formulas via one of the deduction rules.

2.2 The Soundness Theorem

This section is to prove the soundness theorem by induction on the length of proofs. Before giving the proof, we give the following lemmas:

Lemma2.1. Each axiom schema is valid.

Proof. As for the axiom schema $L1, L2, L3$, we do not check their validity and two references are [1] and [3].

($L5$) By the definition 2.3, it is easy to prove it.

($L6$) By the definition 2.3, it is easy to prove it.

($L7_1$) Since the accessibility relation R_1 for $\Box_1$ is an equivalence relation, it follows that $\Box_1\varphi \rightarrow \Box_1\Box_1\varphi$ is valid.

($L7_2$) Since the accessibility relation R_2 for $\Box_2$ is an equivalence relation, it follows that $\Box_2\varphi \rightarrow \Box_2\Box_2\varphi$ is valid.   □

Lemma2.2. The deduction rules preserve validity.

Proof. We prove that (N) preserves the validity. Since for any model $\langle W, R_1, R_2, I \rangle$ based on any frame $\langle W, R_1, R_2 \rangle$ and any $w \in W$, $M, w \vDash \varphi$.

Let w_1 be any possible world. Since R_1 and R_2 are the equivalence relations on W, we can obtain that for any $w_1' \in W$ if $w_1 R_1 w_1'$ then $M, w_1' \vDash \varphi$, and for any $w_1'' \in W$ if $w_1 R_2 w_1''$ then $M, w_1'' \vDash \varphi$. It follows that $M, w_1 \vDash \Box\varphi$. Since for any $w_1 \in W$ $M, w_1 \vDash \Box\varphi$, it follows that $\vDash \Box\varphi$.   □

Theorem 2.1(The Soundness Theorem). For any set of formulas Γ and formula φ, if $\Gamma \vdash \varphi$, then $\Gamma \vDash \varphi$.

Proof. For any set of formulas Γ and formula φ, since $\Gamma \vdash \varphi$, φ is the last member of a sequence which is a deduction from Γ. So we can use induction on the number of the sequence to prove this theorem as follows:

For the base step, the sequence has only one formula, namely φ. Then φ must be an axiom of $^{S5_1 \wedge S5_2}$ or a member of Γ, and then $\Gamma \vDash \varphi$.

Now suppose that the sequence contains n formulas,where n>1,and suppose as induction hypothesis that $\Gamma \vDash \alpha$ follows from $\Gamma \vdash \alpha$,which sequence is fewer than n members. There are the following cases:

Case a. φ is an axiom of $^{S5_1 \wedge S5_2}$ or a member of Γ, then we have $\Gamma \vDash \varphi$

Case b. φ is obtained by modus ponens rule from a formula ψ and a formula $\psi \rightarrow \varphi$ in the sequence. So by induction hypothesis, it obtains that $\Gamma \vDash \psi$ and $\Gamma \vDash \psi \rightarrow \varphi$, then it follows that $\Gamma \vDash \varphi$.

Case c. $\varphi = \Box\psi$ is obtained by the inference rule N from ψ. So by induction hypothesis, it follows that $\Gamma \vDash \psi$. Since if for any model $\langle W, R_1, R_2, I \rangle$ based on any frame $\langle W, R_1, R_2 \rangle$ and any $w \in W$ $M, w \vDash \psi$ then for any model $\langle W, R_1, R_2, I \rangle$ based on any frame $\langle W, R_1, R_2 \rangle$ and any $w \in W$ $M, w \vDash \Box\psi$, it follows that $\Gamma \vDash \Box\psi$.   □

2.3 The Completeness Theorem

The completeness theorem is to be proved in this section. The proof method of the complete theorem is similar to the classical canonical model method ([1]). We shall construct two relations on W and prove whether the two relations are equivalence relations or not. And one relation corresponds to the accessibility relation for $\square_1$, while the other corresponds to the accessibility relation for $\square_2$.

Definition 2.6. Γ is consistent iff there is no finite set $\{\varphi_1,...,\varphi_n\} \subseteq \Gamma$ such that:

$$\vdash \neg(\varphi_1 \wedge ... \wedge \varphi_n).$$

By definition 2.6, we can prove that Γ is inconsistent iff there is some formula φ such that $\Gamma \vdash \varphi$ and $\Gamma \vdash \neg\varphi$.

Lemma 2.3. Suppose that Σ is a consistent set of formulas. Then there is a maximal consistent set of formulas Σ^* such that $\Sigma \subseteq \Sigma^*$.

In constructing a model in which the possible worlds are maximal consistent sets of formulas we will have to specify when one world is accessible from another(that model, in this paper, is also called canonical model). Thereby, the accessibility relation R_1 and R_2, in the canonical model, is defined as follows:

Definition 2.7. For any two distinct maximal consistent sets Σ_1^*, Σ_2^*, we define two binary relations R_1 and R_2 on W as follows:

(1) We shall say that $\Sigma_1^* R_1 \Sigma_2^*$ iff Σ_1^* and Σ_2^* satisfy the following condition: For any formula φ if $\square\varphi \in \Sigma_1^*$ then $\varphi \in \Sigma_2^*$ (written:$S^-(\Sigma_1^*) = \{\varphi : \square\varphi \in \Sigma_1^*\}$).

(2) We shall say that $\Sigma_1^* R_2 \Sigma_2^*$ iff Σ_1^* and Σ_2^* satisfy the following condition: For any formula φ if $\square\varphi \in \Sigma_1^*$ then $\square\varphi \in \Sigma_2^*$ (written: $S(\Sigma_1^*) = \{\square\varphi : \square\varphi \in \Sigma_1^*\}$).

Lemma 2.4. Let $\Gamma^* = \{\Sigma_0^*, \Sigma_1^*, ...\}$ be the set of all maximal consistent sets. If for any $i, j \in \mathbb{N}$ we define $\Sigma_i^* R_1 \Sigma_j^*$ iff $S^-(\Sigma_i^*) \subseteq \Sigma_j^*$, and $\Sigma_i^* R_2 \Sigma_j^*$ iff $S(\Sigma_i^*) \subseteq \Sigma_j^*$, then both the relation R_1 and R_2 are equivalence relations on W.

Proof. (1) In order to prove that R_1 is an equivalence relation, we shall prove the following three conditions:

1) For any $i \in N$, if $\Sigma_i^* \in \Gamma^*$ then $\Sigma_i^* R_1 \Sigma_i^*$.

2) For any $\Sigma_i^*, \Sigma_j^* \in \Gamma^*$, if $\Sigma_i^* R_1 \Sigma_j^*$ then $\Sigma_j^* R_1 \Sigma_i^*$.

3) For any $\Sigma_1^*, \Sigma_2^*, \Sigma_3^* \in \Gamma^*$, if $\Sigma_1^* R_1 \Sigma_2^*$ and $\Sigma_2^* R_1 \Sigma_3^*$ then $\Sigma_1^* R_1 \Sigma_3^*$.

1). For any $i \in N$, we shall prove $S^-(\Sigma_i^*) \subseteq \Sigma_i^*$. For any formula φ, if $\square\varphi \in \Sigma_i^*$ then $\varphi \in S^-(\Sigma_i^*)$. Since $\square\varphi \to \varphi(L5)$ and $\square\varphi \in \Sigma_i^*$, $\varphi \in \Sigma_i^*$. It follows that $S^-(\Sigma_i^*) \subseteq \Sigma_i^*$.

2). We shall prove that if $S^-(\Sigma_i^*) \subseteq \Sigma_j^*$ then $S^-(\Sigma_j^*) \subseteq \Sigma_i^*$, that is to say, we shall prove that for any formula β if $\square\beta \in \Sigma_j^*$ then $\beta \in \Sigma_i^*$. Suppose $\beta \notin \Sigma_i^*$. $\neg\beta \in \Sigma_i^*$. By $L6$ and $\neg\beta \in \Sigma_i^*$ it follows that $\square\Diamond\neg\beta \in \Sigma_i^*$.

Since $S^-(\Sigma_i^*) \subseteq \Sigma_j^*$, it follows that $\Diamond\neg\beta \in \Sigma_j^*$, that is, $\neg\square\beta \in \Sigma_j^*$. Since $\neg\square\beta \in \Sigma_j^*$ and $\square\beta \in \Sigma_j^*$, Σ_j^* is not consistent, which is a contradiction to the hypothesis of this lemma.

3). We need to prove that if $S^-(\Sigma_1^*) \subseteq \Sigma_2^*$ and $S^-(\Sigma_2^*) \subseteq \Sigma_3^*$ then $S^-(\Sigma_1^*) \subseteq \Sigma_3^*$, that is to say, for any formula β, if $\square\beta \in \Sigma_1^*$ then $\beta \in \Sigma_3^*$. We can prove it by $L5$ and $L7_1$.

What we need to explain is that R_1 is the accessibility relation for $\Box_1$ and in this case for any formula φ $\Box\varphi \equiv \Box_1\varphi$. Thereby, we can use the axiom schema $L7_1$.

(2) Now, we prove that the relation R_2 is an equivalence relation. the following three conditions shall be proved:

1) For any $i \in N$, if $\Sigma_i^* \in \Gamma^*$ then $\Sigma_i^* R_2 \Sigma_i^*$.
2) For any $\Sigma_i^*, \Sigma_j^* \in \Gamma^*$, if $\Sigma_i^* R_2 \Sigma_j^*$ then $\Sigma_j^* R_2 \Sigma_i^*$.
3) For any $\Sigma_1^*, \Sigma_2^*, \Sigma_3^* \in \Gamma^*$, if $\Sigma_1^* R_2 \Sigma_2^*$ and $\Sigma_2^* R_2 \Sigma_3^*$ then $\Sigma_1^* R_2 \Sigma_3^*$.

It is easy to prove the item 1)-3). We omit the proof procedures. $\quad\square$

Lemma 2.5. Let Γ be any consistent set of formulas containing $\neg\Box\psi$, then $S^-(\Gamma) \cup \{\neg\psi\}$ is consistent and $S(\Gamma) \cup \{\neg\psi\}$ is consistent, where $S^-(\Gamma) = \{\varphi : \Box\varphi \in \Gamma\}$ and $S(\Gamma) = \{\Box\varphi : \Box\varphi \in \Gamma\}$.

Proof. (1) We shall prove that $S^-(\Gamma) \cup \{\neg\psi\}$ is consistent as follows:

Suppose that $S^-(\Gamma) \cup \{\neg\psi\}$ is not consistent. Then there exists some finite subset $\{\varphi_1, ..., \varphi_n\} \cup \{\neg\psi\}$ of $S^-(\Gamma) \cup \{\neg\psi\}$ such that $\vdash \neg(\varphi_1 \wedge ... \wedge \varphi_n \wedge \neg\psi)$. Then

$$
\begin{aligned}
\vdash (\varphi_1 \wedge ... \wedge \varphi_n) \to \neg\psi \ &\text{iff} \vdash \Box(\varphi_1 \wedge ... \wedge \varphi_n) \to \psi \\
&\text{iff} \vdash \Box(\varphi_1 \wedge ... \wedge \varphi_n) \to \Box\psi \\
&\text{iff} \vdash (\Box\varphi_1 \wedge ... \wedge \Box\varphi_n) \to \Box\psi \\
&\text{iff} \vdash \neg(\Box\varphi_1 \wedge ... \wedge \Box\varphi_n \wedge \neg\Box\psi)
\end{aligned}
$$

Thereby, $\{\Box\varphi_1, ..., \Box\varphi_n\} \cup \{\neg\Box\psi\}$ is not consistent. By the definition of $S^-(\Gamma)$ it follows that $\{\Box\varphi_1, ..., \Box\varphi_n\} \cup \{\neg\Box\psi\}$ is a subset of Γ. Thereby, Γ is not consistent, which is a contradiction to the hypothesis of this lemma.

(2) We shall prove that $S(\Gamma) \cup \{\neg\psi\}$ is consistent as follows. Suppose that $S(\Gamma) \cup \{\neg\psi\}$ is not consistent. Then there exists some finite subset $\{\Box\varphi_1, ..., \Box\varphi_n\} \cup \{\neg\psi\}$ of $S(\Gamma) \cup \{\neg\psi\}$ such that $\vdash \neg(\Box\varphi_1 \wedge ... \wedge \Box\varphi_n \wedge \neg\psi)$. Then

$$
\begin{aligned}
\vdash (\Box\varphi_1 \wedge ... \wedge \Box\varphi_n) \to \neg\psi \ &\text{iff} \vdash \Box(\Box\varphi_1 \wedge ... \wedge \Box\varphi_n) \to \psi \\
&\text{iff} \vdash \Box(\Box\varphi_1 \wedge ... \wedge \Box\varphi_n) \to \Box\psi \\
&\text{iff} \vdash (\Box\varphi_1 \wedge ... \wedge \Box\varphi_n) \to \Box\psi \\
&\text{iff} \vdash \neg(\Box\Box\varphi_1 \wedge ... \wedge \Box\Box\varphi_n \wedge \neg\Box\psi)
\end{aligned}
$$

Thereby, $\{\Box\Box\varphi_1, ..., \Box\Box\varphi_n\} \cup \{\neg\Box\psi\}$ is not consistent. For any $\Box\varphi_i$, by $L7_2(\Box\varphi \equiv \Box_2\varphi)$ it follows that $\Box\Box\varphi \in \Gamma$, where i=1,...,n. Thereby, $\{\Box\Box\varphi_1, ..., \Box\Box\varphi_n\} \cup \{\neg\Box\psi\}$ is a subset of Γ. Then Γ is not consistent, which is a contradiction to the hypothesis of this lemma.

What we need to explain is that R_2 is the accessibility relation for $\Box_2$ and in this case for any formula φ $\Box\varphi \equiv \Box_2\varphi$. Thereby, we can use the axiom schema $L7_2$. $\quad\square$

The canonical model for $S5_1 \wedge S5_2$, M, is like any other model, a quadruple $\langle W, R_1, R_2, I \rangle$. W is the set of all sets of maximal consistent sets of formulas. I.e. $w \in W$ iff w is a maximal consistent set of formulas. If w and w' are both in W, then wR_1w' iff $S^-(w) \subseteq w'$. And if w and w' are both in W, then wR_2w'

iff $S(w) \subseteq w'$. For any propositional variable p $I(p) \subseteq W$, and for any $w \in I(p)$ p is true in w iff $p \in w$. For any other formula this has to be proved as follows:

Lemma 2.6. Let $M = \langle W, R_1, R_2, I \rangle$ be the canonical model for $^{S5_1} \wedge ^{S5_2}$. Then for any formula φ and any world w, $M, w \vDash \varphi$ iff $\varphi \in w$.

Proof. We prove the lemma by induction on the structure of formulas.

Case a. $\varphi := p$: By definition, this lemma holds.

Case b. $\varphi := \neg\alpha$:

$$M, w \vDash \neg\alpha \text{ iff } M, w \nvDash \alpha$$
$$\text{iff } \alpha \notin w$$
$$\text{iff } \neg\alpha \in w$$

Case c. $\varphi := \alpha \to \beta$:

$$\alpha \to \beta \notin w \text{ iff } \neg(\alpha \to \beta) \in w$$
$$\text{iff } \alpha \in w \text{ and } \neg\beta \in w$$
$$\text{iff } M, w \vDash \alpha \text{ and } M, w \nvDash \beta$$
$$\text{iff } M, w \nvDash \alpha \to \beta$$

Case d. $\varphi := \Box\alpha$:

($\Leftarrow$) **Subcase d.1.** $\Box\alpha \in w$: By the definition of R_1, R_2, we have the following two cases:

(1) For any $w' \in W$, if wR_1w' then $\alpha \in w'$. $\alpha \in w'$ iff $M, w' \vDash \alpha$ by induction hypothesis. Then $M, w \vDash \Box\alpha$ because for any w' if wR_1w' and $M, w' \vDash \alpha$.

(2) For any $w' \in W$, if wR_2w' then $\Box\alpha \in w'$. Since $\Box\alpha \in w'$ and $\Box\alpha \to \alpha$, $\alpha \in w'$. $\alpha \in w'$ iff $M, w' \vDash \alpha$ by induction hypothesis. Then $M, w \vDash \Box\alpha$ because for any w' if wR_2w' and $M, w' \vDash \alpha$.

Subcase d.2. $\neg\Box\alpha \in w$: By the lemma 2.5 , both $S^-(w) \cup \{\neg\alpha\}$ and $S(w) \cup \{\neg\alpha\}$ are consistent. By the lemma 2.3, we can enlarge $S^-(w) \cup \{\neg\alpha\}$ and $S(w) \cup \{\neg\alpha\}$ into maximal consistent sets of formulas w_1 and w_2, respectively. Thereby, there exist w_1, w_2 such that $S^-(w) \subseteq w_1$ and $S(w) \subseteq w_2$.

Since $S^-(w) \subseteq w_1$ and $S(w) \subseteq w_2$, by the definition of R_1 and R_2, it follows that wR_1w_1 and wR_2w_2. For $\neg\alpha \in w_1$ and $\neg\alpha \in w_2$, it follows that $M, w_1 \models \neg\alpha$ and $M, w_2 \models \neg\alpha$ by induction hypothesis. So there exist w_1, w_2 such that wR_1w_1 and wR_2w_2 and $M, w_1 \models \neg\alpha$ and $M, w_2 \models \neg\alpha$. Therefore, by the definition 2.3, it follows that $M, w \nvDash \Box\alpha$.

($\Rightarrow$) **Subcase d.3.** $M, w \vDash \Box\alpha$: By the definition 2.3, we have:

(1) For any $w' \in W$, if wR_1w' then $M, w' \vDash \alpha$; and

(2) For any $w' \in W$, if wR_2w' then $M, w' \vDash \alpha$.

From (1), we have:

(3) For any $w' \in W$, if wR_1w' then $\alpha \in w'$ by induction hypothesis.

From (2),we have:

(4) For any $w' \in W$, if wR_2w' then $\alpha \in w'$ by induction hypothesis.

Assume $\neg\Box\alpha \in w$. Since w is a maximal consistent set of formulas, it follows that there exists $w_1, w_2 \in W$ such that wR_1w_1 and wR_2w_2 and $\neg\alpha \in w_1$ and $\neg\alpha \in w_2$ by the lemma 2.5. So $\neg\alpha \in w_1$ is a contradiction to (3), and $\neg\alpha \in w_1$ and $\neg\alpha \in w_2$ is also a contradiction to (4). Thereby, $\Box\alpha \in w$ for w is a maximal consistent set of formulas.

Subcase d.4. $M, w \nvDash \Box\alpha$: By the definition 2.3, we have:

(1) There exists $w_1 \in W$ such that wR_1w_1 and $M, w_1 \nvDash \alpha$; or

(2) There exists $w_2 \in W$ such that wR_2w_2 and $M, w_2 \nvDash \alpha$.

From (1), by induction hypothesis, $\neg\alpha \in w_1$; and from (2), $\neg\alpha \in w_2$. Suppose $\Box\alpha \in w$, by the definition of R_1 and R_2, it follows that:

(3) For any $w' \in W$ if wR_1w' then $\alpha \in w'$; and

(4) For any $w' \in W$ if wR_2w' then $\Box\alpha \in w'$.

From (4) and $L5$, we have:

(5) For any $w' \in W$ if wR_2w' then $\alpha \in w'$.

It follows that (3) is in contradiction with wR_1w_1 and $\neg\alpha \in w_1$. And (5) also is in contradiction with wR_2w_2 and $\alpha \in w_2$. Thereby, $\neg\Box\alpha \in w$ for that w is a maximal consistent set of formulas.   □

Theorem 2.2 (The Completeness Theorem). For any set of formulas Γ and formula φ, if $\Gamma \vDash \varphi$, then $\Gamma \vdash \varphi$.

Proof. Suppose $\Gamma \nvdash \varphi$, so $\Gamma \cup \{\neg\varphi\}$ is consistent. Then there is a maximal consistent set of formulas w such that $\Gamma \cup \{\neg\varphi\} \subseteq w$ by the lemma 2.3.

Let $M = \langle W, R_1, R_2, I \rangle$ be a canonical model, where W is the set of all sets of maximal consistent sets of formulas. I.e. $w \in W$ iff w is a maximal consistent set of formulas. If w and w' are both in W, then wR_1w' iff $S^-(w) \subseteq w'$. And if w and w' are both in W, then wR_2w' iff $S(w) \subseteq w'$. By the lemma 2.4 it follows that R_1 and R_2 are the equivalence relations on W.

For any formula $\alpha \in \Gamma \cup \{\neg\varphi\}$, $\alpha \in w$. By the theorem 2.6, $M, w \vDash \alpha$. So $M, w \vDash \Sigma$ and $v \vDash \neg\varphi$, which is a contradiction to $\Sigma \vDash \varphi$.

So for any set of formulas Γ and formula φ, if $\Gamma \vDash \varphi$, then $\Gamma \vdash \varphi$.   □

3 Conclusion

In this paper, we present an axiomatic system for a modality $\Box\varphi \equiv \Box_1\varphi \wedge \Box_2\varphi$, and prove that the axiomatic system for $\Box$ is sound and complete. The axiomatic system for $\Box$ is different from $S5$. What we need to point out is that $L7_1$ and $L7_2$ are not the axiom schemas for $\Box$, which is needed when proving the completeness theorem. An interesting problem is to give a sound and complete axiomatic system for the modality corresponding to the accessibility relation $R = R_1 \cap R_2$, where the equivalence relation R_i is the accessibility relation for $\Box_i, i = 1, 2$.

Acknowledgement. This work is supported by National Natural Science Foundation of China under grant No. 91224006,61035004,61173063,61203284,61363047, and the ministry of science and technology project under grant No. 201303107.

References

[1] Hughes, G.E., Cresswell, M.J.: A new introduction to modal logic. Burns & Oates (1996)

[2] Lin, T.Y., Liu, Q.: First-order rough logic i: approximate reasoning via rough sets. Fundamenta Informaticae 27(2), 137–153 (1996)

[3] Ebbinghaus, H.-D., Flum, J., Thomas, W.: Mathematical logic. Springer (1994)

[4] Carnielli, W.A., Pizzi, C., Bueno-Soler, J.: Modalities and multimodalities, vol. 12. Springer (2008)

[5] Corsi, G., Orlandelli, E.: Free quantified epistemic logics. Studia Logica 101(6), 1159–1183 (2013)

[6] Blanco, R., Casado, G.d.M., Requeno, J.I., Colom, J.M.: Temporal logics for phylogenetic analysis via model checking. In: 2010 IEEE International Conference on Bioinformatics and Biomedicine Workshops (BIBMW), pp. 152–157. IEEE (2010)

[7] Proietti, C.: Intuitionistic epistemic logic, kripke models and fitch's paradox. Journal of philosophical logic 41(5), 877–900 (2012)

[8] van Benthem, J., Minică, Ş.: Toward a dynamic logic of questions. In: Logic, Rationality, and Interaction, pp. 27–41. Springer (2009)

[9] Sietsma, F., van Eijck, J.: Action emulation between canonical models. In: Proceedings of the 10th Conference on Logic and the Foundations of Game and Decision Theory, p. 6 (2012)

Verification of Branch-Time Property Based on Dynamic Description Logic

Yaoguang Wang, Liang Chang, Fengying Li, and Tianlong Gu

Guangxi Key Laboratory of Trusted Software,
Guilin University of Electronic Technology,Guilin 541004,China
wangyguang@qq.com, {changl,lfy,cctlgu}@guet.edu.cn

Abstract. The dynamic description logic DDL provides formalism for describing dynamic system in the semantic Web environment Model checking is a formal verification method based on state transition system. In this paper, we bring dynamic description logic into model checking. Firstly, state transition systems considered in model checking are modeled as complex actions in dynamic description logic. Secondly, a kind of temporal description logic DL-CTL is introduced to specify temporal properties on state transition systems, where DL-CTL is a DL-based extension of propositional branch-time temporal logic CTL. Finally, verification algorithm is presented with the help of reasoning mechanisms provided by description logic.

Keywords: dynamic description logic, verification, temporal description logic, action theory.

1 Introduction

Model checking [1] is a formal verification method widely used in recent years, which is based on state transition system. However, the traditional model checking has some limitations. Firstly, it does not consider what makes state change. Secondly, it just uses temporal logic based on proposition logic to specify the property, which limits the scope of specifying properties. So, researchers begin to combine the action theory [2] to verification problem and consider the action make the state change.

Verification problem with action theory has been addressed by some researchers. In [3], the author puts forward a method of verifying temporal properties based on infinite sequence of Golog program and checks whether the execution of program sequence can satisfy temporal properties. In [4], the author aims at the fully automated verification of non-terminating Golog programs and uses an extension of situation calculus by constructing the first-order temporal logic CTL^*. However, the problems both of them consider are all undecidable. For this reason, [5] begins to use the action theory based on decidable description logic to check whether there is an execution sequence of action can satisfy the temporal property specified in linear temporal description logic DL-LTL [6].

However, [5] just considers the atom action and only the linear-time properties can be verified. Based on this limitation, we consider the action in dynamic description logic that contains action constructors like sequence, choice, iterator or test action. For

Z. Shi et al. (Eds.): IIP 2014, IFIP AICT 432, pp. 161–170, 2014.

the reason that the action in DDL contains choice action constructor, it becomes possible that the action can make the state change in branch structure. The verification problem becomes whether there is a model generated by an execution complex action that meets temporal formula specified in branch temporal description logic DL-CTL. Instead of considering the actual execution sequences of actions, we consider execution complex action sequences accepted by a given non- deterministic finite automaton NFA. If a NFA is an abstraction of the action, i.e. all possible execution sequences of the action are accepted by NFA, then any property that holds in all the actions accepted by NFA is also a property that is satisfied by any execution of the actions. Therefore, we not only add the action in dynamic description logic to verification problem, but also can verify the branch-time properties.

2   Preliminaries

2.1   Temporal Description Logic DL-CTL

Description logics [5-7] are a well-known family of formalisms to represent the knowledge, which offers the considerable expressive power going far beyond the propositional logic and the reasoning is still decidable. DL-CTL is the temporal extension to description logics, which extends the propositional branch-time logic (CTL) by allowing for the use of axioms of the basic description logic in place of propositional letters. The properties in this paper will be expressed in DL-CTL.

The concepts in DL-CTL is similar with DL, which are inductively defined a set N_C of concept name, a set N_R of role name, and a set N_I of individual name. The concept construction is the same as those do in DL and the formula is constructed with the temporal operators in CTL. At the same time, the propositional letters in CTL are replaced by ABox assertions of description logic.

Definition 1. DL-CTL formula is defined as follows :

$$\phi, \psi ::= C \sqsubseteq D | C(p) | R(p, q) | \neg \phi | \phi \wedge \psi | EX\phi | AF\phi | E(\phi U\psi)$$

where p, $q \in N_I$, $R \in N_R$, C,D are concept name. We can also introduce the formula such as false, true, $\phi \vee \psi$, $\phi \rightarrow \psi$, $AX \phi$, $EF \phi$, $EG \phi$, $AG \phi$, $A(\phi U\psi)$, $E(\phi R\psi)$, $A(\phi R\psi)$.

The semantic of DL-CTL is similar with CTL and it is based on the structure, in which their states are organized by branch structure. However, the different from CTL is that the state in DL-CTL is not mapped to a set of propositional letters but mapped to DL interpretations. Thus, the state change in state transition system can be viewed as an interpretation change in DL-CTL.

Definition 2. DL-CTL structure is a tetrad $M = (S, T, \Delta, I)$:

(1) S is a set of all states;

(2) $T \subseteq S \times S$ is binary relation of state which means the transition between two states;

(3) Δ is the interpretation domain;

(4) For every state $s \in S$, the function I gives s a DL interpretation $I(s) = (\Delta, \cdot^{I(s)})$ and the interpretation function $\cdot^{I(s)}$ must meet the following conditions:

(i) For every concept $C_i \in N_C$, there is $C_i^{I(s)} \subseteq \Delta$;

(ii) For every role name $R_i \in N_R$, there is $R_i^{I(s)} \subseteq \Delta \times \Delta$;

(iii) For every individual name $p_i \in N_I$ there is $p_i^{I(s)} \in \Delta$, and for every state $s \in S$ there is $p_i^{I(s)} = p_i^{I(s')}$.

Definition 3. A DL-CTL structure is M=(S, T, Δ, I), the semantic of concept and formula in DL-CTL are inductively defined as follows.

Firstly, for every states $s \in S$, a concept C is interpreted $C^{I(s)}$, which is a subset of Δ.

(1) $(\neg C)^{I(s)} := \Delta \backslash C^{I(s)}$;

(2) $(C \sqcup D)^{I(s)} := C^{I(s)} \cup D^{I(s)}$;

(3) $(\forall R.C)^{I(s)} := \{x \mid \text{for every } y \in \Delta: \text{if}(x, y) \in R^{I(s)}, \text{then } y \in C^{I(s)}\}$.

Secondly, for every states $s \in S$, M, $s \models \phi$ means ϕ holds at state s of the structure M.

(4) $(M, s) \models C \sqsubseteq D$ iff $C^{I(s)} \subseteq D^{I(s)}$;

(5) $(M, s) \models C(p)$ iff $p^{I(s)} \in C^{I(s)}$;

(6) $(M, s) \models R(p, q)$ iff $(p^{I(s)}, q^{I(s)}) \in R^{I(s)}$;

(7) $(M, s) \models \neg \phi$ iff $(M, s) \not\models \phi$;

(8) $(M, s) \models \phi \wedge \psi$ iff $(M, s) \models \phi$ and $(M, s) \models \psi$;

(9) $(M, s) \models EX \phi$ iff there is a state s' and (s, s') $\in T$ and $(M, s') \models \phi$;

(10) $(M, s) \models AF \phi$ iff for every path starting with s, there is always a state s' so that sT*s' and $(M, s') \models \phi$;

(11) $(M, s) \models E(\phi U \psi)$ iff there exists a path starting with s, and there exists an integer $k \geq 0$ so that $(M, s+k) \models \psi$ and for every $0 \leq i < k$, $(M, s+i) \models \phi$.

In this paper, we consider that the state transition is caused by the application of action. For the reason that every state in DL-CTL structure is mapped to a DL interpretation, we can say that the change of the interpretation of one state to the next is also caused by action.

2.2 Dynamic Description Logic DDL

The action theory based on description logic is inherited the advantage of the action theory and the description logic, which not only has the more expressive power but also makes the reasoning tasks decidable. Dynamic description logic (DDL) is a kind of action theory based on description logic, which is proposed by Chang. L et al [8], who has introduced complex action to the action theory based on description logic.

In this paper, we consider the action theory based on dynamic description logic and the basic definitions of action will be given below.

Definition 4. Let T is an acyclic TBox. An action for T is generated below.
$$\pi, \pi' = \alpha | \varphi? | \pi \cup \pi' | \pi, \pi' | \pi^*$$
where α is an atom action and φ is an assertion. $\varphi?$, $\pi \cup \pi'$, π, π', π^* are respectively called test action, choice action, sequential action and iterated action.

The complex action is composed by these actions sequence. For example, $(\varphi?)$, $a \cup b$, $(c, d)^*$ is the complex action and a, b, c, d are respectively the atom action and φ is an assertion.

An atom action is a triple (**pre, occ, post**), and the detail description of atom action can be found in [7]. In this paper, we just consider the action without **occ** for convenience.

We say that α is executable in an interpretation I if I is a model of **pre**. If the execution of action can change the interpretation I to I′, we can say that α makes I⇒I′.

In the existing researches, action is considered as the cause of the state transition, which also means that action is the binary relation between states. For the reason that the action in dynamic description logic contains choice constructor with which the action can make the state change in branch-time state transition system, it becomes possible to verify the branch-time property based on dynamic description logic. Next, some necessary definitions corresponding to this case will be given below.

Definition 5. For every complex action π, let Σ be the set of the atom action or test action in π. Every element in Σ can be view as a word and let L(π) be the minimal set of string that defined by the following rules:

(1) If π is an atom action or test action then, L (π) = π;

(2) L (π∪π′) =L (π)∪L(π′);

(3) L (π, π′) = {$l_1 l_2$|l_1∈ L(π) and l_2∈ L(π′)};

(4) L($π^*$)=L($π^0$)∪L($π^1$)∪L($π^2$)…,where L($π^0$) is an empty string and for every i ≥1,there is L($π^i$)=L($π^{i-1}$,π).

So, each string in L(π) corresponds one of the action execution sequence of π.

Definition 6. For a complex action π and one of its action execution sequence l_i, where l_i ∈L(π), we use |l_i| to denote the length of action execution sequence, l_i(j) to denote the j-th atom or test action in i-th action execution sequence.

For example, for a complex action π=a,b,c,d, L(π)={l_1}.That is to say, l_1 is the only action execution sequence and l_1=abcd; For another complex actionπ=(a,b)∪(c,d,e), L(π)={l_1,l_2}, l_1 and l_2 are the two action execution sequences of π and l_1=ab, l_2=cde, l_1(1)=a, l_2(2)=d.

Definition 7. Let T be an acyclic TBox, A an ABox, and π a complex action for T. For the interpretation of a state s_0, there is I(s_0)⊨A, then for an execution action sequence l_i of π and a path starting from s_0, if l_i(j) is executable in I(s_j) and it makes I(s_j)⇒ I(s_{j+1}), then, we call this path is a path generated by l_i. A DL-CTL structure generated by π is a tree structure where s_0 is the root and it contains all paths generated by each corresponding execution action sequence of π.

In this paper, for verification problem based on action, we consider whether there is a model M generated by an execution complex action that meets the property specified in DL-CTL. According to automaton theory, every action π can be constructed to a non-deterministic finite automaton (NFA). Instead of considering the actual execution sequences of actions, we abstract the complex action to non-deterministic finite state automaton (NFA) and just consider the action accepted by a NFA.

Definition 8. A= (Q, Σ, δ, q_0, F) is a non-deterministic finite automaton and Q is a set of state, Σ is the alphabet, δ: Q× Σ →2^Q is a transition function, q_0∈Q is initial state and F⊆Q is the final state set. We can say that A is a NFA for Σ and Σ is the set of the atom action or test action. The language accepted by A is L (A), which also can be treated as L (π).

According to the theory of formalism, the verification problem can be considered as satisfiability problem, which asks whether there is a complex action π accepted by NFA that satisfies the property specified in DL-CTL. The formal definition of this problem is shown in Definition 9 and an actual example will be given to this problem in the following section.

Definition 9. Let $\mathcal{T}$ be an acyclic TBox, $\mathcal{A}$ an ABox, and Σ a finite set of action for $\mathcal{T}$. $\mathcal{B}$ is a NFA for the alphabet Σ, and φ a DL-CTL formula. φ is satisfiable w.r.t $\mathcal{T}$, $\mathcal{A}$, and $\mathcal{B}$ if there is a DL-CTL structure M generated by π from $\mathcal{A}$ w.r.t $\mathcal{T}$ such that M, $s_0 \models \varphi$.

2.3   An Example

An example of buying a book will be given to show the problem we have discussed above. Assume the fact is: Jim wants to buy book A and B in bookstore, if the bookstore does not have the two books, it has to order them and then Jim can buy the two books at the same time.

According to the fact, the property can be described like this: though the bookstore does not have the two books, Jim can eventually get them. Obviously, if Jim wants to buy the two books, the bookstores must have the two books; if not, the bookstore must order the two books. If the bookstores just order one of them, Jim still can't buy the two books. Based on these facts, we give a formal definition of this case.

Firstly, we give a basic definition and symbol in this case, where the concept set $N_C=\{$student, book, instore$\}$, the role set $N_R=\{$bought, has$\}$, the individual name set $N_I=\{$Jim,book_a,book_b$\}$,The action set:$\{$buyBook_a,buyBook_b,order_a,order_b$\}$.

Based on these definitions, the background knowledge can be described below.

$$student \equiv person \sqcap \exists has.books$$

Then, we define the actions given above.

$buyBook\_a \equiv (\{$student (Jim), book (book_a), instore (book_a)$\}$, $\{\neg$ instore (book_a), bought (Jim, book_a)$\}$);
$buyBook\_a \equiv (\{$student (Jim), book (book_b), instore (book_b)$\}$, $\{\neg$ instore(book_b), bought (Jim, book_b)$\}$);
$order\_a \equiv (\{$book (book_a), $\neg$instore (book_a)$\}$, $\{$instore (book_a)$\}$);
$order\_b \equiv (\{$book (book_b), $\neg$instore (book_b)$\}$, $\{$instore (book_b)$\}$);

The property described in this example can be specified in following DL-CTL formula φ.

$$EF(\neg instore(book\_a) \wedge \neg instore(book\_b) \rightarrow EF(bought(Jim,book\_a) \wedge bought(Jim,book\_b)))$$

From the fact we know that if Jim want to buy the two books, the bookstore must have the two books; if not, the bookstore must order them, and Jim can go to buy the book and the get the book. That is to say, as long as the action *order_a, order_b, buyBook_a, buyBook_b* exists in complex action π, the property φ will be satisfied.

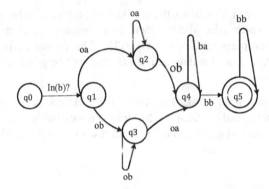

Fig. 1. A non-deterministic finite automaton $B_{buyBook}$

The NFA $B_{buyBook}$ is depicted in Figure 1. The alphabet of $B_{buyBook}$ is Σ, which consists of the actions defined above. For the convenience of describing the actions, the actions *instore (book)?, buyBook_a, buyBook_b, order_a, order_b* are abbreviated with *In(b)?, ba ,bb ,oa ,ob.*

For action $\pi=In(b)?(((oa\cup ob)(oa)^*,(ob),(ba)^*,(bb)^*$, it is easy to check that one of its action execution sequence $In(b)oa,oa(oa...)ob,ba,(bb...)\in L(B_{buyBook})$ can generate a branch structure that can satisfy the property that Jim eventually get the two books described above.

3   Verification Algorithm Based on DDL

In this paper, we consider a restricted situation that action contains cyclic sequence of action, which is similar to the case given in [5]. We solve this problem defined above by the reduction from the satisfiability problem of DL-CTL formula to the consistency problem introduced in [7]. The detail algorithm is given in Algorithm 1 and the detail construction of each step will be given in the following section.

Algorithm 1. *Given an ABox $\mathcal{A}$, a TBox $\mathcal{T}$, an action π and a formula ϕ, whether there is a model M generated by π that meets ϕ can be decided by the following steps:*
Step 1. *Construct an acyclic TBox $\mathcal{T}_{red}$, and an ABox $\mathcal{A}_{red}$ from $\mathcal{A}$, $\mathcal{T}$, π and ϕ;*
Step 2. *Construct an ABox $\mathcal{A}_{pre}$ from π;*
Step 3. *Using tableau rules to compute an ABox $\mathcal{A}\phi$ from ϕ;*
Step 4. *Using the reasoning services provided by description logic to decide whether $\mathcal{A}_{red}\cup \mathcal{A}_{pre}\cup \mathcal{A}\phi$ is consistent w.r.t $\mathcal{T}_{red}$, if $\mathcal{A}_{red}\cup \mathcal{A}_{pre}\cup \mathcal{A}\phi$ is consistent w.r.t $\mathcal{T}_{red}$, return 'yes', else return 'no' .*

3.1   Construction of T_{red}

Firstly, we assume that there are no DL-CTL negation signs in φ, which is similar with the method in [10] when dealing with LTL negation signs. It allows DL-CTL signs occur only in front of ABox assertion rather than temporal operator.

Secondly, we refer to the method for solving the projection problem in [10] to the finite sequence of atom action to construct T_{red} and A_{red}.

In order to define T_{red}, we define T_N, which contains all individual names in the input.

$$T_N = \{N \equiv \sqcup \{a\}, \text{ for all } a \in N_I\} \tag{1}$$

Let Sub be the set of the subconcepts in the input. For every $C \in Sub$, if C is not a defined concept name of T, then there is a concept definition of T_C^i and T_{Sub}^i. Moreover, T_{Sub}^i contains only those concept definitions. The concept definition of T_C^i can be found in [10].

Now, according to the method given in [5], we are ready to assemble T_{red}:

$$T_{red} = T_N \cup (\cup_i^{m+2n-1} T_{Sub}^i) \cup \{T_A^i \equiv T_E^i | A \equiv E \in T, \text{ for } i \leq m+2n-1\} \tag{2}$$

TBox T_N and T_{Sub}^i can ensure that the interpretations of concept and role names remain unchanged by actions on the anonymous objects and the last part of T_{red} is to make sure that T is satisfied no matter how actions change an interpretation.

3.2   Construction of A_{red} and A_{pre}

A_{red} is an ABox which record the changes by actions on the named objects. For every ABox assertion ϕ, we define $\phi^{(i)}$. If $\phi = C(a)$, $\phi^{(i)} = T_C^i$ and if $\phi = r(a,b)$, $\phi^{(i)} = r^{(i)}(a,b)$.

In order to meet the semantic of action, we need to get the pre-definitions of A_{post}^i, A_{min}^i, A_{ini}, which can be found in [10], and then we can get ABox A_{red} :

$$A_{red} = A_{ini} \cup (\cup_i^{m+2n-1} A_{post}^i) \cup (\cup_i^{m+2n-1} A_{min}^i) \tag{3}$$

In [5], we know that from every model of A_{red} and T_{red}, we can construct the cruial part of a DL-CTL structure generated by π from A and T. We can also see that any finite sequence $I(s_0)....I(s_{m+2n-1})$ satisfy the property stated in the above items can be extended to an DL-CTL structure generated by $\pi = \pi_1 \cdots \pi_m (\pi_1' \cdots \pi_n')^\pi$ from A w.r.t T by setting $I(s_{m+kn+i}) = I(s_{m+n+i})$ for all $k \geq 2$ and $0 \leq i < n$.

To enforce the excitability of the execution action sequence l_i, which $l_i(j)$ is the j-th atom or test action in $l_i(j)$, we define ABox A_{pre} , which is similar with the method given in [5] dealing with the atom actions.

$$A_{pre} = \cup \{\gamma^{(j)} | \gamma \in pre_j, \text{ for } 0 \leq j \leq m+2n-1\} \tag{4}$$

where pre_j is the set of pre-conditions of l_i , the excution action sequence of π.

3.3   Construction of $\mathcal{A}\phi$

In order to ensure that the DL-CTL formula φ is satisfied, we generate additional ABox $\mathcal{A}\phi$ by applying a non-deterministic tableau algorithm. We have time-stamped copied $\phi^{(i)}$ for every subformula φ of ϕ, which means that ϕ holds at time point i.

Different from the method for solving the semantic the temporal operator of LTL given in [10], we use the approach given below to solve the problem of DL-CTL with these tableau rules. The tableau algorithm starts with an initial set $S=\{\phi^{(0)}\}$, and then modifies this set by tableau rules until there are no more rules to apply, part of the tableau rules are described below and other rules can be inferred by these rules.

(1) $\neg\neg$rule: If $(\neg\neg\phi)^{(i)} \in S$ and $(\phi)^{(i)} \notin S$, let $(\phi)^{(i)} \in S$;

(2) $\wedge$ rule: If $(\phi_1\wedge\phi_2)^{(i)} \in S$ and $\{(\phi_1)^{(i)}, (\phi_2)^{(i)}\} \notin S$, let $\{(\phi_1)^{(i)}, (\phi_2)^{(i)}\} \in S$;

(3) $\neg\wedge$ rule: If $\neg(\phi_1\wedge\phi_2)^{(i)} \in S$ and $(\neg\phi_1)^{(i)} \notin S$, let$(\neg\phi_1)^{(i)}\in S$; If$\neg(\phi_1\wedge\phi_2)^{(i)} \in S$ and $(\neg\phi_2)^{(i)} \notin S$,let $(\neg\phi_2)^{(i)}\in S$;

(4) $\neg$EX rule: If $(\neg EX\phi_1)^{(i)} \in S$ and $(AX\neg\phi_1)^{(i)} \notin S$,let $(AX\neg\phi_1)^{(i)} \in S$, and let $(\neg\phi_1)^{(i+1)} \in S$ in all possible branches at the next time point ,then remove $(AX\neg\phi_1)^{(i)}, (\neg EX\phi_1)^{(i)}$;

(5) $\neg$AX rule: If $(\neg AX\phi_1)^{(i)} \in S$ and $(EX\neg\phi_1)^{(i)} \notin S$, let $(EX\neg\phi_1)^{(i)} \in S$, and let $(\neg\phi_1)^{(i)}\in S$ in next time point in all branches at i+1, and all the branches are respectively recorded by $l_1, l_2...$

(6) EG rule: If $(EG\phi)^{(i)} \in S$ and $(\phi)^{(i)} \notin S$,and use $(\phi)^{(i)}, (\phi)^{(i+1)}\cdots$to lable all the states aftertimeiandlet$\{(\phi)^{(i)}, (\phi)^{(i+1)}\cdots(\phi)^{(j)}\} \in S$ $(j\leq m+2n-1)$, then remove$(EG\phi)^{(i)}$ and all the possible branches are recorded by $l_1, l_2...$

(7) AG rule: If $(AG\varphi)^{(i)} \in S$ and $(\varphi)^{(i)} \notin S$, use $(\phi)^{(i)}, (\phi)^{(i+1)}\cdots$to lable all the states after time i and let $\{(\phi)^{(i)}, (\phi)^{(i+1)}\cdots(\phi)^{(j)}\} \in S$ $(j\leq m+2n-1)$,then remove$(EG\phi)^{(i)}$.

(8) EU rule: If $(E(\phi_1 U\phi_2))^{(i)} \in S$,for i there are two conditions: $i\leq m+n$ and $i>m+n$. When $i>m+n$, use $(\phi)^{(i)},\cdots(\phi)^{(k-1)}, (\phi)^{(k)}$ to lable all the states after time i and make $\{(\phi)^{(i)}...(\phi)^{(k-1)} (\phi)^{(k)}\} \in S$, then remove $(E(\phi_1 U\phi_2))^{(i)}$, when $i\leq m+n$,use $(\phi)^{(i)},\cdots(\phi)^{(m+2n-1)}, (\phi)^{(m+n)}, \cdots (\phi)^{(k-1)}, (\phi)^{(k)}$ to lable all the states after time i and make$\{(\phi)^{(i)},\cdots(\phi)^{(m+2n-1)}, (\phi)^{(m+n)},\cdots(\phi)^{(k-1)}, (\phi)^{(k)}\} \in S$, then remove $(E(\phi_1 U\phi_2))^{(i)}$. For the two conditions, use $l_1, l_2...$to record the existing branches.

(9) AU rule: If $(A(\phi_1 U\phi_2))^{(i)} \in S$,for i there are two conditions: $i\leq m+n$ and $i>m+n$. When $i>m+n$, use $(\phi)^{(i)},\cdots(\phi)^{(k-1)}, (\phi)^{(k)}$ to lable all the states after time i and make $\{(\phi)^{(i)}\cdots (\phi)^{(k-1)} (\phi)^{(k)}\} \in S$, then remove $(A(\phi_1 U\phi_2))^{(i)}$,when $i \leq m+n$, use $(\phi)^{(i)},\cdots(\phi)^{(m+2n-1)}, (\phi)^{(m+n)},\cdots(\phi)^{(k-1)}, (\phi)^{(k)}$ to lable all the states after time i and make$\{(\phi)^{(i)},\cdots(\phi)^{(m+2n-1)}, (\phi)^{(m+n)},\cdots(\phi)^{(k-1)}, (\phi)^{(k)}\} \in S$,then remove $(A(\phi_1 U\phi_2))^{(i)}$.

(10) $\neg$EU rule: If $(\neg E(\phi_1 U\phi_2))^{(i)} \in S$ and at the same time $(\neg\phi_1)^{(i)}\in S, (\neg\phi_2)^{(i)} \notin S$, let $\{(\neg\phi_2)^{(i)}, (\neg\phi_2)^{(i)}\phi_1\}\in S$; If$(\neg E(\phi_1 U\phi_2))^{(i)} \in S$ and $(\neg\phi_2)^{(i)} \notin S$,at the same time, $(AX\neg E(\phi_1 U\phi_2))^{(i)} \notin S$, let $\{(\neg\phi_2)^{(i)}, (AX\neg E(\phi_1 U\phi_2))^{(i)}\}\in S$,then,use the rule AX and EU defined above.

(11) $\neg$AU rule: If $(\neg A(n_1 U n_2))^{(i)} \in S$ and $(\neg n_1)^{(i)} \notin S, (\neg n_2)^{(i)} \notin S$,let $\{(\neg n_2)^{(i)}, (\neg n_2)^{(i)}n\}\in S$; If $(\neg A(n_1 U n_2))^{(i)}\in S$ and $(\neg n_2)^{(i)} \notin S, (EX\neg A(n_1 U n_2))^{(i)} \notin S$, let $\{(\neg n_2)^{(i)}, (EX\neg A(\phi_1 U\phi_2))^{(i)}\}\in S$, then,use the rule EX and AU defined above .

(12) ¬EG rule: If $(¬EGn_1)^{(i)} \in S$ and $(¬\phi_1)^{(i)} \notin S$, let $(¬\phi_1)^{(i)} \in S$; If $(¬EG\phi_1)^{(i)} \in S$ and $(AX¬EG\phi_1)^{(i)} \notin S$, let $(AX¬EG\phi_1)^{(i)} \in S$, then, use the rule AX and EG defined above and use l_1, l_2...to record the existing branches.

(13) ¬AG rule: If $(¬AG\phi_1)^{(i)} \in S$ and $(¬\phi_1)^{(i)} \notin S$, let $(¬\phi_1)^{(i)} \in S$; If $(¬AG\phi_1)^{(i)} \in S$ and $(EX¬AG\phi_1)^{(i)} \notin S$, let$(EX¬AG\phi_1)^{(i)} \in S$, then use the rule EX and AG defined above.

It can be shown that the tableau rules always terminate with a finite set S, which contains only time-stamped DL-assersions and the final S is an ABox. Since there exists the branch time operator, it will generate not only one ABox, depending by the choices made in the rules. We say that $A\phi$ is induced by ϕ w.r.t π if it is one of the ABoxes produced by applying the above rules to $\{\phi^{(0)}\}$.

In this case, we introduce the verification problem based on dynamic description logic and conisder its dual, the satisfiability problem, which is introduced in Definition 7. Finally, we reduce this problem to consistency of an ABox w.r.t an acyclic Tbox.

Theorem 1. The DL-CTL formula ϕ is satisfiable w.r.t T, A and an NFA B iff there is an ABox $A\phi$ induced by ϕ w.r.t π such that $A_{red} \cup A_{pre} \cup A\phi$ is consisitent with T_{red}.

This theorem can be proved with a similar process presented in [10]. Due to space limitations, we omitted the proof here.

4   Conclusions

Traditional verification technology is based on transition system and the property is specified in propositional logic, which limits the scope of describing the property. So, the temporal description logic DL-LTL is put forward to specify the temporal property in [6]. Based on DL-LTL, F.Baader considers runtime verification problem in [9], which observes changes to the state without knowing how they are caused.

In order to explore the cause of state transition, [5] assumes the action can make the state change and then combine the decidable action theory based on description logic to decide whether there is an infinite execution atom actions that can satisfy the linear property specified in DL-LTL. At last, a specific approach is given for solving this problem.

However, [5] just considers the atom actions and just can verify the liner property. For such limitations, we consider complex action based on dynamic description logic and use the temporal description logic DL-CTL to specify the properties. For the verification problem whether the property specified in DL-CTL holds in a model generated by a complex action, we consider its dual, the satisfiability problem whether there is an execution of complex action that can satisfy the property. For the convenience of this problem, we abstract complex action to a non-deterministic finite state automaton (NFA) and consider whether there is a complex action accepted by NFA that satisfy the temporal property. Finally, we reduce this verification problem to consistency problem in description logic and give an approach to it.

In this paper, we abstract the action to a non-deterministic finite state automaton and every possible execution actions are accepted by this NFA. A future work will consider the actual action rather than its abstraction.

Acknowledgements. This work is supported by the National Natural Science Foundation of China (Nos. 61363030, 61262030), the Natural Science Foundation of Guangxi Province (No.2012GXNSFBA053169, 2012GXNSFAA053220) and the Science Foundation of Guangxi Key Laboratory of Trusted Software.

References

1. Clarke, E.M., Grumberg, O., Peled, D.A.: Model Checking. MIT Press (1999)
2. Reiter, R.: Knowledge in action: logical foundations for describingand implementing dynamical systems. MIT Press, Cambridge (2001)
3. Giacomo, G.D., Ternovskaia, E., Reiter, R.: Non-terminating processes in the situation calculus. In: Proceedings of the AAAI 1997 Workshop on Robots, Softwoods, Immobots: Theories of Action, Planning and Control (1997)
4. Claβen, J., Lakemeyer, G.: A Logic for non-terminating Golog programs. In: Proceedings KR 2008, pp. 589–599. AAAI Press (2008)
5. Baader, F., Liu, H.K.: ul Mehdi, A.: Verifying properties of infinite sequences of description logic actions. In: Proceedings of ECAI 2010 (2010)
6. Baader, F., Ghilardi, S., Lutz, C.: LTL over description logic axioms. In: Proceedings of KR 2008, pp. 684–694. AAAI Press, Cambridge (2008)
7. Baader, F., Lutz, C., Sattler, U., Wolter, F.: Integrating description logics and action formalisms: First results. In: Proceedings AAAI 2005 (2005)
8. Chang, L., Shi, Z.Z., Gu, T.L., Zhao, L.: A family of dynamic description logics for representing and reasoning about actions. Journal of Automated Reasoning 49(1), 19–70 (2012)
9. Baader, F., Bauer, A., Lippmann, M.: Runtime verification using a temporal description logic. In: Ghilardi, S., Sebastiani, R. (eds.) FroCoS 2009. LNCS, vol. 5749, pp. 149–164. Springer, Heidelberg (2009)
10. Baader, F., Liu, H.K.: Intergrate Action Formalisms into Linear Temporal Description Logic. LTCS-Report 09-03, TU Dresden, Germany (2009)

Dynamic Description Logic Based on DL-Lite

Na Zhang, Liang Chang, Zhoubo Xu, and Tianlong Gu

Guangxi Key Laboratory of Trusted Software, Guilin University of Electronic Technology,
Guilin 541004, China
zhnamengluo@163.com, {changl,xzbli_11,cctlgu}@guet.edu.cn

Abstract. Description logics offer considerable expressive power for describing knowledge about static application domains while reasoning is still decidable. The dynamic description logic DDL is a family of dynamic extensions of description logics for representing and reasoning about knowledge of dynamic application domains. In order to provide effective reasoning mechanisms, systems of DDL investigated in the literatures assume that there is no general concept inclusion(GCI) contained in the knowledge base. In this paper, we build a system of dynamic description logic based on the tractable description logic DL-Lite$_R^{pr}$, in such a way that all the knowledge described by DL-Lite$_R^{pr}$ is supported by our system. A decision algorithm is provided for our system DDL-Lite$_R^{pr}$.Termination and correctness of the algorithm are proved.

Keywords: description logic, dynamic description logic, action theory, satisfiability, tableau algorithm.

1   Introduction

With the rapid development of the Semantic Web, description logics [1] are playing an important role in it, which are recommended by W3C as the basis of Web Ontology Language OWL [2]. About static application domains, they provide considerable expressive power and decidable reasoning mechanisms [3]. But they can't directly deal with knowledge of dynamic application domains which are characterized by actions.

For this limitation, Shi et al. [4] put forward a dynamic description logic DDL based on a combination of description logic ALC, dynamic logic and action theory. Based on DDL, a family of dynamic description logics named DDL(X$^@$)[5] was proposed for representing and reasoning about actions, where the minimal change semantics[6] were used to define the semantics of atomic action definitions. But all the current decision algorithms for dynamic description logics are all restricted to requiring that TBoxes of description logics don't include GCIs any more. The DL-Lite family [7] is a family of DLs tailored to capture conceptual modeling constructs while keeping reasoning.

Based on [5], in this paper, we first of all propose a dynamic description logic DDL-Lite$_R^{pr}$ and give its syntax and semantics. And we adopt model-based semantics $\mathcal{L}_{\subseteq}^a$[8] of ABox update to define the semantics of atomic action definitions. Then a tableau decision algorithm which supports GCIs for DDL-Lite$_R^{pr}$ is given. Finally, the termination and correctness of the algorithm are proved.

Z. Shi et al. (Eds.): IIP 2014, IFIP AICT 432, pp. 171–177, 2014.

2   Dynamic Description Logic DDL-Lite$_R^{pr}$

DDL-Lite$_R^{pr}$ primitive symbols include a set N_C of concept names, a set N_R of role names, a set N_I of individual names and a set N_A of action names. With the help of a set of constructors, starting from these symbols, roles, concepts, formulas and actions can be inductively constructed respectively.

Definition 1. Roles of DDL-Lite$_R^{pr}$ are formed according to the following syntax rule:
$$R ::= P \mid P^- \text{ where } P \in N_R.$$

Definition 2. Concepts of DDL-Lite$_R^{pr}$ are formed according to the following syntax rule:
$$C, C' ::= A_i \mid \neg C \mid \exists R \text{ where } A_i \in N_C, R \text{ is a role.}$$

The form of $C_1 \sqsubseteq C_2$ is called a GCI short for a general concept inclusion assertion, where C_1 and C_2 are any concept. The form of $R_1 \sqsubseteq R_2$ is called a role inclusion assertion where R_1 and R_2 are any role. Each finite set $\mathcal{T}$ of GCIs and role inclusion assertions is called a TBox of DDL-Lite$_R^{pr}$, where disjointness that involves roles is forbidden.

Definition 3. Formulas of DDL-Lite$_R^{pr}$ are formed according to the following syntax rule:
$$\varphi, \varphi' ::= C(p) \mid R(p, q) \mid <\pi> \varphi \mid \neg \varphi \text{ where } p, q \in N_I, R \in N_R, C \text{ is a concept, } \pi \text{ is an action.}$$

Formulas of the form $C(p)$, $R(p, q)$, $<\pi>\varphi$ and $\neg\varphi$ are respectively called concept assertion, role assertion, diamond assertion and negation formula. Concept assertions, role assertions, negations of concept assertions, and negations of role assertions are all called ABox assertions. A finite set of ABox assertions is called an ABox of DDL-Lite$_R^{pr}$. For any ABox $\mathcal{A}$, we use $\mathcal{A}^-$ to denote the set $\{\neg\varphi \mid \varphi \in \mathcal{A}\}$.

Definition 4. With respect to a TBox $\mathcal{T}$, an atomic action definition of DDL-Lite$_R^{pr}$ is of the form $\alpha \equiv (P, E)$, where,

(1) $\alpha \in N_A$ is an atomic action name;
(2) P is a finite set of ABox assertions for describing the pre-conditions of the action;
(3) E is a finite set of ABox assertions for describing the post-conditions.

For each finite set $\mathcal{Ac}$ of atomic action definitions, if no action name occurs on the left-hand sides for more than once, then we call $\mathcal{Ac}$ an ActBox of DDL-Lite$_R^{pr}$.

Definition 5. With respect to a TBox $\mathcal{T}$, an ActBox $\mathcal{Ac}$, actions of DDL-Lite$_R^{pr}$ are formed according to the following syntax rule:
$$\pi, \pi' ::= \alpha \mid \varphi? \mid \pi \cup \pi' \mid \pi; \pi' \mid \pi^* \text{ where } \alpha \text{ is an atomic action, } \varphi \text{ is a formula.}$$

Actions of the form α, $\varphi?$, $\pi \cup \pi'$, $\pi; \pi'$ and π^* are respectively called atomic action, test action, choice action, sequential action and iterated action.

Definition 6. A DDL-Lite$_R^{pr}$ -model is of the form $M = (W, T, \Delta, I)$, where,
1) W is a non-empty finite set composed of states;
2) T is a function that maps every action name $\alpha \in N_A$ to a binary relation $T(\alpha) \subseteq W \times W$;
3) Δ is a non-empty set made up of individuals;

4) I is a function which associates every state $w \in W$ a DL-interpretation $I(w) = (\Delta, \cdot^{I(w)})$, where the function $\cdot^{I(w)}$

— maps each concept name $C_i \in N_C$ to a set $C_i^{I(w)} \subseteq \Delta$,
— maps each role name $R_i \in N_R$ to a binary relation $R_i^{I(w)} \subseteq \Delta \times \Delta$, and
— maps each individual name $p_i \in N_I$ to an individual $p_i^{I(w)} \in \Delta$, with the constraints that $p_i^{I(w)} = p_i^{I(w')}$ for any state $w' \in W$.

Definition 7. Let $\mathcal{T}, \mathcal{A}c$ be a TBox, an ActBox respectively. $M = (W, T, \Delta, I)$ is a model of DDL-Lite$_R^{pr}$. The semantics of roles, concepts, formulas and actions of DDL-Lite$_R^{pr}$ are defined inductively as follows.

Firstly, for any state $w \in W$, each role R is interpreted as a binary relation $R^{I(w)} \subseteq \Delta \times \Delta$ and each concept C is interpreted as a set $C^{I(w)} \subseteq \Delta$. The semantics of roles and concepts of DDL-Lite$_R^{pr}$ are defined inductively as follows:

1. $(R^-)^{I(w)} = \{(y, x) \mid x \in \Delta, y \in \Delta \text{ and } (x, y) \in R^{I(w)}\}$;
2. $(\neg C)^{I(w)} = \Delta \setminus C^{I(w)}$, where "$\setminus$" is the set difference operator;
3. $(\exists R)^{I(w)} = \{x \in \Delta \mid \text{there is some } y \in \Delta \text{ such that } (x, y) \in R^{I(w)}\}$.

Secondly, for any state $w \in W$, the satisfaction-relation $(M, w) \models \varphi$ for any formula φ is defined inductively as follows:

4. $(M, w) \models C(p)$ iff $p^I \in C^{I(w)}$;
5. $(M, w) \models R(p, q)$ iff $(p^I, q^I) \in R^{I(w)}$;
6. $(M, w) \models \neg \varphi$ iff it is not the case that $(M, w) \models \varphi$;
7. $(M, w) \models <\pi>\varphi$ iff some state $w' \in W$ exists with $(w, w') \in T(\pi)$ and $(M, w') \models \varphi$;
8. $(M, w) \models [\pi] \varphi$ iff for every state $w' \in W$: if $(w, w') \in T(\pi)$ then $(M, w') \models \varphi$.

Finally, each action π is interpreted as a binary relation $T(\pi) \subseteq W \times W$ according to the following definitions:

9. $T(\varphi?) = \{(w, w) \mid (M, w) \models \varphi\}$;
10. $T(\pi \cup \pi') = T(\pi) \cup T(\pi')$;
11. $T(\pi; \pi') = \{(w, w') \mid \text{there is some state } w'' \in W \text{ such that } (w, w'') \in T(\pi) \text{ and } (w'', w') \in T(\pi')\}$;
12. $T(\pi^*) = $ reflexive transitive closure of $T(\pi)$.

A model M satisfies a TBox $\mathcal{T}$, denoted by $M \models \mathcal{T}$, if and only if for every state $w \in W$, $C_1^{I(w)} \subseteq C_2^{I(w)}$ for every concept inclusion assertion $C_1 \sqsubseteq C_2 \in \mathcal{T}$ and $R_1^{I(w)} \subseteq R_2^{I(w)}$ for every role inclusion assertion $R_1 \sqsubseteq R_2 \in \mathcal{T}$. A state w of a model M satisfies an ABox $\mathcal{A}$, denoted by $(M, w) \models \mathcal{A}$, if and only if $(M, w) \models \varphi$ for every ABox assertion $\varphi \in \mathcal{A}$.

According to $\mathcal{A}c$, any atomic action α is specified by some atomic action definition $\alpha \equiv (P, E)$. Then w.r.t. a TBox $\mathcal{T}$, a model M satisfies an atomic action definition $\alpha \equiv (P, E)$, in symbols $M \models_\mathcal{T} \alpha \equiv (P, E)$, if and only if $M \models \mathcal{T}$ and $T(\alpha) = \{(w, w') \mid ① (M, w) \models P, ② (M, w') \models E, ③ \text{for every state } w'' \in W, \text{if } (M, w'') \models E, \text{ then } dist_\sqsubseteq^d (I(w), I(w')) \subseteq dist_\sqsubseteq^d (I(w), I(w''))\}$.

The distance function between interpretations is specified in details as follows.

Firstly, given two interpretations $I(w) = (\Delta, \cdot^{I(w)})$ and $I(w') = (\Delta, \cdot^{I(w')})$, the distance between them is denoted by $dist_\sqsubseteq^d (I(w), I(w'))$. Then $dist_\sqsubseteq^d (I(w), I(w')) = I(w) \ominus I(w')$

$$= \bigcup_{C \in N_C} ((C^{I(w)} - C^{I(w')}) \cup (C^{I(w')} - C^{I(w)})) \cup \bigcup_{R \in N_R} ((R^{I(w)} - R^{I(w')}) \cup (R^{I(w')} - R^{I(w)}))$$

where $\ominus$ is the set symmetric difference operator. Distances under $\mathrm{dist}^d_{\sqsubseteq}$ are compared by set inclusion.

Definition 8. A model M satisfies an ActBox $\mathcal{A}c$ w.r.t. a TBox $\mathcal{T}$, in symbols $M \vDash_{\mathcal{T}} \mathcal{A}c$, if and only if $M \vDash_{\mathcal{T}} \alpha \equiv (P, E)$ for every atomic action definition $\alpha \equiv (P, E) \in \mathcal{A}c$.

Definition 9. A formula φ is satisfiable w.r.t. a TBox $\mathcal{T}$ and an ActBox $\mathcal{A}c$ if and only if there is a model $M = (W, T, \Delta, I)$ and a state $w \in W$ such that $M \vDash \mathcal{T}, M \vDash_{\mathcal{T}} \mathcal{A}c$ and $(M, w) \vDash \varphi$.

3   Tableau Decision Algorithm for DDL-Lite$_R^{\mathrm{pr}}$

Let $\mathcal{T}, \mathcal{A}c$ be a TBox and an ActBox respectively. Let φ be a DDL-Lite$_R^{\mathrm{pr}}$ -formula which is defined w.r.t. $\mathcal{A}c$. For the convenience of presentation, we firstly transform the formula φ into a normal form $\mathrm{nf}(\varphi)$ according to the following steps.

(1) Replace every occurrence of atomic actions with their atomic action definitions. Let φ' be the resulted formula.

(2) Transform φ' into an equivalent one in negation normal form by pushing negations inwards according to the following equivalences:

$$\neg (<\pi>\psi) = [\pi] \neg\psi \qquad \neg ([\pi] \psi) = <\pi> \neg \psi \qquad \neg \neg\psi = \psi$$

A full closure of an ABox $\mathcal{A}$ w.r.t. TBox $\mathcal{T}$, denoted by $fcl_{\mathcal{T}}(\mathcal{A})$, is the set of all ABox assertions f such that $\mathcal{A} /\!\!= _{\mathcal{T}} f$.

Next, we introduce some definitions involved in the algorithm as follows.

Given a TBox $\mathcal{T}$, a prefix $\sigma.\mu$ with a sequential action σ and a set μ of ABox assertions is constructed according to the following syntax rule:

$$\sigma.\mu ::= (\varnothing, \varnothing).\varnothing \mid \sigma; (P, E). (\mu\backslash (fcl_{\mathcal{T}}(E))^-)\cup E$$

where $(\varnothing, \varnothing)$ and (P, E) are atomic actions, $\sigma;(P, E)$ is a sequential action.

We also use $\sigma_0.\mu_0$ to denote the prefix $(\varnothing, \varnothing).\varnothing$ and call it the initial prefix. A prefixed formula is a pair $\sigma.\mu:\varphi$, where $\sigma.\mu$ is a prefix and φ is a formula.

A branch $\mathcal{B}$ is a union of a set $\mathcal{B}_{PF}$ of prefixed formulas and a set $\mathcal{B}_E$ of eventuality records which is of the form $X \equiv <\pi^*>\varphi$.

A branch $\mathcal{B}$ is completed if and only if it can't be expanded by any tableau expansion rule.

An eventuality record $X \equiv <\pi^*>\varphi$ is fulfilled in a branch $\mathcal{B}$ if and only if there is a prefix $\sigma.\mu$ such that both $\sigma.\mu: X \in \mathcal{B}$, and $\sigma.\mu:\varphi \in \mathcal{B}$.

A branch $\mathcal{B}$ is ignorable if and only if it is completed but contains some eventuality record $X \equiv <\pi^*>\varphi$ which is not fulfilled.

A branch $\mathcal{B}$ is contradictory if and only if there is some prefix $\sigma.\mu$ and some formula φ such that both $\sigma.\mu:\varphi \in \mathcal{B}$ and $\sigma.\mu:\neg\varphi \in \mathcal{B}$.

For any branch $\mathcal{B}$, $\{\psi/\sigma_0.\mu_0:\psi \in \mathcal{B}$ and ψ is an ABox assertion $\}$ is denoted by $IV_{\mathcal{B}}$.

Tableau expansion rules on inverse roles, non-atomic actions are the same with [5].

¬atom$_{<>}$-rule	If $\sigma.\mu:\neg<(P, E)>\varphi \in \mathcal{B}$, $\{\sigma.\mu:\psi^\neg \mid \psi \in P\} \cap \mathcal{B}=\varnothing$, and there is a prefix $\sigma'.\mu'$ with both $\mu'=(\mu\backslash(fcl_T(E))^\neg)\cup E$ and $\sigma'.\mu':\neg\varphi\notin \mathcal{B}$, then set either $\mathcal{B}:= \mathcal{B}\cup\{\sigma'.\mu':\neg\varphi\}$ or $\mathcal{B}:= \mathcal{B}\cup\{\sigma.\mu:\neg\psi\}$ for some $\psi \in P$.
atom$_{<>}$-rule	If $\sigma.\mu:<(P, E)>\varphi \in \mathcal{B}$, and if $\{\sigma.\mu: \psi \mid \psi \in P\} \subsetneq \mathcal{B}$, or no prefix $\sigma'.\mu'$ exists with both $\mu'=(\mu\backslash(fcl_T(E))^\neg)\cup E$ and $\sigma'.\mu':\varphi \in \mathcal{B}$, then: if there is no prefix $\sigma'.\mu'$ with $\mu'=(\mu\backslash(fcl_T(E))^\neg)\cup E$, then introduce a prefix $\sigma''.\mu'' := \sigma;(P, E).(\mu\backslash(fcl_T(E))^\neg)\cup E$ and set $\mathcal{B}:= \mathcal{B}\cup \{\sigma.\mu: \psi \mid \psi \in P\}\cup\{\sigma''.\mu'':\varphi\}\cup \{\sigma''.\mu'': \phi \mid \phi \in \mu''\}$, else find a prefix $\sigma'.\mu'$ with $\mu'=(\mu\backslash(fcl_T(E))^\neg)\cup E$, and set $\mathcal{B}:= \mathcal{B}\cup \{\sigma.\mu: \psi \mid \psi \in P\}\cup\{\sigma'.\mu':\varphi\}\cup \{\sigma'.\mu': \phi \mid \phi \in \mu'\}$.

Fig. 1. Tableau expansion rules on atomic actions

B-rule	If $\sigma.\mu:\varphi \in \mathcal{B}$, and φ is an ABox assertion, then : 1. $\mathcal{A}=\{\varphi\}$; 2. if $\{\sigma_0.\mu_0: \phi \mid \phi \in \mathcal{A}^{\text{Regress}(\sigma.\mu)}\} \subsetneq \mathcal{B}$, then any $\psi \in \mathcal{A}^{\text{Regress}(\sigma.\mu)}$ and $\sigma_0.\mu_0:\psi \notin \mathcal{B}$ set $\mathcal{B}:= \mathcal{B}\cup\{\sigma_0.\mu_0: \psi\}$.

Fig. 2. Tableau expansion rules on ABox assertions

For B-rule, ABox $\mathcal{A}^{\text{Regress}(\sigma.\mu)}$ is constructed according to the following Algorithm 1.

Algorithm 1. Given a regress setting $\mathsf{K}=(T, \mathcal{A})$ and $\mu^\neg$, given that $\mathcal{A}$ and $\mu^\neg$ w.r.t. T are satisfiable respectively, construct an ABox $\mathcal{A}'$ by the following steps.

Step 1. Let $\mathcal{A}'$ is an empty set and S is equivalent to $fcl_T(\mathcal{A})$.
Step 2. If S is not an empty set, do the following operations:
    1. Choose some ψ of S and at the same time remove ψ from S.
    2. If a union of the two sets $\{\psi\}$, $fcl_T(\mu^\neg)$ is satisfiable, then add ψ to $\mathcal{A}'$.
    3. Repeat Step 2 until the condition is unallowed.
The construction process ensures the following Lemma 3 and $\mathcal{A}^{\text{Regress}(\sigma.\mu)}$ is a union of $\mathcal{A}'$, $\mu^\neg$.

Finally, the tableau decision algorithm is described as follows.

Algorithm 2. The satisfiability of a formula φ w.r.t. a TBox T and an ActBox $\mathcal{A}c$ is decided according to the following steps:

Step1. Construct a branch $\mathcal{B}:=\{\sigma_0.\mu_0: \text{nf}(\varphi)\}$.
Step2. If all the tableau expansion rules are applied to $\mathcal{B}$ in random order so that they yield a completed branch $\mathcal{B}'$, and $\mathcal{B}'$ is neither contradictory nor ignorable, and $IV_{\mathcal{B}'}$ of $\mathcal{B}'$ is consistent w.r.t. T, then the algorithm returns *"TRUE"*, else returns *"FALSE"*.

Theorem 1. Algorithm 2 terminates.

Firstly, the number of branches which will be investigated by Algorithm 2 is finite. Then, for each branch $\mathcal{B}$ investigated by Algorithm 2, the number of ABox assertions contained in the initial view $IV_{\mathcal{B}}$ is also finite. Finally, the consistency of $IV_{\mathcal{B}}$ w.r.t. T can be decided with terminable procedures.

To demonstrate the correctness of Algorithm 2, some notations are introduced as follows.

Firstly, with reference to Algorithm AlignAlg($(T, \mathcal{A})$, $\mathcal{N}$)[8], Algorithm 1 has the following properties.

Lemma 1. Let T, A, N be a TBox, an ABox and an ABox, respectively. A, N w.r.t. T are consistent, if A^N is an ABox which is constructed according to Algorithm 1 under A and the update information N, then we have $A^N = fcl_T(A)\text{-}(fcl_T(N))^\neg \cup N$.

Lemma 2. Let T be a TBox, $\sigma.\mu$ is a prefix w.r.t. T, where $\sigma = (\varnothing, \varnothing)$; (P_1, E_1); (P_2, E_2); ...; (P_n, E_n). For any ABox A, if A w.r.t. T is consistent, then there must be $((A^{E_1})^{E_2}\ldots)^{E_n} = A^\mu$.

Secondly, we introduce branch-model mappings to act as bridges between branches and models.

Let T, Ac, B and $M = (W, T, \Delta, I)$ be a TBox, an ActBox, a branch and a model respectively. A branch-model mapping δ w.r.t. T, B and M is a function from $\sum$ of prefixes occurring in B to states of M, satisfying that for each pair of prefixes $\sigma.\mu$ and $\sigma'.\mu'$ occurring in B : if there is an atomic action (P, E), $\sigma'=\sigma;(P, E)$ and $\mu' = (\mu\backslash(fcl_T(E))^\neg)\cup E$, then:
(1) $M \vDash T$; (2) $(M, \delta(\sigma.\mu)) \vDash P$; (3) $(M, \delta(\sigma'.\mu')) \vDash E$; (4) for any $\sigma''.\mu'' \in \sum$, if $(M, \delta(\sigma''.\mu'')) \vDash E$, then $\text{dist}^a_\not\subseteq (I(\delta(\sigma.\mu)), I(\delta(\sigma'.\mu'))) \subseteq \text{dist}^a_\not\subseteq (I(\delta(\sigma.\mu)), I(\delta(\sigma''.\mu'')))$.

By means of a branch-model mapping, we can get the following property.

Lemma 3. Let $A^{Regress(\sigma.\mu)}$ be a ABox constructed by the regression operator in the B-rule. Let $M = (W, T, \Delta, I)$ be a model with $M \models T$. Then, for any branch-model mapping δ w.r.t. T, B and M, we have $(M, \delta(\sigma.\mu)) \vDash A$, iff $(M, \delta(\sigma_0.\mu_0)) \vDash A^{Regress(\sigma.\mu)}$.

Finally, it is easy to prove the correctness of Algorithm 2.

Theorem 2. Algorithm 2 returns "φ is satisfiable w.r.t. T and Ac" if and only if φ w.r.t. T and Ac is satisfiable.

4   Conclusion

Dynamic description logic DDL-Lite$_R^{pr}$ not only inherits the feature of representing and reasoning about knowledge of dynamic application domains as the same with the existing dynamic description logics, but also at the same time its TBoxes are composed of GCIs. We broaden knowledge offered by TBoxes of dynamic description logics and provide a reliable algorithm foundation for the practical application of a dynamic description logic with GCIs. Further work is to optimize the algorithm and develop a corresponding dynamic description logic reasoning machine.

Acknowledgements. This work is supported by the National Natural Science Foundation of China (Nos. 61363030, 61100025, 61262030), the Natural Science Foundation of Guangxi Province (No.2012GXNSFBA053169) and the Science Foundation of Guangxi Key Laboratory of Trusted Software.

References

1. Baader, F., Calvanese, D., McGuinness, D., Nardi, D., Patel-Schneider, P.F.: The Description Logic Handbook: Theory, Implementation and Applications. Cambridge University Press, Cambridge (2003)

2. Horrocks, I., Patel-Schneider, P.F., Harmelen, F.V.: From SHIQ and RDF to OWL: the making of a web ontology language. J. Web Semant. 1(1), 7–26 (2003)
3. Baader, F., Sattler, U.: An Overview of Tableau Algorithms for Description Logics. Studia Logica 69, 5–40 (2001)
4. Shi, Z.Z., Dong, M.K., Jiang, Y.C., Zhang, H.J.: A logical foundation for the semantic Web. Science in China, Ser. F: Information Sciences 48(2), 161–178 (2005)
5. Chang, L., Shi, Z.Z., Gu, T.L., Zhao, L.Z.: A family of dynamic description logics for representing and reasoning about actions. Journal of Automated Reasoning 49(1), 19–70 (2012)
6. Baader, F., Lutz, C., Miličić, M., Sattler, U., Wolter, F.: Integrating description logics and action formalisms: first results. In: Proc. of the 12th Nat. Conf. on Artificial Intelligence (AAAI 2005), pp. 572–577. AAAI Press / MIT Press (2005)
7. Artale, A., Calvanese, D., Kontchakov, R., Zakharyaschev, M.: The DL-Lite family and relations. J. of Artificial Intelligence Research 36, 1–69 (2009)
8. Kharlamov, E., Zheleznyakov, D.: Capturing instance level ontology evolution for DL-lite. In: Aroyo, L., Welty, C., Alani, H., Taylor, J., Bernstein, A., Kagal, L., Noy, N., Blomqvist, E. (eds.) ISWC 2011, Part I. LNCS, vol. 7031, pp. 321–337. Springer, Heidelberg (2011)

Formalizing the Matrix Inversion Based on the Adjugate Matrix in HOL4

Liming Li[1], Zhiping Shi[1]*, Yong Guan[1], Jie Zhang[2], and Hongxing Wei[3]

[1] Beijing Key Laboratory of Electronic System Reliability Technology,
Capital Normal University, Beijing 100048, China
liliminga@126.com, shizhiping@gmail.com
[2] College of Information Science & Technology,
Beijing University of Chemical Technology, Beijing 100029, China
[3] School of Mechanical Engineering and Automation, Beihang University,
Beijing 100083, China

Abstract. This paper presents the formalization of the matrix inversion based on the adjugate matrix in the HOL4 system. It is very complex and difficult to formalize the adjugate matrix, which is composed of matrix cofactors. Because HOL4 is based on a simple type theory, it is difficult to formally express the sub-matrices and cofactors of an n-by-n matrix. In this paper, special n-by-n matrices are constructed to replace the $(n-1)$-by-$(n-1)$ sub-matrices, in order to compute the cofactors, thereby, making it possible to formally construct aadjugate matrices. The Laplace's formula is proven and the matrix inversion based on the adjugate matrix is then inferred in HOL4. The paper also presents formal proofs of properties of the invertible matrix.

Keywords: Formal verification, Theorem proving, Matrix inversion, HOL4, Adjugate matrix.

1 Introduction

Matrix theory is widely applied in many areas, and a great deal of research and development is currently being conducted on vector, matrix, and space transformation in some theorem provers. Yatsuka Nakamura et al. presented the formalization of a matrix of real elements, and Nobuyuki Tamura et al. described the determinant and inverse of a matrix of real elements in Mizar [4,5]. Ioana Pasca formalized interval matrices and verified the conditions for their regularity in the COQ system [7,8]. Steven Obua described in [6] how the vector and matrix can be formalized in Isabelle/HOL. John Harrision formalized the real vector type $\mathbb{R}^N$ for a variable N and verified many properties about the real matrix and Euclidean space in HOL-light [1,2]. To the best of our knowledge, there is no explicit form of the inverse matrix in existing theorem provers.

We think that there are two main difficulties in formalizing the matrix inversion. First, it is not easy to describe $(n-1)$-ary space based on a simple

---

* Corresponding author.

Z. Shi et al. (Eds.): IIP 2014, IFIP AICT 432, pp. 178–186, 2014.

type theory [9,10], even though sets with an $n - 1$ size can be generalized using the complement of one element subset [11,12]. More importantly, $\mathbb{R}^{(n-1)\times(n-1)}$ and $\mathbb{R}^{n\times n}$ are different types, so it becomes very difficult to define the minor using the determinant of the sub-matrix. Accordingly, it is hard to explicate the inverse matrix using an analytic solution.

Indeed, it is very difficult to describe an $(n - 1)$-ary square matrix in HOL4. However, the cofactors of a matrix can be defined using the determinant of an n-by-n rather than an $(n - 1)$-by-$(n - 1)$ matrix. Therefore, we constructed an n-by-n matrix, the determinant of which is equal to the cofactor. With such a method, the problem of the expression of the cofactor is solved. And then, the explicit form of the inverse matrix can be formalized using the adjugate matrix.

In this paper, we start from definitions of invertibility and the inverse matrix, and prove some important properties about the inverse matrix. Then, we systematically give formal definitions of the cofactor and the adjugate matrix, and formal proof of Laplace's formula. Finally, we formalize the matrix inversion and prove some major theorems about it.

2　Formalization of the Invertible Matrix

2.1　Definition of Invertibility and the Inverse Matrix

Nonsingular linear transformation and its inverse transform have been applied in many important areas. The matrix is the operator of the space transform. General linear groups are such groups, the objects of which are all nonsingular matrices of a given size and the operation is matrix multiplication, since every element in a group has to be invertible. The multiplicative inverse of a nonsingular matrix is the inverse matrix.

In linear algebra an n-by-n (square) matrix $\mathbf{A}$ is called invertible (it is also called non-singular or nondegenerate) if there exists an n-by-n matrix $\mathbf{A}'$ such that

$$\mathbf{A}\mathbf{A}' = \mathbf{A}'\mathbf{A} = \mathbf{E}_n \tag{1}$$

where $\mathbf{E}_n$ denotes the n-by-n identity matrix and the multiplication used is ordinary matrix multiplication. If this is the case, then the matrix $\mathbf{A}'$ is uniquely determined by $\mathbf{A}$ and is called the inverse of $\mathbf{A}$, denoted by $\mathbf{A}^{-1}$.

Here, two definitions are involved. One is the invertibility of the matrix, and the other is the inverse matrix. Non-square matrices (m-by-n matrices for which $m \neq n$) have no inverses. However, in some cases, such an m-by-n matrix may have a left inverse or right inverse. Without loss of generality, the definitions are defined as followed in HOL4.

Definition 1. *invertible_def:*

$$invertible(\mathbf{A} : \mathbb{R}^{m\times n}) := \exists \mathbf{A}'(: \mathbb{R}^{n\times m}).\ \mathbf{A}\mathbf{A}' = \mathbf{E} \wedge \mathbf{A}'\mathbf{A} = \mathbf{E}$$

where the type of $\mathbf{A}$ is $\mathbb{R}^{m\times n}$ denotes matrix $\mathbf{A}$ is the m-by-n real matrix, similar to the following. The first $\mathbf{E}$ denotes the m-by-m identity matrix and the second $\mathbf{E}$ denotes the n-by-n identity matrix, the same as below.

Definition 2. *MATRIX_INV_DEF:*

$$(\mathbf{A} : R^{m \times n})^{-1} := \varepsilon \mathbf{A}'(: \mathbb{R}^{n \times m}).\ \mathbf{A}\mathbf{A}' = \mathbf{E} \wedge \mathbf{A}'\mathbf{A} = \mathbf{E}$$

there is an ε-term in the definition, $\varepsilon \mathbf{A}'(: \mathbb{R}^{n \times m}).\ \mathbf{A}\mathbf{A}' = \mathbf{E} \wedge \mathbf{A}'\mathbf{A} = \mathbf{E}$ denotes an $\mathbf{A}'$ such that $\mathbf{A}\mathbf{A}' = \mathbf{E} \wedge \mathbf{A}'\mathbf{A} = \mathbf{E}$.

2.2 Formalization and Proof of Important Properties

In accordance with the definition of the invertibility of a matrix, it is easy to prove the following two theorems.

Theorem 1. *MATRIX_INV:*

$$\forall \mathbf{A}(: \mathbb{R}^{n \times n}).\ invertible(\mathbf{A}) \Rightarrow \mathbf{A}\mathbf{A}^{-1} = \mathbf{E} \wedge \mathbf{A}^{-1}\mathbf{A} = \mathbf{E}$$

Theorem 2. *INVERTIBLE_MATRIX_INV:*

$$\forall \mathbf{A}(: \mathbb{R}^{m \times n}).\ invertible(\mathbf{A}) \Rightarrow invertible(\mathbf{A}^{-1})$$

Furthermore, the following properties hold for an invertible matrix $\mathbf{A}$.

Theorem 3. *MATRIX_RMUL_EQ:*

$$\forall \mathbf{A}(: \mathbb{R}^{m \times n})\ (\mathbf{X}\ \mathbf{Y})(: \mathbb{R}^{n \times p}).\ invertible(\mathbf{A}) \Rightarrow (\mathbf{X} = \mathbf{Y} \Leftrightarrow \mathbf{A}\mathbf{X} = \mathbf{A}\mathbf{Y})$$

Theorem 4. *MATRIX_LMUL_EQ:*

$$\forall \mathbf{A}(: \mathbb{R}^{m \times n})\ (\mathbf{X}\ \mathbf{Y})(: \mathbb{R}^{p \times m}).\ invertible(\mathbf{A}) \Rightarrow (\mathbf{X} = \mathbf{Y} \Leftrightarrow \mathbf{X}\mathbf{A} = \mathbf{Y}\mathbf{A})$$

The above theorems are succinct and symmetrical, but they are less conveniently expended in the practical proof. They can be changed into the following two theorems:

Theorem 5. *MATRIX_EQ_LMUL_IMP:*

$$\forall \mathbf{A}(: R^{m \times n})\ (\mathbf{X}\ \mathbf{Y})(: \mathbb{R}^{n \times p}).\ invertible(\mathbf{A}) \wedge \mathbf{A}\mathbf{X} = \mathbf{A}\mathbf{Y} \Rightarrow \mathbf{X} = \mathbf{Y}$$

Theorem 6. *MATRIX_EQ_RMUL_IMP:*

$$\forall \mathbf{A}(: \mathbb{R}^{m \times n})\ (\mathbf{X}\ \mathbf{Y})(: \mathbb{R}^{p \times m}).\ invertible(\mathbf{A}) \wedge \mathbf{X}\mathbf{A} = \mathbf{Y}\mathbf{A} \Rightarrow \mathbf{X} = \mathbf{Y}$$

Thus, it can be proved that the inverse of an invertible matrix's inverse is itself.

Theorem 7. *MATRIX_INV_INV:*

$$\forall \mathbf{A}(: \mathbb{R}^{m \times n}).\ invertible(\mathbf{A}) \Rightarrow (\mathbf{A}^{-1})^{-1} = \mathbf{A}$$

The nature of an invertible transformation can be described as follows:

Theorem 8. *MATRIX_INV_TRAN_UNIQ:*

$$\forall \mathbf{A}(: \mathbb{R}^{m \times n}) \ (\mathbf{x} \ \mathbf{y})(: \mathbb{R}^n). \ invertible(\mathbf{A}) \wedge \mathbf{A}\mathbf{x} = \mathbf{y} \Rightarrow \mathbf{x} = \mathbf{A}^{-1}\mathbf{y}$$

here, the type of $\mathbf{x}$ and $\mathbf{y}$ is $\mathbb{R}^n$ denotes they are both n-ary real vectors. The following theorems can be proven:

Theorem 9. *MATRIX_MUL_LINV_UNIQ:*

$$\forall \mathbf{A}(: \mathbb{R}^{m \times n}) \ (\mathbf{X} \ \mathbf{Y})(: \mathbb{R}^{n \times p}). \ invertible(\mathbf{A}) \wedge \mathbf{A}\mathbf{X} = \mathbf{Y} \Rightarrow \mathbf{X} = \mathbf{A}^{-1}\mathbf{Y}$$

Theorem 10. *MATRIX_MUL_RINV_UNIQ:*

$$\forall \mathbf{A}(: \mathbb{R}^{m \times n}) \ (\mathbf{X} \ \mathbf{Y})(: \mathbb{R}^{p \times m}). \ invertible(\mathbf{A}) \wedge \mathbf{X}\mathbf{A} = \mathbf{Y} \Rightarrow \mathbf{X} = \mathbf{Y}\mathbf{A}^{-1}$$

Similarly, the following theorems hold for the square matrix $\mathbf{A}$:

Theorem 11. *MATRIX_LINV_UNIQ:*

$$\forall (\mathbf{A} \ \mathbf{B})(: \mathbb{R}^{n \times n}). \ \mathbf{A}\mathbf{B} = \mathbf{E} \Rightarrow \mathbf{A} = \mathbf{B}^{-1}$$

Theorem 12. *MATRIX_RINV_UNIQ:*

$$\forall (\mathbf{A} \ \mathbf{B})(: \mathbb{R}^{n \times n}). \ \mathbf{A}\mathbf{B} = \mathbf{E} \Rightarrow \mathbf{B} = \mathbf{A}^{-1}$$

The above theorems are all related to the inverse matrix, but do not indicate the explicit form of the inverse matrix. However, in practical applications it is very necessary to explicate the inverse matrix. For example, for non-singular linear transformations, their inverse transforms are always needed. In order to solve such problems, the explicit form of the inverse matrix must be formalized. There are many methods for achieving matrix inversion and an analytical solution is applied in this paper.

3   Formalizing the Explicit Form of the Inverse Matrix

Writing the adjugate matrix is an efficient way of calculating the inverse of small matrices. To determine the inverse, we calculate a matrix of cofactors:

$$\mathbf{A}^{-1} = \frac{\begin{bmatrix} \mathbf{A}_{00} & \mathbf{A}_{10} & \cdots & \mathbf{A}_{(n-1)0} \\ \mathbf{A}_{01} & \mathbf{A}_{11} & \cdots & \mathbf{A}_{(n-1)1} \\ \vdots & \vdots & \ddots & \vdots \\ \mathbf{A}_{0n} & \mathbf{A}_{1n} & \cdots & \mathbf{A}_{(n-1)(n-1)} \end{bmatrix}}{|\mathbf{A}|} \tag{2}$$

where $|\mathbf{A}|$ is the determinant of $\mathbf{A}$, and $\mathbf{A}_{ij}$ is the cofactor of the corresponding subscript elements of matrix $\mathbf{A}$.

3.1   Formalizing Determinant

The determinant of an n-by-n matrix $\mathbf{A}$ is defined with the **Leibniz** formula.

Definition 3. *DET_DEF*:

$$DET(\mathbf{A} : \mathbb{R}^{n \times n}) := \sum_{p \in S_n} SIGN(p) \prod_{i=0}^{n-1} a_{i,p(i)}$$

It is denoted as $|\mathbf{A}|$. Here, the sum is computed over all permutations p of the set $\{0, 1, \cdots, n-1\}$. The signature of a permutation p is denoted as $SIGN(p)$ and defined as $+1$ if p is even and -1 if p is odd. $a_{i,p(i)}$ is the i-th row and the $p(i)$-th column element of the matrix $\mathbf{A}$.

Cramer's rule is an explicit formula for the solution of a system of linear equations with as many equations as unknowns. The formula is valid whenever the system has a unique solution. It expresses the solution in terms of the determinants of the (square) coefficient matrix and of matrices obtained from it by replacing one column by the vector of the right-hand sides of the equations. Its formalization is as follows:

Theorem 13. *CRAMER*:

$$\forall \mathbf{A} : \mathbb{R}^{n \times n} \ \mathbf{x} \ \mathbf{b}.$$
$$|\mathbf{A}| \neq 0 \Rightarrow$$
$$\mathbf{A}\mathbf{x} = \mathbf{b} \Leftrightarrow \mathbf{x} = \frac{[|\mathbf{b}, a_1, \cdots, a_{n-1}|, \cdots, |a_0, \cdots, a_{j-1}, \mathbf{b}, a_{j+1}, \cdots, a_{n-1}|, \cdots, |a_0, \cdots, a_{n-2}, \mathbf{b}|]}{|A|}$$

here, $\mathbf{a}_i$ is the the i-th column vector of the matrix $\mathbf{A}$. This theorem is presented with some ellipsis for comprehensibility. Its formal description in HOL4 is as follows.

```
!A:real['n]['n]  x   b.
   ~(DET(A) = &0)
     ==> ((A ** x = b) <=>
     (x =
FCP k. DET(FCP i j. if j = k then b ' i else A ' i ' j) / DET(A)))
```

3.2   Formalizing the Cofactor and the Adjugate Matrix

The Difficulty of Defining Minor. In classical mathematical theory, the minor M_{ij} is defined as the determinant of the $(n-1) \times (n-1)$-matrix that results from $\mathbf{A}$ by removing the i-th row and the j-th column. The expression $(-1)^{i+j} M_{ij}$ is known as cofactor A_{ij}. However, with this definition it is very difficult to formalize the minor in a theorem prover. For example, in HOL4 an n-ary real vector type $\mathbb{R}^n$ is formalized with $real[:' n]$, where $'n$ is a type variable, supposing that $dimindex(:' n) = n$, where n is a num type and $:' n$ is a set [3]. Although $dimindex(sub1(:' n)) = n-1$, it is still difficult to define an $(n-1)$-ary real vector type $\mathbb{R}^{n-1}$ in this way. Even if it could be expressed, $\mathbb{R}^n$ and $\mathbb{R}^{n-1}$ are two different types, and so are $\mathbb{R}^{(n-1)\times(n-1)}$ and $\mathbb{R}^{n\times n}$. The type of determinant that was previously defined is $\mathbb{R}^{n\times n} \to \mathbb{R}$, but the type of the expected minor should be $\mathbb{R}^{(n-1)\times(n-1)} \to \mathbb{R}$. Therefore, it is also hard to define a minor with the determinant.

Defining the Cofactor by Contributing the n-by-n Matrix. To Define the cofactor with the determinant of $\mathbb{R}^{n\times n} \to \mathbb{R}$, it is essential to construct a matrix with $\mathbb{R}^{n\times n}$ that is the same size as the original matrix $\mathbf{A}$. The following key point is to construct a matrix whose determinant is equal to the cofactor. When **Cramer**'s rule is formalized, $\mathbf{a}_j$ is replaced with $\mathbf{b}$ to obtain a new matrix. Similarly, the cofactor A_{ij} can be expressed by the determinant of an n-by-n matrix by replacing $\mathbf{a}_j$ with a standard basis $\mathbf{e}_i$ in matrix $\mathbf{A}$. In order to simplify the proof, a more concise matrix can be used to define the cofactor.

Definition 4. *COFACTOR_DEF:*

$$A_{ij} := DET \begin{bmatrix} a_{00} & \cdots & a_{0(j-1)} & 0 & a_{0(j+1)} & \cdots & a_{0(n-1)} \\ \vdots & \ddots & \vdots & \vdots & \vdots & \ddots & \vdots \\ a_{(i-1)0} & \cdots & a_{(i-1)(j-1)} & 0 & a_{(i-1)(j+1)} & \cdots & a_{(i-1)(n-1)} \\ 0 & \cdots & 0 & 1 & 0 & \cdots & 0 \\ a_{(i+1)0} & \cdots & a_{(i+1)(j-1)} & 0 & a_{(i+1)(j+1)} & \cdots & a_{(i+1)(n-1)} \\ \vdots & \ddots & \vdots & \vdots & \vdots & \ddots & \vdots \\ a_{(n-1)0} & \cdots & a_{(n-1)(j-1)} & 0 & a_{(n-1)(j+1)} & \cdots & a_{(n-1)(n-1)} \end{bmatrix}$$

The type of definition above is $\mathbb{R}^{n\times n} \to \mathbb{N} \to \mathbb{N} \to \mathbb{R}$, which matches with the determinant defined previously. Therefore, the cofactor of the corresponding subscripts can be expressed, and its formal definition is described below:

```
val COFACTOR_DEF = Define
 '(COFACTOR:real['n]['n]-> num -> num -> real) A i j =
    DET ((FCP k l. if k = i then (if l = j then &1 else &0) else
           (if l = j then &0 else A ' k ' l)):real['n]['n])';
```

Formalizing the adjugate matrix. The transpose of the matrix of cofactors is known as the adjugate matrix, i.e.,

Definition 5. *ADJUGATE_MATRIX_DEF:*
$$ADJUGATE\_MATRIX(\mathbf{A} : \mathbb{R}^{n\times n}) := \{\mathbf{A}_{ij}\}^T$$

noted as $\mathbf{A}^*$. In HOL4, it can be formally defined in the following way:

```
val ADJUGATE_MATRIX_DEF = Define
 '(ADJUGATE_MATRIX:real['n]['n]-> real['n]['n]) A =
                    TRANSP(FCP i j. COFACTOR A i j)';
```

3.3 Formalizing and Proving Laplace's Formula

The determinant of matrix $\mathbf{A}$ is expanded along an arbitrary row or column as follows:

Theorem 14. *LAPLACE_ROW:*

$$\forall \mathbf{A} : \mathbb{R}^{n \times n} \ k. \ k < n \Rightarrow |\mathbf{A}| = \sum_{j=0}^{n-1} a_{kj} A_{kj}$$

Theorem 15. *LAPLACE_COLUMN:*

$$\forall \mathbf{A} : \mathbb{R}^{n \times n} \ k. \ k < n \Rightarrow |\mathbf{A}| = \sum_{i=0}^{n-1} a_{ik} A_{ik}$$

Likewise, these corollaries can be proven.

Corollary 1. *LAPLACE_ROW_COROLLARY:*

$$\forall A(: \mathbb{R}^{n \times n}) \ i \ j. \ i < n \wedge j < n \wedge i \neq j \Rightarrow \sum_{k=0}^{n-1} a_{ik} A_{jk} = 0$$

Corollary 2. *LAPLACE_COLUMN_COROLLARY:*

$$\forall A(: \mathbb{R}^{n \times n}) \ i \ j. \ i < n \wedge j < n \wedge i \neq j \Rightarrow \sum_{k=0}^{n-1} a_{ki} A_{kj} = 0$$

3.4 Proving the Explicit Form of the Inverse Matrix

Using the definition of matrix multiplication, the following theorems can be proven.

Corollary 3. *LAPLACE_COROLLARY_LMUL:*

$$\forall A(: \mathbb{R}^{n \times n}). \ \mathbf{A}\mathbf{A}^* = |\mathbf{A}|\mathbf{E}$$

Corollary 4. *LAPLACE_COROLLARY_RMUL:*

$$\forall A(: \mathbb{R}^{n \times n}). \ \mathbf{A}^*\mathbf{A} = |\mathbf{A}|\mathbf{E}$$

A square matrix $\mathbf{A}$ is invertible if and only if its determinant is not 0.

Theorem 16. *INVERTIBLE_DET_NZ:*

$$\forall \mathbf{A}(: \mathbb{R}^{n \times n}). \ invertible(\mathbf{A}) \Leftrightarrow |\mathbf{A}| \neq 0$$

This theorem can be proven based on the definition of the inverse matrix and the properties of the determinant mentioned before.

Hence, the inversion of the invertible matrix can be explicated as follows:

Theorem 17. *MATRIX_INV_EXPLICIT:*

$$\forall \mathbf{A}(: \mathbb{R}^{n \times n}). \ invertible(\mathbf{A}) \Rightarrow \mathbf{A}^{-1} = \frac{\mathbf{A}^*}{|\mathbf{A}|}$$

Here is the formal proof in HOL4:

Moving the antecedent of the above goal into the assumptions and doing left multiplication with matrix $\mathbf{A}$ at both sides of the equation, the following form is obtained.

$$\frac{\mathbf{A}\mathbf{A}^{-1} = \frac{\mathbf{A}\mathbf{A}^*}{|\mathbf{A}|}}{invertible(\mathbf{A})} \tag{3}$$

It can be proven by rewriting *LAPLACE_COROLLARY_LMUL* and *MATRIX_INV*.

4   Inverse Transforming and Solving the Matrix Equation

After formalizing the matrix inversion, some questions about the inverse matrix can be expressed using its explicit form. As a consequence, the inverse of the nonsingular transformation can be formalized.

Theorem 18. *TRAN_INV_EXPLICIT:*

$$\forall \mathbf{A}(: \mathbb{R}^{n \times n}) \ \mathbf{x} \ \mathbf{y}(: \mathbb{R}^n). invertible(\mathbf{A}) \wedge \mathbf{A}\mathbf{x} = \mathbf{y} \Rightarrow \mathbf{x} = \frac{\mathbf{A}^*}{|\mathbf{A}|}\mathbf{y}$$

Solving the matrix equation with an invertible coefficient matrix is widely applied in robot, real-time simulations, and MIMO wireless communication. It can be represented in formal form as below:

Theorem 19. *MATRIX_MUL_LINV_EXPLICIT:*

$$\forall \mathbf{A}(: \mathbb{R}^{n \times n}) \ (\mathbf{X} \ \mathbf{Y})(: \mathbb{R}^{n \times m}). invertible(\mathbf{A}) \wedge \mathbf{A}\mathbf{X} = \mathbf{Y} \Rightarrow \mathbf{X} = \frac{\mathbf{A}^*}{|\mathbf{A}|}\mathbf{Y}$$

It can be proven using *MATRIX_INV_EXPLICIT* and *MATRIX_MUL_LINV_UNIQ*.

5   Conclusions

To solve the problem of formalizing the cofactor, we proposed a method using the determinant of an n-by-n matrix to express the cofactor. Consequently, we formally described the explicit form of an inverse matrix using the adjugate matrix method, and proved some of the important properties. Our work enriched theories of HOL4 and is expected to extend the scope of application of HOL4.

Acknowledgements. First and foremost we thank Prof. Shengzhen Jin for the guidance and encouragement that he gave us. We also thank Dr. John Harrison for his many good suggestions.

This work was supported by the International S&T Cooperation Program of China (2010DFB10930, 2011DFG13000); the National Natural Science Foundation of China (60873006, 61070049, 61170304, 61104035); the Beijing Natural Science Foundation and S&R Key Program of BMEC(4122017, KZ201210028036). Support was also received from the Open Project section of State Key Laboratory of Computer Architecture and the Trusted Software section of the Guangxi Key Laboratory.

References

1. Harrison, J.V.: A HOL theory of euclidean space. In: Hurd, J., Melham, T. (eds.) TPHOLs 2005. LNCS, vol. 3603, pp. 114–129. Springer, Heidelberg (2005)
2. Robert, C., Solovay, M., Arthan, R.D., Harrison, J.: Some new results on decidability for elementary algebra and geometry. ArXiV preprint 0904.3482, Submitted to the Annals of Pure and Applied Logic (2009)
3. Slind, K., Norrish, M.: A brief overview of HOL4. In: Mohamed, O.A., Muñoz, C., Tahar, S. (eds.) TPHOLs 2008. LNCS, vol. 5170, pp. 28–32. Springer, Heidelberg (2008)
4. Nakamura, Y., Tamura, N., Chang, W.: A Theory of Matrices of Real Elements. J. Formalized Mathematics 14(1), 21–28 (2006)
5. Tamura, N., Nakamura, Y.: Determinant and Inverse of Matrices of Real Elements. J. Formalized Mathematics 15(3), 127–136 (2007)
6. Obua, S.: Proving bounds for real linear programs in isabelle/HOL. In: Hurd, J., Melham, T. (eds.) TPHOLs 2005. LNCS, vol. 3603, pp. 227–244. Springer, Heidelberg (2005)
7. Pasca, I.: Formal Proofs for Theoretical Properties of Newton's Method. Rapport de recherché INRIA Sophia Antipolis, 28 pages (February 2010)
8. Paşca, I.: Formally verified conditions for regularity of interval matrices. In: Autexier, S., Calmet, J., Delahaye, D., Ion, P.D.F., Rideau, L., Rioboo, R., Sexton, A.P. (eds.) AISC 2010. LNCS, vol. 6167, pp. 219–233. Springer, Heidelberg (2010)
9. Andrews, P.B.: An Introduction to Mathematical Logic and Type Theory: To Truth Through Proof. Academic Press (1986)
10. Church, A.: A formulation of the Simple Theory of Types. Journal of Symbolic Logic 5, 56–68 (1940)
11. Diaconescu, R.: Axiom of choice and complementation. Proceedings of the American Mathematical Society 51, 176–178 (1975)
12. Gordon, M.J.C.: Representing a logic in the LCF metalanguage. In: Néel, D. (ed.) Tools and notions for program construction: an advanced course, pp. 163–185. Cambridge University Press (1982)

A Heuristic Approach to Acquisition of Minimum Decision Rule Sets in Decision Systems

Zuqiang Meng, Liang Jiang, Hongyan Chang, and Yuansheng Zhang

College of Computer, Electronics and Information, Guangxi University Nanning 530004, China
mengzuqiang@163.com

Abstract. In rough set theory, not too much work pays attention to the acquisition of decision rules and to the uses of the obtained rule set as classifier to predict data. In fact, rough set theory also can be applied to train data and create classifiers and then complete data prediction. This paper systematically studies the problem of acquisition of decision rules in decision systems. The main outcomes of this research are as follows: (1) the specific definition of minimum rule set is given, and such a minimum rule set can be used as a classifier to predict new data; (2) a new approach to finding out all minimum rule sets for a decision system, Algorithm 1, is proposed based on discrimination function, but with relatively low execution efficiency; (3) By improving Algorithm 1, a heuristic approach to computing a special minimum rule set, Algorithm 3, is proposed, which works far more efficiently than Algorithm 1. The outcomes can form the foundation for applying rough set theory to data classification and offer a new resolution to data classification.

Keywords: Heuristic approach, minimum decision rule set, rule acquisition, classifier, data classification.

1 Introduction

Rough set theory[1] is a powerful mathematical tool to deal with insufficient, incomplete or vague information. Nowadays the majority of work in rough set theory focuses on attribute reduction[2]. Attribute reduction is in fact feature selection and has been applied to address many practical problems[3-6]. After attribute reduction, data sets can be used for a variety of applications and data classification is an important application. However, not too much work pays attention to data classification with rough set theory. In rough set theory, most methods of data classification are to use rough set theory to select features and then use other tools to train data and create classifier[7]. In fact, rough set theory also can be used to extract decision rules from decision systems. Although some scholars have made contribution to this work[8-10], there is still a lack of systematic foundation for rule acquisition in rough set theory. This paper systematically studies the problem of acquisition of decision rules in decision systems. First, the specific definition of minimum rule sets is given, and then a new approach to finding out all minimum rule sets for a decision system is proposed based on discrimination function, and finally a heuristic approach

Z. Shi et al. (Eds.): IIP 2014, IFIP AICT 432, pp. 187–196, 2014.

to computing a special minimum rule set is proposed. Each of the two proposed approaches has its own advantages and disadvantages. The work provides a relatively complete solution for extracting decision rules from decision systems and using rough set theory to classify data.

The rest of the paper is organized as follows. Section 2 reviews some basic concepts linked to Information system and decision logic. Section 3 gives the definition of minimum rule set. Section 4 proposes a general approach to the acquisition of all minimum rule sets in a given decision system. Section 5 then proposes a heuristic approach to the acquisition of a minimum rule set. Section 6 gives an example to show how to use the heuristic approach. Section 7 finally concludes the paper.

2   Information System and Decision Logic

An **information system** is usually expressed in the following form: $IS = (U, A, \{V_a\}, f_a)_{a \in A}$, where U is a nonempty finite set of objects, standing for a given universe; A is a nonempty finite set of attributes; V_a is a value set(domain) of attribute a; f_a is called information function, i.e., $f_a : U \rightarrow V_a$, which denotes the value of function f on attribute a; If V_a and f_a are obvious, then $(U, A, \{V_a\}, f_a)_{a \in A}$ can be denoted as (U, A) for short.

For any $B \subseteq A$, subset B determines a binary relation, denoted as $TR(B)$, which is defined as follows: $TR(B) = \{(x,y) \mid f_a(x) = f_a(y)$ for any $a \in B$ and any $x, y \in U\}$.

It is easy to prove that $TR(B)$ is reflexive, symmetric, transitive, and thereby is an equivalence relation on U. Thus, equivalence relation $TR(B)$ can divide the universe U into several disjoint subsets, which are known as equivalence classes. Suppose that $X_1, X_2, \ldots, X_n$ are all equivalence classes induced by attribute set B, then $\{X_1, X_2, \ldots, X_n\}$ is a partition $U/TR(B)$ of the universe U, denoted by $U/TR(B) = \{X_1, X_2, \ldots, X_n\}$.

Extraction of decision rules is in fact the problem of extracting description of granules (equivalence classes), which are expressed with decision logic.

Let $IS(B) = <U, B, \{V_a\}, f_a>_{a \in B}$, $B \subseteq A$. Then decision logic language $DL(B)$ with respect to B is defined as following:

(1) (a, v) is an atomic formula, where $a \in B$, $v \in V_a$,

(2) an atomic formula is a formula in $DL(B)$;

(3) if φ is a formula, then $\sim\varphi$ is also a formula in $DL(B)$;

(4) if both φ and ψ are formulae, then $\varphi \vee \psi$, $\varphi \wedge \psi$, $\varphi \rightarrow \psi$, $\varphi \equiv \psi$ are all formulae;

(5) only the formulae obtained according to the above Steps (1) to (4) are formulae in $DL(B)$.

If φ is a simple conjunction, which consists only of atomic formulae and connectives $\wedge$, then φ is called a **basic formula**.

The following definition gives the relationship between formulae in $DL(B)$ and granules $IS(B)$:

For any $s \in U$, the relationship between s and formulae in $DL(B)$ is defined as following:

(1) $s \models (a, v)$ iff $f_a(s) = v$

(2) $s \models \sim\varphi$ iff not $s \models \varphi$

(3) $s \models \varphi \wedge \psi$ iff $s \models \varphi$ and $s \models \psi$

(4) $s \models \varphi \vee \psi$ iff $s \models \varphi$ or $s \models \psi$

(5) $s \models \varphi \rightarrow \psi$ iff $s \models \sim\varphi \vee \psi$

(6) $s \models \varphi \equiv \psi$ iff $s \models \varphi \rightarrow \psi$ and $s \models \psi \rightarrow \varphi$.

For formula φ, if $s \models \varphi$, then we say that the object s satisfies formula φ. Let $m(\varphi)$ = $\{ s \mid s \models \varphi \}$, that is, $m(\varphi)$ is the set of all those objects that satisfy formula φ; For subset $g \subseteq U$, if $m(\varphi) = g$, then g is said to be **descriptive** and φ is a **description** of g, denoted by $DES(g)$. It is easy to find that for the same subset $g \subseteq U$, there are possibly many different descriptions: $\varphi_1, \varphi_2, \ldots, \varphi_m$, such that $m(\varphi_1) = m(\varphi_2) = \ldots = m(\varphi_m) = g$. This is one of the reasons why extracting all decision rules from a decision system is a NP-hard problem.

3   Minimum Rule Set in Decision System

We note that some principles of attribute reduction can also be applied to extracting decision rules from decision systems and then to acquiring minimum rule set, which can be used as a classifier.

Decision system is a special case of information system, which can be regarded to be generated by partitioning attribute set A into two disjoint subsets, or by adding some attributes to A.

A **decision system** is usually denoted by $DS = (U, C \cup D, \{V_a\}, f_a)_{a \in A = C \cup D}$,where U, V_a and f_a have the same meanings as that in the above section; C is a nonempty finite set of attributes, called condition attribute set; D is a nonempty finite set of attributes, called decision attribute set, and $C \cap D = \varnothing$; The 4-triple $(U, C \cup D, \{V_a\}, f_a)_{a \in A = C \cup D}$ is usually denoted as $(U, C \cup D)$ for short, namely, $DS = (U, C \cup D)$.

As mentioned above, we can consider that decision system $(U, C \cup D)$ consists of two information systems: (U, C) and (U, D).

Suppose that $\phi \in DL(C)$ and $\phi \in DL(D)$. Implication form $\phi \rightarrow \psi$ is said to be a **decision rule** in decision system $(U, C \cup D)$. If both ϕ and ψ are basic formula, then $\phi \rightarrow \psi$ is called **basic decision rule**.

Decision rule has two important measuring indices, confidence and support, which are defined as following:

$$conf(\phi \rightarrow \psi) = \frac{\mid m(\phi) \cap m(\psi) \mid}{\mid m(\phi) \mid}\text{[11]} \qquad (1)$$

$$sup(\phi \rightarrow \psi) = \frac{|m(\phi) \cap m(\psi)|}{|U|} \qquad [2]$$

where $conf(\phi \rightarrow \psi)$ and $sup(\phi \rightarrow \psi)$ are confidence and support of decision rule $\phi \rightarrow \psi$, respectively.

In general, we usually find those decision rules whose confidence is equal to 1. Obviously, if $m(\phi) \subseteq m(\psi)$, then $conf(\phi \rightarrow \psi) = 1$. Of course, it is not enough to guarantee that confidence is equal to 1. In fact, the larger the confidence, the better it is, although it is small than 1 with certainty. That is, extracting decision rules is the procedure of finding those rules whose confidence is equal to 1 and whose support is as large as possible. In order to illustrate this procedure, we further introduce some related concepts.

Definition 1. For decision system $(U, C \cup D)$, $B \subseteq C$ and $x \in U$, let $[x]_B = \{y \in U \mid (x, y) \in TR(B)\}$, then $[x]_B$ is said to be an equivalence class with respect to attribute subset B.

Property 1. For decision system $(U, C \cup D)$, $B_1 \subseteq B_2 \subseteq C$ and any $x \in U$, $[x]_{B2} \subseteq [x]_{B1}$.

Proof. It is straightforward.

Property 1 shows that with the decrease of attribute subset, corresponding equivalence classes become larger and larger, thus increase corresponding rule's support. That is, by deleting some attributes from C, we can increase rule's support according to formula (2), which is the foundation of finding decision rules in this paper.

In fact, by deleting some attributes from condition attribute set, every object in the universe U can induce at least one decision rule. Thus, all objects from U can induce a lot of decision rules and then form a very large decision rule set. Obviously, many rules are redundant and should be removed from rule set. Therefore, one basic problem is that what is the criterion of a minimum rule set?

Definition 2. For decision system $DS = (U, C \cup D)$, if rule $\varphi \rightarrow \psi$ is true in $DL(C \cup D)$, i.e., for any $x \in U$ $x \models \varphi \rightarrow \psi$, then rule $\varphi \rightarrow \psi$ is said to be **consistent** in DS, denoted by $\models_{DS} \varphi \rightarrow \psi$, if there exists at least object $x \in U$ such that $x \models \varphi \wedge \psi$, then rule $\varphi \rightarrow \psi$ is said to be **satisfiable** in DS.

Consistency and satisfiability are the basic properties that must be satisfied by decision rules. In addition, an efficient and effective decision rule set should be as small as possible, that is, it should be minimized.

Definition 3. For decision system $(U, C \cup D)$, object $x \in U$ and decision rule r: $\varphi \rightarrow \psi$, if $x \models r$, then it is said that rule r **covers** object x, or object x is **covered** by rule r; let coverage(r) denote the set of all objects that are **covered** by rule r and coverage$^{-1}(x)$ the set of all rules that cover object x, that is:

coverage$(r) = \{x \in U \mid x \models r\}$,
coverage$^{-1}(x) = \{r' \mid r'$ cover x, i.e., $x \in$ coverage$(r')\}$.

Definition 4. For decision system $(U, C \cup D)$, decision rules r_1 and r_2, if coverage(r_1) $\subseteq$ coverage(r_2), then it is said that r_2 **functionally cover** r_1, denoted by $r_1 \leq r_2$.

Obviously, those rules that are functionally covered by other rules should be removed out from rule set.

Definition 5. For decision system $(U, C \cup D)$ and $x \in U$, $[x]_B$ is said to be **maximized** if $[x]_B \subseteq [x]_C$ and for any $B' \subset B$ $[x]_B \not\subseteq [x]_D$, where $B \subseteq C$; rule $DES([s]_B) \to DES([s]_D)$ is said to be **reduced** if $[x]_B$ is maximized.

A decision rule set $\wp$ is said to be minimal if it satisfies the following properties:

(1) any rule from $\wp$ should be consistent;

(2) any rule from $\wp$ should be satisfiable;

(3) any rule from $\wp$ should be reduced;

(4) for any two rules $r_1, r_2 \in \wp$, neither $r_1 \leq r_2$ nor $r_2 \leq r_1$.

When a decision rule set satisfies all the four properties, each rule is effective and the whole rule set is minimal and then matching efficiency can be improved greatly.

4   A General Approach to Acquisition of All Minimum Rule Sets

Suppose that φ be a formula and $B = \{a_1, a_2, ..., a_m\}$ be a condition attribute subset. Let $set(\varphi)$ be the set of all attributes that appear in formula φ ,and $\wedge B$ and $\vee B$ respectively be the simple conjunction and the simple disjunction that consists of all the attributes appear in B with $\wedge$ as connective, i.e., $\wedge B = a_1 \wedge a_2 \wedge ... \wedge a_m$. For example, let $\varphi = a \wedge b \wedge c$ then $set(\varphi) = \{a, b, c\}$; let $B = \{a, b, c\}$ then $\vee B = a \vee b \vee c$ and $\wedge B = a \wedge b \wedge c$.

Definition 6. For decision system $(U, C \cup D)$, let $\alpha([x]_C, x') = \vee \{a \in C \mid x' \notin [x]_{\{a\}}\}$, where $x, x' \in U$; suppose that $U - [x]_D = \{x'_1, x'_2, ..., x'_m\}$, then the discrimination function of $[x]_C$ is defined as following:

$$f([x]_C) = \alpha([x]_C, x'_1) \wedge \alpha([x]_C, x'_2) \wedge ... \wedge \alpha([x]_C, x'_m).$$

The discrimination function is a conjunctive normal form. By using absorption law and distribution law, it can be converted to be a disjunctive normal form, which consists of several simple conjunctions. Each simple conjunction corresponds to a decision rule. Without loss of generality, suppose that $f([x]_C)$ is converted to disjunctive normal form $\rho_1 \vee \rho_2 \vee ... \vee \rho_{m'}$, where ρ_i is a simple conjunction, $i=1,2,...,$ m'. Then all the rules that induced by equivalence class $[x]_C$ are as follows:

$r_1: DES([x]_{set(\rho_1)}) \to DES([x]_D),$

$r_2: DES([x]_{set(\rho_2)}) \to DES([x]_D),$

$r_{m'}: DES([x]_{set(\rho_{m'})}) \to DES([x]_D).$

Let $R([x]_C)$ be the set of all rules that induced by $[x]_C$, i.e., $R([x]_C) = \{r_1, r_2, ..., r_{m'}\}$. Suppose that decision class $[x]_D$ is an union of several equivalence classes: $[x_1]_C$, $[x_2]_C$, ..., $[x_p]_C$, i.e., $[x]_D = [x_1]_C \cup [x_2]_C \cup ... \cup [x_p]_C$. Let $R([x]_D) = R([x_1]_C) \cup R([x_2]_C) \cup ... \cup R([x_p]_C)$. Obviously, some induced rules in $R([x]_D)$ are redundant and should be removed from rule set. Or in other words, $R([x]_D)$ needs to be further reduced.

Definition 7. For decision system $(U, C \cup D)$ and decision class $[x]_D$, suppose $\wp$ is a subset of rule set $R([x]_D)$, i.e., $\wp \subseteq R([x]_D)$, then $\wp$ is a **reduct** of $R([x]_D)$ if any rule in $\wp$ is consistent, satisfiable, reduced and for any two rules $r_1, r_2 \in \wp$, neither $r_1 \leq r_2$ nor $r_2 \leq r_1$.

Similar to attribute reduction, $R([x]_D)$ possibly has more than one reduct. We can also use discrimination function to find all reducts of $R([x]_D)$.

Suppose $[x]_D = \{x_1, x_2, ..., x_h\}$. Let $f([x]_D) = [\vee \text{coverage}^{-1}(x_1)] \wedge [\vee \text{coverage}^{-1}(x_2)] \wedge ... \wedge [\vee \text{coverage}^{-1}(x_h)]$. Obviously, $f([x]_D)$ is a conjunctive normal form. Similarly, it can be converted to a disjunctive normal form by using absorption law and distribution law, in which each simple conjunction corresponds to a reduct of $R([x]_D)$. Assume that $f([x]_D)$ is converted to disjunctive normal form $\beta_1 \vee \beta_2 \vee ... \vee \beta_{h'}$. Then all reducts of $R([x]_D)$ are as follows: $set(\beta_1), set(\beta_2), ..., set(\beta_{h'})$.

It is not difficult to prove that for any $r \in set(\beta_i)$, r is consistent, satisfiable, reduced and for any two rules $r_1, r_2 \in set(\beta_i)$, neither $r_1 \leq r_2$ nor $r_2 \leq r_1$, $i \in \{1, 2, ..., h'\}$. Therefore, $set(\beta_i)$ is a reduct of $R([x]_D)$.

Again, we know that that if $[x_1]_D \neq [x_2]_D$ then $[x_1]_D \cap [x_2]_D = \varnothing$ and then $R([x_1]_D) \cap R([x_2]_D) = \varnothing$. Hence for any two different decision classes $[x_1]_D$ and $[x_2]_D$, we may concurrently calculate all reducts of $[x_1]_D$ and $[x_2]_D$.

According to analysis above, we can give a complete solution to acquisition of all minimum rule sets in decision system $DS = (U, C \cup D)$, which is called Algorithm 1.

Algorithm 1. For calculating all minimum rule sets.

(1) compute all different equivalence classes: $[x_1]_D, [x_2]_D, ..., [x_n]_D$, and let $DC(DS) = \{[x_1]_D, [x_2]_D, ..., [x_n]_D\}$;

(2) for decision class $[x_i]_D \in DC(DS)$, find out all equivalence classes whose union is equal to $[x_i]_D$, and suppose the set of all such equivalence classes is $EC([x_i]_D)$;

(3) for each $[x']_C \in EC([x_i]_D)$, compute $R([x']_C)$ using discrimination function and then obtain $R([x_i]_D) = \cup \{R([x']_C) \mid [x']_C \in EC([x_i]_D)\}$;

(4) find all reducts of $R([x_i]_D)$ using discrimination function, and suppose the set of all reducts of $R([x_i]_D)$ is $RED[R([x_i]_D)]$;

(5) compute the set of all minimum rule sets: $\{ \bigcup_{i=1}^{n} \wp_i \mid \wp_i \in RED[R([x_i]_D)],$ where $i \in \{1,2,...,n\}\}$.

Steps (2)-(4) can be performed concurrently, with which we can improve the efficiency of the approach. The biggest advantage of this approach lies in that it can find out all minimum rule sets for a decision system. However, we know that in steps

(3) and (4) discrimination function is used, where a conjunctive normal form is converted to be a disjunctive normal form using absorption law and distribution law. This conversion involves problem of combination explosion and is a NP-hard problem. Therefore, the approach is suitable for large data set. Furthermore, it is unnecessary to find all minimum rule sets for a decision system in real applications. In fact, a special minimum rule set is more desired in most cases. Therefore, the time-consuming steps, conversion of conjunctive normal form to disjunctive normal form, should be abandoned in real applications.

5   A Heuristic Approach to Acquisition of a Minimum Rule Set

Suppose that $f(t_1, t_2, ..., t_r)$ is a conjunctive normal form, where $t_1, t_2, ..., t_r$ are r different items that occur in the formula. They can stand for attributes or rules. Further, suppose that after using absorption law: $f(t_1, t_2, ..., t_r) = \gamma_1 \wedge \gamma_2 \wedge ... \wedge \gamma_k$, where γ_i is a simple disjunction, and for any different simple disjunctions γ_i and γ_j, neither $set(\gamma_i) \subseteq set(\gamma_j)$ nor $set(\gamma_j) \subseteq set(\gamma_i)$, $i,j \in \{1,2,...,k\}$. Now we introduce how to obtain a special simple conjunction of $f(t_1, t_2, ..., t_r)$, which corresponds to a reduct of attribute set or rule set.

Let us sort items $t_1, t_2, ..., t_r$ in an ascending order by their occurring frequencies in formula $f(t_1, t_2, ..., t_r)$ and suppose the ascending order is $<t'_1, t'_2, ..., t'_r>$, where t'_r occurs most frequently in formula $f(t_1, t_2, ..., t_r)$ and therefore it is viewed as the most important item. That is, the finally obtained simple conjunction should contain some important items like $..., t_{r-1}, t_r$ and remove some unimportant items like $t_1, t_2, ...$ as possible. The approach to obtaining a special simple conjunction of $f(t_1, t_2, ..., t_r)$ is called Algorithm 2, which is described as follows.

Algorithm 2. For calculating a special simple conjunction (called a reduct of $f(t_1, t_2, ..., t_r)$ for simplicity).
  (1) let $f' = f(t_1, t_2, ..., t_r)$
  (2) for $i = 1$ to r do
        {
  (3)   for each γ_j in f', if there is no such γ_j in f' that $set(\gamma_j) = \{t'_i\}$ then
             remove item t_i from γ_j for each γ_j, denoted by $\gamma_j = \gamma_j - \{t_i\}$;
  (4)   apply absorption law to f';
        }
  (5) $set(f')$ is reduct.

Take $f(a_1, a_2, a_3, a_4, a_5) = (a_2 \vee a_3) \wedge (a_3 \vee a_4 \vee a_5) \wedge (a_1 \vee a_3) \wedge (a_2 \vee a_4) \wedge (a_1 \vee a_4 \vee a_5)$ for example. It can be observed that the occurring frequencies of a_1, a_2, a_5, a_3, a_4 are 2, 2, 2, 3, 3, respectively. According to Algorithm 2, the reduction steps are as follows:

  (1) let $f' = (a_2 \vee a_3) \wedge (a_3 \vee a_4 \vee a_5) \wedge (a_1 \vee a_3) \wedge (a_2 \vee a_4) \wedge (a_1 \vee a_4 \vee a_5)$;
  (2) remove a_1 from f': $f' = (a_2 \vee a_3) \wedge (a_3 \vee a_4 \vee a_5) \wedge (a_3) \wedge (a_2 \vee a_4) \wedge (a_4 \vee a_5)$;
  (3) after applying absorption law: $f' = (a_3) \wedge (a_2 \vee a_4) \wedge (a_4 \vee a_5)$;

(4) remove a_2 from f': $f' = (a_3) \wedge (a_4) \wedge (a_4 \vee a_5)$;
(5) after applying absorption law: $f' = (a_3) \wedge (a_4)$;
(6) as a result $set(f') = \{a_3, a_4\}$.

By using Algorithm 2, we can modify Algorithm 1 to a heuristic approach, which is used to acquire a special minimum rule set and is described as follows:

Algorithm 3. For acquiring a special minimum rule set.

(1)   compute all different equivalence classes: $[x_1]_D$, $[x_2]_D$, ..., $[x_n]_D$, and let $DC(DS) = \{[x_1]_D, [x_2]_D, ..., [x_n]_D\}$;

(2)   for decision class $[x_i]_D \in DC(DS)$, find out all equivalence classes whose union is equal to $[x_i]_D$, and suppose the set of all such equivalence classes is $EC([x_i]_D)$;

(3)   for each $[x']_C \in EC([x_i]_D)$, sort attributes occurring $f([x']_C)$ in an ascending order by their occurring frequencies, and then use Algorithm 2 to obtain a reduct of $f([x']_C)$, with which we create a rule set, denoted as $R''([x']_C)$, and finally let $R''([x_i]_D) = \cup \{R''([x']_C) \mid [x']_C \in EC([x_i]_D)\}$;

(4)   sort rules occurring in $f([x]_D)$ in an ascending order by their occurring frequencies, and then use Algorithm 2 to obtain a reduct of $f([x]_D)$, with which we obtain a rule set, denoted as $RED''[R''([x_i]_D)]$;

(5)   return a minimum rule set: $\cup \{RED''[R''([x_i]_D)] \mid [x_i]_D \in DC(DS)\}$.

6   Example Analysis

Consider decision system $DS = (U, C \cup D)$, where $U = \{x_1, x_2, ..., x_6\}$, $C = \{a_1, a_2, a_3\}$ and $D = \{d\}$, presented in Table 1.

Table 1. A decision system

U	a_1	a_2	a_3	d
x_1	2	3	2	1
x_2	1	1	1	2
x_3	3	3	3	3
x_4	2	2	2	3
x_5	2	2	1	2
x_6	2	3	3	1

According to Algorithm 3, steps to obtain a special minimum rule set are as follows:

(1) by computing, $DC(DS) = \{\{x_1, x_6\}, \{x_2, x_5\}, \{x_3, x_4\}\} = \{[x_1]_D, [x_2]_D, [x_3]_D\}$;

(2) for decision class $[x_3]_D \in DC(DS)$, $EC([x_3]_D) = \{\{x_3\}, \{x_4\}\} = \{[x_3]_C, [x_4]_C\}$; $f([x_3]_C) = (a_1 \vee a_3) \wedge (a_1 \vee a_2 \vee a_3) \wedge (a_1 \vee a_2 \vee a_3) \wedge a_1$, and through the statistics the ascending order is $<a_2, a_3, a_1>$, and then by using Algorithm 2, the obtained reduct of $f([x_3]_C)$ is $\{a_1\}$; so we have $R''([x_3]_C) = \{(a_1, 3) \rightarrow (d, 3)\}$, and similarly, we have $R''([x_4]_C) = \{(a_2, 2) \wedge (a_3, 2) \rightarrow (d, 3)\}$; finally, $R''([x_3]_D) = R''([x_3]_C) \cup R''([x_4]_C) = \{(a_1, 3) \rightarrow (d, 3), (a_1, 2) \wedge (a_1, 2) \rightarrow (d, 3)\}$;

(3) by the similar way, we can compute $R''([x_1]_D)$ and $R''([x_2]_D)$. All the results are as follows:

$R''([x_1]_D) = \{r_1: (a_2, 3) \wedge (a_3, 2) \rightarrow (d, 1),\quad r_2: (a_1, 2) \wedge (a_3, 3) \rightarrow (d, 1)\}$

$R''([x_2]_D) = \{r_3: (a_3, 1) \rightarrow (d, 2),\quad r_4: (a_3, 1) \rightarrow (d, 2)\}$

$R''([x_3]_D) = \{r_5: (a_1, 3) \rightarrow (d, 3),\quad r_6: (a_2, 2) \wedge (a_3, 2) \rightarrow (d, 3)\}$

(4) by computing, we have: coverage$^{-1}(x_1) = \{r_1\}$, coverage$^{-1}(x_2) = \{r_3, r_4\}$, coverage$^{-1}(x_3) = \{r_5\}$, coverage$^{-1}(x_4) = \{r_6\}$, coverage$^{-1}(x_5) = \{r_3, r_4\}$, coverage$^{-1}(x_6) = \{r_2\}$. Take $RED''[R''([x_2]_D)]$ for example. Because $f([x_2]_D) = (r_3 \vee r_4) \wedge (r_3 \vee r_4)$ and the ascending order is $<r_3, r_4>$, so it can be found that $\{r_4\}$ is a reduct of $f([x_2]_D)$ and therefore $RED''[R''([x_2]_D)] = \{r_4\}$; by the same way, $RED''[R''([x_1]_D)] = \{r_1, r_2\}$, $RED''[R''([x_3]_D)] = \{r_5, r_6\}$;

(5) the finally obtained minimum rule set is $RED''[R''([x_1]_D)] \cup RED''[R''([x_2]_D)] \cup RED''[R''([x_3]_D)] = \{r_1, r_2\} \cup \{r_4\} \cup \{r_5, r_6\} = \{r_1, r_2, r_4, r_5, r_6\}$.

That is, $\{r_1, r_2, r_4, r_5, r_6\}$ is a special minimum rule set that obtained by Algorithm 3. We can prove that any rule in the rule set is consistent, satisfiable, reduced and for any two rules r_1, r_2 in the rule set, neither $r_1 \leq r_2$ nor $r_2 \leq r_1$. So it is a minimum rule set.

7   Conclusion

In rough set theory, many scholars pay more attention to attribute reduction, and not too much work pays attention to the acquisition of decision rules with rough set theory. Most methods use rough set theory to reduce data in the first place, and then use other algorithms, such as CART, SVM[7], to create classifier based on the reduced data and use it to classify or predict new data. In fact, rough set theory can also be used to extract decision rules from data sets. By creating minimum rule sets, which is used as classifiers, we can also achieve better data prediction. In this paper, we propose two important algorithms, Algorithm 1 and Algorithm 3. Algorithm 1 can find out all minimum rule sets for a given data set theoretically. But it applies discrimination function to complete its computing, which is a time-consuming operation, so its disadvantage is low efficiency, especially for large data sets. By improving Algorithm 1, a heuristic approach to computing a special minimum rule set, Algorithm 3, is proposed. Using occurring frequency of attributes or rules as heuristic information, Algorithm 3 can rapidly obtain a special minimum rule set. It works far more efficiently than Algorithm 1 and has high application value.

Acknowledgement. This work is supported by the National Natural Science Foundation of China (No. 61363027) and the Natural Science Foundation of Guangxi Province, China (No. 2012GXNSFAA053225).

References

[1] Pawlak, Z.: Rough Sets. International Journal of Computer and Information Sciences 11(5), 341–356 (1982)

[2] Li, M., Shang, C.X., Feng, S.Z., Fan, J.P.: Quick Attribute Reduction in Inconsistent Decision Tables. Information Sciences 254, 155–180 (2014)

[3] Liu, J., Hu, Q., Yu, D.: A Weighted Rough Set Based Method Developed for Class Imbalance Learning. Information Sciences 4, 1235–1256 (2008)

[4] Meng, Z.Q., Shi, Z.Z.: A Fast Approach to Attribute Reduction in Incomplete Decision Systems with Tolerance Relation-based Rough Sets. Information Sciences 179, 2774–2793 (2009)

[5] Meng, Z.Q., Shi, Z.Z.: Extended Rough Set-based Attribute Reduction in Inconsistent Incomplete Decision Systems. Information Sciences 204, 44–66 (2012)

[6] Yang, H.H., Liu, T.C., Lin, Y.T.: Applying rough sets to prevent customer complaints for IC packaging foundry. Expert Systems with Applications 32, 151–156 (2007)

[7] Hu, Q.H., Liu, J.F., Yu, D.: Mixed Feature Selection Based on Granulation and Approximation. Knowledge-Based Systems 21(4), 294–304 (2008)

[8] Chang, L.Y., Wang, G.Y., Wu, Y.: An Approach for Attribute Reduction and Rule Generation Based on Rough Set Theory. Journal of Software 10(11), 1206–1211 (1999)

[9] Shao, E.X., Zhang, L.S., Yang, Y.Z., Li, F., Yang, H.,, J.X.: Method of Rule Extraction Based on Rough Set Theory. Computer Science 38(1), 232–235 (2011)

[10] Du, Y., Hu, Q., Zhu, P., Ma, P.: Rule Learning for Classification Based on Neighborhood Covering Reduction. Information Sciences 181 (2011)

[11] Pawlak, Z., Skowron, A.: Rough Membership Functions. In: Yager, R.R., Fedrizzi, M., Kacprzyk, J. (eds.) Advances in the Dempster–Shafer Theory of Evidence, pp. 251–271. John Wiley and Sons (1994)

Cooperative Decision Algorithm for Time Critical Assignment without Explicit Communication

Yulin Zhang, Yang Xu, and Haixiao Hu

School of Computer Science and Engineering,
University of Electronic Science and Technology of China
xuyang@uestc.edu.cn

Abstract. Decentralized time critical assignment is popular in the domains of military and emergency response. Considering the large communication delay, unpredictable environment and constraints, to make a rational decision in time critical context toward their common goal, agents have to jointly allocate multiple tasks based on their current status. This process can be built as a Markov Decision Model when they can share their status freely. However, it is computationally infeasible when explicit communication becomes a severe bottleneck. In this paper, by modelling the decision process as a Partially Observable Markov Decision Process for each agent and building heuristic approaches to estimate the state and reduce the action space, we apply greedy policies in our heuristic algorithms to quickly respond to time critical requirement.

Keywords: Multi-agent systems, task assignment, time critical.

1 Introduction

With the progress of research in distributed artificial intelligence, cooperative agents have been applied in dynamic and complex application domains such as disaster response[1], military[2] and business organizations[3] in recent years. In those domains, task allocation becomes a basic problem that needs to be solved efficiently [3]. The basic task allocation problem is organized as follows: Given a set of agents and a set of tasks, where each agent obtains some payoff for each task, find a one-to-one assignment of agents to tasks to maximize the total payoff of all agents. This problem can be solved (near) optimally in polynomial time by centralized algorithms [4,5], and decentralized algorithms [3,6]. In all of these works, it is assumed that communication is available and the environment is static. However, in real domain, the communication is not available and the environment including the status of the agents and the tasks may change. For example, when a group of robots are deployed to carry out time critical tasks for Urban Search and Rescue (USAR), communication infrastructure is most likely destroyed in disaster and communication may be restricted for each rescue robot. Another example, when missiles are launched to attack hostile targets,

Z. Shi et al. (Eds.): IIP 2014, IFIP AICT 432, pp. 197–206, 2014.

the missiles have to keep silent due to security reasons. In both of these domains, the environment is dynamically changing and necessary cooperative replanning or reallocation is required. Therefore, we summarize those scenarios as a group of time critical agents that they have to cooperate to optimize their common goals without explicit communication.

More specifically, time critical task assignment without explicit communication that we study can be stated as follows: Given a set of agents and a set of tasks, firstly, the capability of each agent is determined by the dynamically changing environment; Secondly, the state of task and agent may change unexpectedly; Thirdly, no explicit communication is allowed; In addition, a payoff can be obtained from a completed task. We need to find the assignment of agents to tasks within critical response time, such that the sum of payoff from all agents is maximized. To improve team performance, team states have to be well maintained from observations. This can be modelled as a Decentralized Partially Observable Markov Decision Process, which is NEXP-complete [7]. And current approaches such as [8,9], try to solve this problem with the help of communication, which is not allowed in our problem.

In this paper, several efforts were made to solve the task assignment problem in time critical context. Firstly, since the capabilities of all agents and the statuses of their tasks are changing with Markovian property, the task assignment process can be built as a Markov Decision Model. However, when communication is not available or unacceptable in time critical context, the team state cannot be accurately observed. In this paper, we model the decision process as a Partially Observable Markov Decision Process for each agent. Secondly, the objective function in our task assignment is a step function, which makes it difficult for the agent to decide whether to jointly assign an uncompleted task to a single agent. We propose an exponential reward function to fit for it. Finally, without communication, there is only a small amount of observation, which makes the problem an open-loop control and toward an NEXP-complete problem [10]. Thus, heuristic approaches are built according to the following observations: 1. Task-related agents may get the same observation with a high probability. 2. When nothing can be referred from others, it's better to assume that their choice stays the same. With these observations, the search space can be dramatically reduced and a greedy policy is applied to help operate in time critical context.

2   Problem Statement

In this section, we provide a time critical assignment problem without explicit communication for cooperative teams.

2.1   Task Assignment Problem

We focus on a cooperative time critical agent team $A = \{a_1, a_2, ..., a_i, ...\}$, such as a group of robots or UAVs, coordinating to carry out a set of tasks $T = \{t_1, t_2, ..., t_j, ...\}$. Each task consists of two parts, $t_j = < \chi_j, \Phi_j >$. χ_j is the payoff to complete t_j, indicating its importance. Φ_j describes t_j's workload, which represents the total amount of work required to complete t_j. Before the task is executed, there is a preparation stage for the agent to prepare for the execution or heading for the target location. During this preparation stage, the potential contribution of the agents to their tasks may change, since their states dynamically change. To obtain the highest payoff with limited agents, we should dynamically assign all the tasks during this stage.

At time t, the potential contribution of a_i to t_j is described as $c(a_i, t_j)^t$, which is dynamically updated as the situation between a_i and t_j changes, according to the domain knowledge. For example, the capability of the rescue robot is related to its remaining energy, holding resources and even distance to its target location. When the robot is of low power, its capability to carry out a long-distance task is very small. If the resource in the robot is going to run out, it is less capable to carry out a resource-consuming mission. After agent a_i is assigned with task $\tau_{a_i}(t)$, all agents assigned with task t_j is summarized as $\Gamma_j(t)$. Thus, if the total contribution $\psi_j(t)$ of all assigned agents for task t_j meets its workload, that is,

$$\psi_j(t) = \sum_{a_i \in \Gamma_j(t)} c(a_i, t_j)^t \geqslant \Phi_j$$

t_j can be completed and $Completed(t_j) = 1$; Otherwise, $Completed(t_j) = 0$.

During the preparation stage, we need to find the best strategy π^* as following to get the highest payoff.

$$\pi^*(t) = argmax_{Jt\_\tau(A,T,t)} \sum_{t_j \in T} \chi_j \cdot Completed(t_j) \tag{1}$$

where $Jt\_\tau(A, T, t)$ is the task allocated for all agents in this team, $Jt\_\tau(A, T, t) = \bigcup_{a_i \in A} \tau_{a_i}(t)$.

2.2 Time Critical Assignment without Explicit Communication

In a time critical system, a tiny delay may lead to disasters. For example, when a UAV in fast mode finds an obstacle in front, the decision about how to change its orbit should be made immediately, or it will be destroyed before it changes its orbit. Therefore, the task assignment should be carried out in a time critical context and no deliberation model should be allowed. In addition, when the time critical system is performing a task far away from its base, communication is extremely unreliable and of long delay. To operate in a time critical context, no explicit communication could be relied on. Without explicit communication, due to the limited sensing capabilities, the agents cannot fully observe the team state but can only get partial observations. Therefore, the agents have to reason from its partial observations to make rational decisions.

3 Cooperative Task Assignment Algorithm

In order to obtain the highest payoff, each agent has to dynamically update its task according to the changing team state, including the capabilities of all agents and the status of each task. Since the task of each agent is only determined by the latest team state rather than the historical ones, the task assignment process can be built as a Markov Decision Model when they can share their status freely. Unfortunately, in time critical context, the agents cannot share the state through unreliable communication with large delay. Each agent can only estimate the team state from the ambient data of its own observation. Due to the limitation of its observation range, the observation obtained by the agent is incomplete and team state is uncertain. Therefore, in this paper, a Partially Observable Markov Decision Process model is built for each agent to reason from its partial observations. Since small amount of observation makes the POMDP problem even more difficult, heuristic approaches are proposed to solve this problem.

3.1 Partially Observable Markov Decision Process Model for Cooperative Task Assignment

The basic decision model for the task assignment process of each agent can be written as a POMDP $< S, Jt\_act, T, \Theta, O, U >$. S is the state space and its value at time t is defined as $s(t)$, Jt_act is the action space of all agents, $T : S \times A \to S$, is the transition function that describes the resulting state $s(t + 1) \in S$ when executing $Jt\_\tau(A, T, t) \in Jt\_act$ in $s(t)$. Θ and O are the observations and their observation function respectively to indicate the possible states under a given observation. $R : S \to \mathbb{R}$ defines the reward of being in a specific state. This model can be applied to each agent.

In this case, the team state $s(t)$ is modelled as the capabilities of all agents and the assigned agents for each task at time $t - 1$:

$$S(t) = \cup_{t_j \in T}(\Gamma_j(t - 1) \cup_{a_i \in A} c(a_i, t_j)^t)$$

After assigning $Jt\_\tau(A, T, t)$ at state $s(t)$, the capability of each agent is transited according to the domain knowledge: $T : c(a_i, t_j)^t \times \tau_{a_i}(t) \to c(a_i, t_j)^{t+1}$. For example, when a rescue robot changes its task, it will change its direction to head for its new target location and improve its capability to accomplish this task in time.

However, each agent cannot fully observe the whole team state but can only get partial observations $\theta(t) \in \Theta$, such as the status of the task, the state of itself and other agents. For example, when the rescue robot is approaching its target location, it can sense the topography of the target area and its remaining energy. From the observation, the belief state $b(t)$ could be derived according to the observation function: $O : \Theta \to B$. For each state $s(t) \in b(t)$, we know the perceptual distribution $Pr(s(t)|\theta(t))$, which describes the likelihood of the team being in the state of $s(t)$ under observation $\theta(t)$. For example, when a robot observes that the traffic to the target location becomes extremely tougher, which will consumes more energy to reach the target location and indicates a capability drop of this agent, others with the same task may also observe the same capability drop with a high probability.

To find the optimal assignment, the key is to maximize the objective function [1]. However, the objective function is a step function, which generates reward only when the task is able to be completed and makes it difficult to decide whether to assign an uncompleted task to a single agent. Hence we propose an exponential reward function to fit for the objective function. And the exponential reward function should satisfy the following two properties: Firstly, for each task t_j, with all the potential agents and their contributions, the potential progress $\psi_j(t)$ of t_j is the sum of all contributions from its assigned agents.

$$\psi_j(t) = \sum_{a_i \in \Gamma_j(t-1)} c(a_i, t_j)^t$$

If $\psi_j(t)$ equals to zero, which means that there is no potential progress for task t_j and the team obtains no reward from task t_j. Secondly, when the task receives enough progress to be completed, the team gets the whole priority of the task as its reward. Considering these two properties of the objective function, the exponential reward function is built as follows:

$$R(\psi_j(t)) = \frac{\chi_j}{e^{\Phi_j} - 1}(e^{\psi_j(t)} - 1)$$

And the utility of $s(t)$ is the total reward from all tasks.

$$U(S(t)) = \sum_{t_j \in T} R(\psi_j(t))$$

In this model, the agents coordinate to find the best strategy π^* to maximize the expected utility.

$$\pi^*(t) = \arg \max_{Jt_{-T}(A,T,t)} \sum_{s(t+1) \in b(t+1)} Pr(s(t+1)|\theta(t+1)) \times U(s(t+1))$$

Without communication, each agent can only get a small amount of observation, which makes this problem an open-loop control and toward an NEXP-complete problem, according to the study of [10].

3.2 Heuristic Approach

In this section, we build heuristic approaches according to the following efforts. Firstly, we assume that an optimal assignment is initialized for each agent offline. Secondly, since agents act in the same environment, task-related agents may obtain the same observation with a high probability. Hence the belief of all task-related agents may change and their tasks may be reassigned. Finally, when we cannot infer anything about others, since the state is Markovian, we can assume that their choice stays the same to keep synchronization of all agents. Thus, the search space can be dramatically reduced and a greedy policy can be applied for each agent in a time critical context.

State Estimation. In a decentralized task assignment domain, several agents may be assigned to the same task. When there is an unexpected accident happening to the task, all agents assigned to the task tend to get the same observation. For example, when a group of robots are assigned to carry out the same task t, if robot α observes that the condition of the task is more difficult than it expected, its capability $\widetilde{c}(\alpha, t)^t$ will drop and other task-related agents may also observe the same disturbance, which drives them to reassign their tasks. Thus, the state can be estimated with all the possible observations from all task-related agents.

$$\widetilde{s}(t) = s(t-1) \cup_{a_i \in \Gamma_t(t-1)} \widetilde{c}(a_i, t)^t$$

where t is the task that causes the disturbance and $\widetilde{c}(a_i, t)^t$ is the estimated capability from the observations.

Action Space Reduction. In state estimation, task-related agents in Γ_t may get the same observation and their states can be easily estimated. But without communication, we know nothing about other agents, which do not share the same task and their states cannot be derived. Hence there are many possible states for these agents, which make it difficult to reason about their states within critical response time. When we know nothing about other agents and what they know, since the state is Markovian, it is reasonable to assume that their state changes regularly and their choices stay the same. That is, the joint action of the agents in $A - \Gamma_t$, whose state cannot be derived, is estimated from that of last assignment.

$$\widetilde{Jt\_\tau}(A - \Gamma_t, T, t) = Jt\_\tau(A - \Gamma_t, T, t - 1)$$

Thus, only the joint action $Jt\_\tau(\Gamma_t, T, t)$ of the task related agents in Γ_t has to be reallocated, which dramatically reduces the action space. With these two parts, the joint action of the team is approximated:

$$Jt\_\tau(A, T, t) = Jt\_\tau(\Gamma_t, T, t) \cup \widetilde{Jt\_\tau}(A - \Gamma_t, T, t)$$

To react in a time critical context, no deliberation model could be used. Therefore, we use greedy policy to find a good joint action $Jt\_\tau(\Gamma_t, T, t)$ for the task-related agents.

Each agent in the team runs algorithm 1. In this algorithm, when the agent gets an observation (line 2), it triggers a task assignment process. For the agents that have different tasks (line 4), it assumes that their tasks keep the same (line 5). According to the domain knowledge, their contributions to the tasks can be updated (line 7). Thus the total potential contribution each task receives could be summarized (line 8). For the task-related agents (line 10), the contribution to each task can be updated from the observation (line 11). To obtain the highest priority and respond within critical time, a greedy policy is conducted to find the task with max reward for each agent. Considering the task may be over completed, the reward has to be calculated in two cases (line 14-19 and line 20-26). When the task is not over completed, the reward is the additional utility after adding current agent (line 15). If the reward is the maximum one, record the task and its reward (line 16-17). When the task is over completed after adding current agent (line 20-21), the reward is the additional utility of the task before it exceeds its priority (line 21). And the same is done to record the maximum reward (line 22-23). Finally, find the task with maximum reward and update its status (line 28).

Algorithm 1 Cooperative Decision Algorithm for Time Critical Assignment

---

1: **for** at time step t **do**
2:     $Observation \leftarrow getObservation$;
3:     init array $\psi(t)$ with zero;
4:     **for** all $\alpha \in A - \Gamma_j(t)$ **do**
5:         $\tau_\alpha(t) \leftarrow \tau_\alpha(t-1)$;
6:         $task \leftarrow \tau_\alpha(t)$;
7:         update $c(\alpha, task)^t$ according to domain knowledge;
8:         $\psi_{task}(t) \leftarrow \psi_{task}(t) + c(\alpha, task)^t$;
9:     **end for**
10:    **for** all $\alpha \in \Gamma_j(t)$ **do**
11:        update $c(\alpha,)^t$ according to domain knowledge from $Observation$
12:        $max_r \leftarrow 0$;
13:        **for** all $task \in T$ **do**
14:            **if** $\psi_{task}(t) + c(\alpha, task)^t \leq \Phi_{task}$ **then**
15:                **if** $R(\psi_{task}(t) + c(\alpha, task)^t) - R(\psi_{task}(t)) > max_r$ **then**
16:                    $max_r \leftarrow R(\psi_{task}(t) + c(\alpha, task)^t) - R(\psi_{task}(t))$;
17:                    $\tau_\alpha(t) \leftarrow task$;
18:                **end if**
19:            **else**
20:                **if** $\psi_{task}(t) \leq \Phi_{task}$ **then**
21:                    **if** $\chi_{task} - R(\psi_{task}(t)) > max_r$ **then**
22:                        $max_r \leftarrow \chi_{task} - R(\psi_{task}(t)$;
23:                        $\tau_\alpha(t) \leftarrow task$;
24:                    **end if**
25:                **end if**
26:            **end if**
27:        **end for**
28:        $\psi_{\tau_\alpha(t)}(t) \leftarrow \psi_{\tau_\alpha(t)}(t) + c(\alpha, \tau_\alpha(t))^t$;
29:    **end for**
30: **end for**

---

4   Experiment Setup and Result

In this section, we present the empirical evaluations of our approach in the cooperative task assignment. In our experiment, a group of agents are deployed from the same base station to carry out several tasks, which takes the agents about 40 time steps to reach the target location. Each task is initialized with different priorities and workloads. For the sake of simplicity, we initiate agents' capabilities for all tasks to 1. The objective is to maximize the priorities of all tasks accomplished by the agent group. Initially, the agents do not have accurate knowledge about all the tasks and the situation may change. When the agent is heading for the target location, its capabilities can be found changed since the situation is different.

In order to investigate the capability of adapting to the dynamics and uncertainties, we compare our approach (*dynamic approach*) with traditional static assignment policy (*static approach*), which initially assigns the optimal policy to each agent offline. Since the agents are initialized with identical capabilities, 0-1 Integer Programming could be used to find the initial optimal policy.

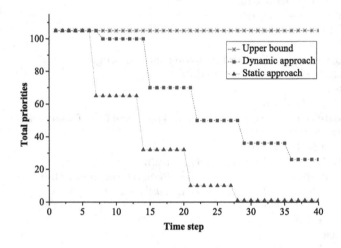

Fig. 1. The experimental results of 20 robots to carry out 20 tasks

To explore the performance of our approach, we initialized 20 agents to carry out 20 tasks with different priorities, and compared our approach with *static approach* and the upper bound. During this experiment, 5 tasks change their resource requirement at different time. Thus, when the agents approach their target locations, they will find their capabilities changed. According to the experimental result in Figure 1, both the total priorities of *dynamic approach* and *static approach* drop when there are unexpected disturbances. However, since *dynamic approach* reasons from the observations and reassigns its task, there is less performance loss in *dynamic approach* than in *static approach*.

In order to investigate the scalability, we vary the number of agents and tasks as 20, 50, 100, and 200 in the experiment in Figure 2(a), and also conduct the experiment with different numbers of disturbances as shown in Figure 2(b). The performance is presented as the ratio of the obtained priorities to the upper bound. From Figure 2(a), when the numbers of agents and tasks scale up, given number of disturbances can only influence a smaller part of tasks. Thus, these disturbances have smaller influences on the team performance when the team

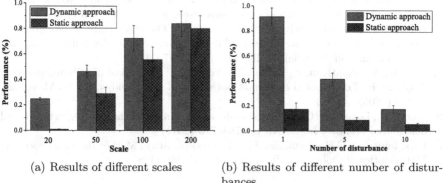

(a) Results of different scales

(b) Results of different number of disturbances

Fig. 2. The experimental results of different scales and disturbances

and tasks scale up. But *dynamic approach* always performs better than *static approach*. When the number of disturbances increases as shown in Figure 2(b), more tasks are influenced and the team performance drops. Because of the reassigning capabilities, *dynamic approach* still outperforms the static one.

5   Conclusion

In this paper, we presented a multi-agent decision algorithm for time critical assignment without communication. In this context, each agent can only obtain a partial observation of the team state and cannot communicate with each other. To reason with partial observations, we built a POMDP model for each agent and proposed heuristic approaches to estimate the state and reduce action space. Our experiments show that our algorithm is feasible to dynamically adjust to disturbances and can be applied in time critical assignment without communication.

References

1. Scerri, P., Kannan, B., Velagapudi, P., Macarthur, K., Stone, P., et al.: Flood disaster mitigation: A real-world challenge problem for multi-agent unmanned surface vehicles. In: Advanced Agent Technology, pp. 252–269 (2012)
2. Day, M.: Multi-agent task negotiation among UAVs to defend against swarm attacks. PhD thesis, Naval Postgraduate School, Monterey, California (2012)
3. Luo, L., Chakraborty, N., Sycara, K.: A distributed algorithm for constrained multi-robot task assignment for grouped tasks. Robotics Institute, p. 1077 (2012)
4. Kuhn, H.: The hungarian method for the assignment problem. Naval Research Logistics 52(1), 7–21 (2005)
5. Burkard, R., Dell'Amico, M., Martello, S.: Assignment problems. Cambridge University Press (2012)

6. Bertsekas, D.P.: The auction algorithm: A distributed relaxation method for the assignment problem. Annals of operations research 14(1), 105–123 (1988)
7. Bernstein, D.S., Zilberstein, S., Immerman, N.: The complexity of decentralized control of Markov decision processes. In: Proceedings of the Sixteenth Conference on Uncertainty in Artificial Intelligence, pp. 32–37 (2000)
8. Nair, R., Varakantham, P., Tambe, M., Yokoo, M.: Networked distributed pomdps: A synthesis of distributed constraint optimization and pomdps. In: AAAI, vol. 20, p. 133 (2005)
9. Zhang, C., Lesser, V.: Coordinated multi-agent reinforcement learning in networked distributed pomdps. In: AAAI, pp. 764–770 (2011)
10. Papadimitriou, C.H., Tsitsiklis, J.N.: The complexity of Markov decision processes. Mathematics of Operations Research, pp. 441–450 (1987)

Using PDDL to Solve Vehicle Routing Problems

Wenjun Cheng[1] and Yuhui Gao[2]

[1] Academy of Military Science, Beijing 100091, China
berrywen@hotmail.com
[2] Beijing Aerospace Control Centre, Beijing 100094, China
michaeldoer@hotmail.com

Abstract. In this paper, we describe a search method based on the language PDDL for solving vehicle routing problems. The Vehicle Routing Problem (VRP) is a classical problem in Operations Research, and there are many different variants of the VRP. This paper describes a new approach to model standard VRP and some variants based on PDDL language, explains how the method constructs model and solves the problem using several PDDL planners, and analyses the planning results of these planners.

Keywords: Artificial Intelligence Planning, Vehicle Scheduling, PDDL, Vehicle Routing.

1 Introduction

The vehicle routing problem is a well-known combinatorial optimization problem in the field of service operations management and logistics [1]. It is described as finding the minimum distance or cost of the combined routes of a number of vehicles that must service a number of customers. The number of feasible solutions for the VRP increases exponentially with the number of customers to be serviced, and almost each of VRP's variants is more complicated than standard VRP. Since there is no known polynomial algorithm that will find the optimal solution, the VRPs is considered NP-hard. For finding solutions of such problems, the use of problem-specific solvers with heuristics is considered a reasonable approach that most of VRP research is based on. However it is realistic to solve VRPs that are structured in the same way using universal planners. This paper describes the methods to model some of the variants of VRP in PDDL. The planners for PDDL are automated, which means that they solve the problem by using heuristics that are pre-developed, meaning that the heuristics do not change depending on the type of problem. So it provides attempt to solve the VRPs through a uniform method.

The remainder of this paper is organized as follows. Section 2 introduces the VRP and its multiple variants; Section 3 describes the features of PDDL language and the main versions of PDDL; Section 4 gives the models of several VRPs based on PDDL; Section 5 provides some efficient PDDL planners to solve the models and compares the planning results. Section 6 concludes with a brief discussion and perspectives.

Z. Shi et al. (Eds.): IIP 2014, IFIP AICT 432, pp. 207–215, 2014.
© IFIP International Federation for Information Processing 2014

2   The Standard VRP and Its Variants

The standard VRP can be described as a weighted graph $G=(V, A, d)$ where $V=\{v_1,v_2,\ldots,v_n\}$ is a set of vertices representing locations with the depot located at v1 and A is the set of arcs. With every $arc(v_i, v_j)$ $(i \neq j)$ is associated a non-negative distance d which can be interpreted as a travel cost or as a travel time. The standard VRP consists of designing a set of least-cost vehicle routes in such a way that

- Each customer is visited only once by a single vehicle, and is served according to their demand.
- Each vehicle must start and end its route at the depot v1, and it is not allowed to visit the depot in between.

The standard VRP can be extended in multiple ways which add some constraints to be satisfied. The most common constraints include:

- Capacity restrictions: each vehicle has finite capacity and each location has a finite demand. So the sum of the demands for the vehicle's tour does not exceed its capacity. Capacity-constrained VRPs will be referred to as CVRPs.
- Total time restrictions: the length of any route may not exceed a prescribed bound. Time-constrained or distance-constrained VRPs will be referred to as DVRPs.
- Time windows restrictions: the service at each customer must start within an associate time window and the vehicle must remain at the customer location during service. The VRP with time windows will be referred to as VRPTW.
- Other restrictions including multiple depots, multiple goods, multiple service modes, multiple transport modes, or different optimization goals. So there are a variety of VRP variants such as multiple depot vehicle routing problem (MDVRP), the vehicle routing problem with backhauls (VRPB), the vehicle routing problem with backhauls and time windows (VRPBTW), and VRP with pickup and delivery (VRPPD).

These VRPs were made by industrial requirements and have been studied since 1995. With the simpler variants of the VRP being solved, the VRP research community pays more attention to rich VRP models and larger size problems recently. We prefer to construct some common models to descript most of VRP variants from simple to complex, and choose a uniform way to solve these models. In the field of artificial intelligence (AI), PDDL is an expressive language which offers to represent many planning knowledge while freeing from the plan search algorithms, and PDDL can satisfy the demand of modelling various kinds of VRP.

3   PDDL and Its Versions

PDDL, the "Planning Domain Definition Language", is a standard language for expressing planning problems and domains is widely used in the planning community [2]. This enables us to effortlessly select the best suited planner for describing the technical and epistemological requirements of planning domains and the capabilities of planners in a uniform way. A planning problem is made of an initial state, a final state and a set of operators; a state is a set of predicates and operators can be seen as a couple of states: preconditions which must be true to allow execution and effects which are true after execution. In PDDL, the specification of the predicates and the operator schemas is separated from the specification of the initial state and the goal condition. The former is typically referred to as the planning domain and the latter as the planning problem, so a number of planning problems can be defined for the same planning domain.

There are dozens of different versions of PDDL and its extension. Every version of PDDL specifies new requirements that should be supported and increases the flexibility and functionality of PDDL language. PDDL1.2 states the basics of the language, such as how the domain and problem file should be composed. One of the most important PDDL versions is 2.1, which specifies the possibility to use numerical variables, plan-metrics, and durative actions which make planning with conditions such as fuel consumption, time constraints and capacity constraints possible [3]. In PDDL versions 2.2, derived predicates and timed initial literals which are useful in modelling VRPs are specified [4]. However, for higher PDDL versions, only a few planners provide effective support, and we cannot use the specifications of these PDDL versions.

4   PDDL Models

4.1   PDDL Model for CVRP

In the problem there are three different types: freights, vehicles, and locations.

Planning problems predicates are objects that are described using Boolean representation. The following predicates represent facts related to the types.

- (at freight ?f, location ?l): the freight ?f is currently at the location ?l.
- (at2 vehicle ?v, location ?l): the vehicle ?v is currently at the location ?l.
- (adjacent location ?l1, location ?l2): the location ?l1 is adjacent to ?l2.
- (in freight ?f, vehicle ?v): the freight ?f is currently loaded by the vehicle ?v.
- (using vehicle ?v): the vehicle ?v is not at the depot.
- (destination freight ?f location ?l): the location ?l is the destination of the freight ?f.

Planning problems functions are objects with states that can be represented numerically (also called numerical fluents). The following predicates represent facts related to the types.

- (totalCost): this descripts the variable that keeps track of the total cost of the solution. It is the key function to solve Multi-criterion and multi-objective optimal problems.
- (weight freight ?f): this descripts the weight of the freight ?f.
- (maxLoad): this descripts the maximum load limit of all vehicles.

- (load vehicle ?v): this descripts how much weight the vehicle ?v is carrying.
- (distance location ?l1 location ?l2): this descripts the distance between ?l1 and ?l2. This function is only available for adjacent locations.

Actions are events that change the state of the problem. The available actions in the domain are defined in terms of their preconditions and effects using the available predicated and the functions of the domain. For example, the follow PDDL statement is an action schema as it appears in the CVRP PDDL domain that represents the actions of the vehicle loading the freight.

```
(:action load
  :parameters (?f - freight ?v - vehicle ?l - location)
  :precondition (
      and (using ?v)
          (at ?f ?l)
          (at2 ?v ?l)
          (>= (maxLoad) (+(load ?v) (weight ?f))))
  :effect (
      and (in ?f ?v)
          (not (at ?f ?l))
          (increase (cargo ?v) (weight ?f))))
)
```

In this problem there are four actions possible:

- useVehicle(vehicle ?v): set the vehicle ?t as available for use.
  precondition: the vehicle ?v is not in use.
  effect: the vehicle ?v is in use.
- load(freight ?f, vehicle ?v, location ?l): loads the freight ?f into the vehicle ?v at the location ?l.
  precondition: the vehicle ?v is in use; it and the freight ?f are both at the location ?l; $maxLoad >= load_v + weight_f$.
  effect: the freight ?f is in the vehicle ?v and is not at the location ?l; $load_v = load_v + weight_f$.
- unload(freight ?f, vehicle ?v, location ?l): unloads the freight ?f from the vehicle ?v at the location ?l.
  precondition: the vehicle ?v is in use and at the location ?l where is the destination of the freight ?f; ?f is in the vehicle ?v.
  effect: the freight ?f is at the location ?l and is not in the vehicle ?v; $load_v = load_v - weight_f$.
- move(vehicle ?v, location ?l_1, location ?l_2): moves the vehicle ?v from ?l_1 to ?l_2.
  precondition: the vehicle ?v is in use and at the location ?l_1; ?l_1 is adjacent to ?l_2.
  effect: the vehicle ?v is at ?l_2 and not at ?l_1.

Inside the problem file, all types of planning objects are defined and initialized, and the planning goal is set. The domain that is described above can have many VRP problem files, which might differ by the number of objects, initial values and goal.

For CVRP PDDL domain, all freights should be at their respective goal location and all vehicles should have returned to their respective origin. The function "totalCost" should be minimized which is the metric of the problem and always equal to the objective function.

Some PDDL features need to be supported in constructing PDDL models for CVRP: typing, fluents, plan metrics. These features are specified before PDDL version 2.1, so the models can be solved by any planner which supports PDDL2.1.

4.2   PDDL Models for DVRP and VRPTW

In the PDDL model for DVRP, it is necessary to convert the transport of vehicles to durative actions. So the PDDL feature "durative actions" is added to the feature list which is must be supported in the PDDL model.

Some correspond changes must be applied to the former PDDL model for CVRP. For example, the follow PDDL statement is a durative action schema that represents the actions of the vehicle loading the freight.

```
(:durative-action load
   :parameters (?f - freight ?v - vehicle ?l - location)
   :duration(= ?duration (loadTime ?f))
   :condition (
     and (over all(using ?v))
         (at start(at ?f ?l))
         (at start(at2 ?v ?l))
         (at start(>= (maxLoad) (+ (load ?v)(weight ?f)))))
   :effect (
     and (at end(in ?f ?v))
         (at start(not (at ?f ?l)))
         (at end(increase (load ?v) (weight ?f))))))
)
(loadTime freight ?f): this function descripts how long
      the freight ?f is loaded or unloaded by vehicles.
```

In VRPTW, there are individual time windows for each customer, and possibly also on the depot. The PDDL feature "timed initial literals" has the possibility to make predicates available according to predetermined times and consequently to make actions feasible at special time. This feature is specified in PDDL 2.2 and only a few planners support it. Only one predicate wants to be added to the domain file of DVRP model to support time windows.

```
(available ?f - freight) : descripts the freight ?f
       available(depend on time windows).
```

The PDDL planning problem generated for above modes is the following (most of the predicates should be self-explanatory):

```
(define (problem VRPP)
  (:domain VRP)
  (:objects f1 - freight f2 - freight
            v1- vehicle v2- vehicle
            l1 - location l2 - location l3 - location)
  (:init
    (adjacent l1 l2) (adjacent l2 l1)
    (adjacent l1 l3) (adjacent l3 l1)
    (= (distance l1 l2) 22) (= (distance l2 l1) 22)
    (= (distance l1 l3) 25) (= (distance l3 l1) 25)
    (= (weight f1) 8) (= (weight f2) 9)
    (= (load v1) 0) (= (load v2)0)
    (at 0 (available f1)) (at 100 (not (available f1)))
    (at 0 (available f2)) (at 100 (not (available f2)))
    (= (maxLoad) 10) (= (totalCost) 0)
    (at f1 l2) (at f2 l3) (at2 v1 l1) (at2 v2 l1)
    (destination f1 l1) (destination f2 l1)
  )
  (:goal
    (and (at f1 l1) (at f2 l1) (at2 v1 l1) (at2 v2 l1))
  )
  (:metric minimize (totalCost))
)
```

4.3   PDDL Models for Other VRPs

For other VRPs, such as MDVRP, VRPB, VRPBTW and VRPPD, the different from above VRPs is focused on multiple vehicle types, multiple service types, route restrictions, and freight constraints. It doesn't need more PDDL features to describe the PDDL model for these VRPs. All we need to do is providing more predicates, functions and actions.

5   Performance of PDDL Planners

The PDDL planners are automated and the heuristics search algorithms embedded in them does not change depending on the type of problem. There are many PDDL planners available on the internet, with certain characteristics that may make them more or less suitable for different kinds of planning problems.

The planners we tried were SAPA [5], LPG-TD [6-8], SGPlan6 [9], TFD [10], POPF2 [11] and DaeYa [12]. All of these planners can solve the DVRP problems because they have the ability to satisfy the "durative-action" feature of PDDL 2.1 and process the problems which contain durative actions. But only SAPA, LPG-TD, POPF2 and SGPlan support the "timed-initial-literals" feature of PDDL 2.2 and can solve the VRPTW problems.

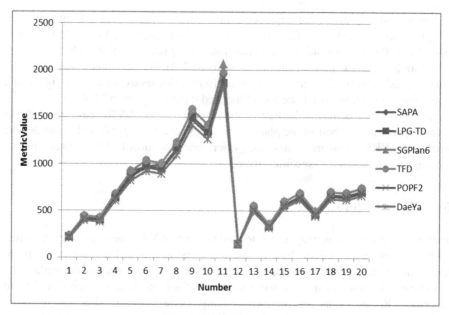

Fig. 1. DVRP Benchmark

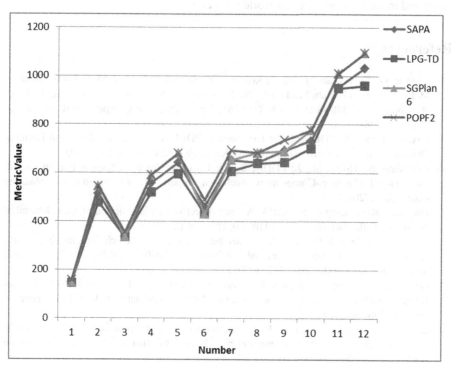

Fig. 2. VRTTW Benchmark

Figure 1 shows the performance of planners on DVRP benchmark instances. 20 DVRP instances used range from between 10 to 30 locations, 10 to 50 freights, and from 10 to 20 vehicles which are auto-generated using the code in Matlab.

Figure 2 shows the performance of planners on VRPTW benchmark instances. 12 VRPTW instances used range from between 10 to 20 locations, 10 to 30 freights, and from 10 to 20 vehicles which are auto-generated using the code in Matlab.

DaeYa planner performs the best of the planners in the DVRP case and LPG-td planner performs the best of the planners in the VRPTW case. We only consider the small-scale VRP problems because the performances of most PDDL planners are not the same level as problem-specific solvers.

6   Conclusions

This paper describes a new approach to model standard VRP and some variants based on PDDL language, explains how the method constructs model and solves the problem using several PDDL planners, and analyses the planning results of these planners. Comparing to problem-specific solvers, this method provides common models to describe VRP problems and uniform solving mode by using different planners. Although the performances of multiple PDDL planners are not the same level as problem-specific solvers, it is realistic to expect planners to solve problems that are structured in the same way as real-world problems.

References

1. Laporte, G.: Fifty Years of Vehicle Routing. Transportation Science 43, 408–416 (2009)
2. McDermott, D., Ghallab, M.: PDDL-The Planning Domain Definition Language. Technical Report, CVC TR-98-003/DCS TR-1165, Yale Center for Computational Vision and Control (1998)
3. Fox, M., Long, D.: PDDL2.1: An Extension to PDDL for Expressing Temporal Planning Domains. Journal of Artificial Intelligence Research (JAIR) 20, 61–124 (2003)
4. Edelkamp, S., Hoffmann, J.: PDDL2.2: The Language for the Classical Part of the 4th International planning Competition. Technical Report, the 4th International planning Competition (2004)
5. Do, M.B., Kambhampati, S.: SAPA: A multi-objective metric temporal planner. Journal of Artificial Intelligence Research (JAIR) 20, 155–194 (2003)
6. Gerevini, A., Serina, I.: LPG: A planner based on a local search for planning graphs with action costs. In: 6th International Conference on Artificial Intelligence Planning and Scheduling (AIPS 2002), pp. 13–22 (2002)
7. Gerevini, A., Saetti, A., Serina, I.: Integrating Planning and Temporal Reasoning for Domains with Durations and Time Windows. In: 19th International Joint Conference on Artificial Intelligence (2005)
8. Gerevini, A., Saetti, A., Serina, I.: An approach to temporal planning and scheduling in domains with predictable exogenous events. Journal of Artificial Intelligence Research (JAIR) 25, 187–231 (2006)

9. Chen, Y., Benjamin, W.: Wah, Chih-Wei Hsu: Temporal planning using subgoal partitioning and resolution in SGPlan. Journal of Artificial Intelligence Research 26, 323–369 (2006)
10. Eyerich, P., Mattmüller, R., Röger, G.: Using the context enhanced additive heuristic for temporal and numeric planning. In: 19th International Conference on Automated Planning and Scheduling (ICAPS), Thessaloniki, Greece (2009)
11. Coles, A.J., Coles, A., Fox, M., Long, D.: Forward chaining partial-order planning. In: 20th International Conference on Automated Planning and Scheduling (ICAPS), Toronto, Ontario, Canada, pp. 42–49 (2010)
12. Dréo, J., Savéant, P., Schoenauer, M., Vidal, V.: Divide-and-Evolve: The marriage of Descartes and Darwin. In: 2011 International Planning Competition (2011)

Automated Localization and Accurate Segmentation of Optic Disc Based on Intensity within a Minimum Enclosing Circle

Ping Jiang[1,2] and Quansheng Dou[1,2]

[1] School of Computer Science and Technology,
Shandong Institute of Business and Technology, 264005 Laishan, Yantai, China
[2] Key Laboratory of Intelligent Information Processing in Universities of Shandong
(Shandong Institute of Business and Technology), 264005 Laishan, Yantai, China
{ccecping,li_dou}@163.com

Abstract. This paper presents a method for automated localization and accurate segmentation of the optic disc. An intensity threshold is determined and select all the pixels whose intensities are greater than the threshold, by erosion the optic disc can be localized. By dilation and region filling, a minimum enclosing circle which can completely hold the optic disc is determined, within the circle, the vessels are eliminated by replacing the darker vessel pixels with brighter pixels. Define the intensity features of the optic disc boundary , and select the pixels according to the features, then the optic disc may be segmented. The experiment shows that compared to the active contour models, this method is more efficient and accurate on the boundary extraction of the optic disc.

Keywords: Automated localization, Accurate segmentation, Intensity threshold, Minimum enclosing circle.

1 Introduction

The fundus images are used for diagnosis by trained clinicians to check for any abnormalities or any change in the retina. The information about the optic disc can be used to examine the severity of some diseases such as glaucoma. Its detection is prerequisite for the segmentation of other normal and pathological features.

A number of studies have reported on automated localization of optic discs; several studies have also reported on the segmentation of optic discs [1-14]. Walter and Klein [1] applied the watershed transformation to the gradient image based on the morphological operations. Lalonde et al.[2] used the Canny edge detector to detect optic disc edge, and by matching the edge map with a circular template the optic disc was segmented. The method of Li and Chutatape[3] is based on an active contour model, and it iteratively matching the landmark points on the disc edge. Osareh et al. [4] and Lowell et al. [5] used the active contour model to find the optimal points based on the external and internal energy of the image. Wong et al. [6] proposed a method based on the level-set technique, and used ellipse to fit the disc boundary. Abramoff et al. [7] employed a pixel classification method using the feature analysis

Z. Shi et al. (Eds.): IIP 2014, IFIP AICT 432, pp. 216–220, 2014.

and nearest neighbor algorithm, then group each pixel to rim, cup or background. Jaspreet Kaur et al. [8] and Siddalingaswamy P. C.[9]   employed iterative thresholding method followed by connected component analysis to automatically localize the approximate center of the optic disc, then the geometric model based implicit active contour is employed to obtain accurate optic disc boundary. Hoover and Goldbaum used a "fuzzy convergence" algorithm to correctly identify the optic disk location in 89% of 81 images [10]. The method of Park et al [11] found the brightest pixels by employing the repeated thresholding technique, then used the roundness of the object to detect optic disk features, and then localized the optic disk by using the Hough transform.

Yet because of fuzzy boundaries, inconsistent image contrast or missed edge features, it is difficult to accurately localize and segment the optic disc. Based on the intensity and shape features, this paper proposed a method for the automated localization and segmentation of the optic disc, without human intervention, the optic disc may be segmented accurately and efficiently by the contour extraction, the contour can then be used to help the doctor for further analysis.

2   Appearance Intensity Threshold Determination for Optic Disc Localization

The optic disc is often the brightest object, ranging from white to yellow, is circular and of reasonably consistent size from patient to patient [12]. So the pixel whose intensity is brighter than the threshold is marked as candidate optic disc pixel. For the retinal images whose sizes are 640×480, the optic discs have more than 1000 but less than 10000 pixels. Clustered those pixels whose intensities are equal when divided by a certain histogram bin width which in this paper is 10, and sort the clusters according to the intensity. Start from the cluster with maximum intensity and check the number of pixels in it, if the number is less than 1000, then go to the next cluster in descending order of intensity until the number of pixels is between 1000 and 10000, and the intensity range of this cluster is determined as the threshold T. Those pixels whose intensities are brighter than the threshold are marked as candidate optic disc with a green color (0xff00ff00) as shown in Figure1.

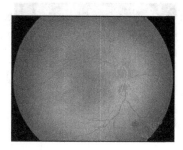

Fig. 1. Retinal image with candidate optic disc marked with green color

As shown in Figure1, there are some regions marked with green color which are not real optic disc, to eliminate the false candidate pixels, morphological erosion is applied on them. Find all the start positions of the candidate regions, structure element *structElem* begins from these positions and tests all the candidate regions so as to find a region which can include it. Figure2 shows an instance of the localization result.

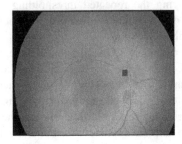

Fig. 2. The result of automatic localization of the optic disc with a rectangle

3   Dilate the Optic Disc Region, Determine the Minimum Enclosing Circle

As shown in Figure2, the green pixels don't occupy the whole disc region, so further steps are needed to gather more optic disc pixels.

3.1   Morphological Dilation

By region grouping, the optic disc pixels are collected together and marked in binary matrix M. The matrix is the same size as the retinal image, if the pixel on the image is colored by green, then the corresponding cell in the matrix is 1, otherwise 0.

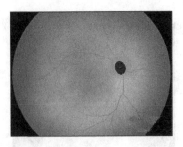

Fig. 3. Dilation result of Figure2

To find more optic disc pixels, double dilation was applied on M , The dilation result was shown in Figure3 with blue color, take all the x and y coordinates of dilated matrix M , and compute the average x and y as the coordinate of the center point $P(x, y)$ of the blue region.

3.2   Determination of Minimum Enclosing Circle

Based on the normal size of optic disk, centered at P , a circle which can totally enclose the optic disc region can be determined. Analyze the intensity distribution of pixels in the circle, and define the intensity features of the pixels on the edge, and mark those pixels having edge features, then the segmentation is done. Two edge features are defined initially: the first is that its intensity is darker than the optic disc localization threshold T ; the second is at least one of its 8 neighbors belongs to the internal optic disc, i.e. the intensity is brighter than the threshold T . According to the two features, some pixels are selected which are used to compute their distances to the center point $pCenter$, and get the average distance $avgD$, set the radius of minimum enclosing circle to be $avgD + d$, where d is the incremental constant used to make the circle include all the optic disc pixels and exclude the false optic disc pixels. So the third feature of the edge pixel is that its distance D to $pCenter$ should satisfy $|D - avgD| \leq d$. D is calculated by equation (1). The determined edge result is shown in Figure4.

$$D = \sqrt{(p.x - pCenter.x)^2 + (p.y - pCenter.y)^2} \qquad (1)$$

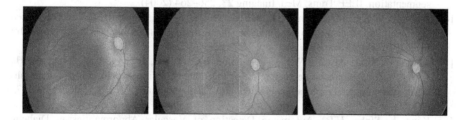

Fig. 4. The optic disc edge results with the third feature added

4   Experiment and Conclusion

The automated localization and segmentation of the optic disc is implemented by JAVA, and it can accurately and efficiently get the contour of the optic disc, yet for the active contour model, to get the optic disc contour, an initial contour need to be drawn first, and because of the noise, vessel intervention etc., normally they cannot get to the accurate boundary, so compared to these methods, our algorithm is more accurate and efficient.

We test our method on DRIVE, DIARETDB1 and the infant retinal images, for segmentation, the method has good performance on the infant images whose optic discs have relatively weak vessel intervention, for those images with strong inside vessels, the method cannot segment all the optic disc, normally a large part the disc, so it needs further improvement which is our future work.

Acknowledgements. This work was partially supported by National Natural Science Foundation of China, under grant No. 61272244, 61175023,61175053 and 61272430, Natural Science Foundation of Shandong Province, under grant No. ZR2013FL022, the Young Foundation, under grant No. 2011QN083.

References

[1] Walter, T., Klein, J.C.: Segmentation of color fundus images of the human retina: detection of the optic disc and the vascular tree using morphological techniques. Int. Symp. Med. Data Anal., 282–287 (2001)

[2] Lalonde, M., Beaulieu, M., Gagnon, L.: Fast and robust optic disc detection using pyramidal decomposition and Hausdorff-based template matching. IEEE Trans. Med. Imaging, 1193–1200 (2001)

[3] Li, H., Chutatape, O.: Automated feature extraction in color retinal images by a model based approach. IEEE Trans. Biomed. Eng. 51, 246–254 (2004)

[4] Osareh, A., Mirmehdi, M., Thomas, B., Markham, R.: Comparison of colour spaces for optic disc localization in retinal images. In: Proc. Int. Conf. Pattern Recog, pp. 743–746 (2002)

[5] Lowell, J., Hunter, A., Steel, D., Basu, A., Ryder, R., Fletcher, E.: Optic nerve head segmentation. IEEE Trans. Med. Imaging 23, 256–264 (2004)

[6] Wong, D.W.K., Liu, J., Lim, J.H., Jia, X., Yin, F., Li, H., Wong, T.Y.: Level-set based automatic cup-to-disc ratio determination using retinal fundus images in ARGALI. In: Proc. IEEE Eng. Med. Biol. Soc., pp. 2266–2269 (2008)

[7] Abramoff, M.D., Alward, W.L.M., Greenlee, E.C., Shuba, L., Kim, C.Y., Fingert, J.H., Kwon, Y.H.: Automated segmentation of the optic disc from stereo color photographs using physiologically plausible features. Invest. Ophthalmol. Vis. Sci. 48, 1665–1673 (2007)

[8] Kaur, J., Sinha, H.P.: Automated Detection of Vascular Abnormalities in Diabetic Retinopathy using Morphological Thresholding. International Journal of Engineering Science & Advanced Technology 2(4), 924–931 (2012)

[9] Sinthanayothin, C., Boyce, J., Cook, H., et al.: Automated localisation of the optic disc, fovea and retinal blood vessels from digital colour fundus images. British Journal of Ophthalmology 83(8), 902 (1999)

[10] Hoover, A., Goldbaum, M.: Locating the optic nerve in a retinal image using the fuzzy convergence of the blood vessels. IEEE Trans Biomed. Eng. 22, 951–958 (2003)

[11] Park, M., Jin, J.S., Luo, S.: Locating the optic disc in retinal images. In: International Conference on Computer Graphics, Imaging and Visualisation, pp. 141–145 (2006)

[12] Jelinek, H.F., Cree, M.J.: Automated Image Detection of Retinal Pathology. CRC Press, Taylor & Francis Group (2009)

An Optimization Scheme for SVAC Audio Encoder

Ruo Shu, Shibao Li, and Xin Pan

China University of Petroleum, Computer and Communication Engineering Institute,
Qingdao 266580, China
{shuruo,lishibao}@upc.edu.cn, langqin12390@126.com

Abstract. Both audio signals and MEL-frequency cepstral coefficients are encoded in SVAC audio encoder. These two independent processing leads to structural redundancies for the encoder design. This paper proposes an optimization scheme using MEL-frequency cepstral coefficients to realize high-frequency reconstruction and removes Bandwidth Extension module of the original encoder. Simulation results prove that the new encoder has considerable improvement on structure, reconstructed high-frequency precision and coding quality.

Keywords: SVAC audio encoder, MEL-frequency ceptral coefficients, bandwidth expansion, linear predictive coding coefficients, quality evaluation.

1 Introduction

Surveillance video and audio coding is designed for application in the field of national security and defense, and is important for maintaining public security and fighting and preventing criminal activities. Chinese government started to establish the national standard of surveillance video and audio coding (SVAC) in 2008, and the standard was approved and published in 2010. At present, the corresponding products are in promotion and have wide application future.

Different from conventional audio encoders mainly used in multi-media fields, SVAC audio encoder has some special technical characteristics for the particularity of applications, such as providing a high recognition rate for voiceprint recognition in back-end after signals are decoded and real-time responses when special events occur, etc. These new techniques lead to complex algorithms, and in addition, combining them with conventional coding technologies simply and directly result in many structural redundancies in SAVC audio encoder. Therefore structure optimization for the encoder needs further research.

This paper proposes an optimization scheme for SVAC audio encoder. The new method reconstructs high-frequency content of signal based on decoded MEL-frequency cepstral coefficients (MFCCs) extracted and coded in front-end, and removes the Bandwidth Extension (BWE) module in original encoder.

2 SVAC Audio Encoder

The overall framework of SVAC audio encoder is shown in Figure1. The coding processing can be divided into two parallel parts: audio coding and characteristic

Z. Shi et al. (Eds.): IIP 2014, IFIP AICT 432, pp. 221–229, 2014.

parameters coding. After sampling rate conversion, input PCM signal in one road is transferred to audio coding module, and can be encoded in different bitrates controlled by the levels of events detected in abnormal events detector. The other road signal is sent to characteristic parameters coding module in which the well-known MFCCs are extracted, quantified and coded in order to prevent the influence of speech coding distortion on voiceprint recognition in back-end.

In audio coding module, input signal is divided into low-frequency band and high-frequency band. Based on the signal type, low-frequency signal is encoded in two switch modes: Algebraic Code Excited Linear Prediction (ACELP) and Transform Audio Coding (TAC), and high-frequency signal is encoded through BWE technology which consumes little bitrates. In characteristic parameters extracting module, MFCCs are extracted, and then quantified and coded in subsequent modules. Here two coding modes are offered for MFCCs: direct coding mode and predictive coding mode (dotted line in Figure 1). The latter one firstly decodes bit stream of the audio encoder and acquires reconstructed signal, then MFCCs of reconstructed signal and original signal are extracted respectively. At last, residual MFCCs are quantified and coded.

It can be seen that the connection between these two road modules lies only in MFCCs' predictive coding mode. From the perspective of encoder design, these two road signal processing (only at different sampling frequencies,) perhaps cause structural redundancies. In view of this, the structure of SVAC audio encoder has potential to be further optimized.

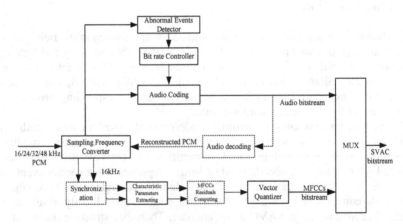

Fig. 1. SVAC audio encoder block diagram

3   SVAC Bandwidth Expansion

As an enhancement technique for audio coding, even under the conditions of limited bit-rates, BWE can present high-frequency content with a very small amount of side information coded in front-end, and use part of low-frequency data combined with the side information to realize high-frequency reconstruction of signal, which further broadens the signal bandwidth of low bit-rate coding, and improves coding quality. BWE technology now has been employed in a variety of audio encoders.

BWE in SVAC audio encoder transmits energy gain factors and 8-order linear predictive coding (LPC) coefficients, and they occupy 16 bits only in one frame after quantization. In audio decoding, high-frequency excitation signal is obtained by gain adjustment of low-frequency excitation signal, and then synthesis filter is designed based on the decoded 8-order LPC coefficients. At last, high-frequency signal is reconstructed and then full-band signal is obtained.

It can be seen that BWE in SVAC audio encoder is based on LPC coefficients, however the encoder also chooses MFCCs as its characteristic parameters. Both of them are widely used speech signal characteristic parameters. Recent studies have shown that the MFCCs have stronger correlation than LPC coefficients between high and low frequency of speech signals and the BWE technology based on MFCCs is superior to the current implementation scheme based on LPC coefficients.

4   Bandwidth Expansion Based on MFCCs

Firstly, we analyze the extraction steps of MFCCs. In order to be unified with SVAC audio encoder, we select speech signals whose frequency spectrums are in the range of 0 kHz ~ 8 kHz. Low-frequency is in 0 kHz ~ 4 kHz range, and high-frequency is in 4 kHz ~ 8 kHz range. Extraction steps are summarized briefly as follows:

a)   *Pre-emphasis:* A single-pole high-pass filter is used to emphasize the high-frequency content.

b)   *Windowing:* A Hamming window is used to mitigate the edge effect of discontinuities between frames.

c)   *Power spectrum estimation:* Fast Fourier Transform(FFT) is applied to obtain the power spectrum;

d)   *Mel-scale filters bank binning:* MEL scale triangular filters are applied. Record the number of low-pass filter with M, and the number of high-pass filter is denoted by N. The output energy of each MEL filter for low-frequency is $X_k \left(0 \leq k \leq M-1\right)$, and the output energy of each MEL filters for high-frequency is $Y_k \left(0 \leq k \leq N-1\right)$;

e)   *Log operation:* The outputs of MEL filters log-energies are obtained.

f)   *Discrete Cosine Transform(DCT):* DCT of the log-energies is applied to obtain the MFCCs as follows:

$$LF: \quad c_n = \sqrt{\frac{2}{M}} \sum_{k=0}^{M-1} (\log X_k) \cos\left(\frac{(2k-1)n\pi}{2M}\right), 0 \leq n \leq M-1$$

$$HF: \quad c_n = \sqrt{\frac{2}{N}} \sum_{k=0}^{N-1} (\log Y_k) \cos\left(\frac{(2k-1)n\pi}{2N}\right), 0 \leq n \leq N-1$$

(1)

Where c_n is the *nth* MFCC.

Details of the derivation of MFCCs-based BWE are described as below.

Seen from the MFCCs extraction process above, two steps involve non-invertible loss of information; discarding phase information in step c) and the many-to-one mapping of the MEL filters in step d). And in the practical application of step f), such as speech recognition, also involves potential loss of information depending on whether the MFCCs vectors are truncated.

Assume that the MFCCs of high-frequency are under well preservation, so by Inverse DCT (IDCT) we can get the complete reconstruction of the energy Y_k of MEL filters. However the evaluation for Y_k in step $d)$ is a many-to-one mapping, so the MEL scale power spectrum cannot be reconstructed precisely through Y_k. Finer cepstral detail can, however, be obtained by interpolating from these log-energies of MFCCs by increasing the resolution of the IDCT as follows:

$$\log \hat{Y}_{k'} = \sqrt{\frac{2}{N}} \sum_{k=0}^{N-1} c_n \cos\left(\frac{(2k'-1)n\pi}{2iN}\right), 0 \leq k' \leq iN - 1 \tag{2}$$

Where i is an interpolation factor, which is decided by the resolution of MEL-scale. For example, when N=6 and the MEL-scale resolution is 1, for the high-frequency ranges in 4 ~ 8 kHz, the calculation for i is:

$$i = \left| \frac{f_{mel}(8kHz) - f_{mel}(4kHz)}{N+1} \right| = 100 \tag{3}$$

Where $f_{mel}()$ is the conversion formula between MEL frequency and the actual linear frequency:

$$f_{mel}(f_{Hz}) = 2595 \log_{10}(1 + f_{Hz}/700) \tag{4}$$

Through exponential transformation of $\hat{Y}_{k'}$ from Mel frequency to linear frequency, we can acquire high-frequency power spectrum, and then with Inverse Fast Fourier Transform (IFFT) we get the autocorrelation coefficients of high-frequency sequence. Furthermore we use Levinson-Durbin iterative algorithm to obtain the solution of Yule-Walker equation, finally, the LPC coefficients are achieved. The generation of high-frequency excitation signal will not be discussed in this paper, and we use the technology defined in SVAC.

5   Improvement on BWE

SVAC audio encoder codes audio signals and MFCCs simultaneously. So obviously MFCCs-based BWE mentioned above can be adopted to replace the original BWE based on LPC coefficients.

For the sampling frequency converter in Figure 1, the output signal transferred to MFCCs coding module is sampled in 16 kHz, and the output signal transferred to audio coding module is sampled in 12.8kHz ~ 38.4kHz range. So bandwidth of encoded audio signal is in 6.4 kHz ~ 19.2 kHz range, and the bandwidth of BWE is half of encoded signals. In SVAC audio coder, there are many coding levels, and Table 1 lists three levels from 1.0 to 1.2. Each level has two internal sampling frequencies.

Table 1. Three coding levels in SVAC audio encoder

Technical parameter	Levels		
	1.0	*1.1*	*1.2*
internal sampling frequency(kHz)	12.8/16	24/25.6	32/38.4
samples per frame	512	512	512
maximum bitrates(bits/s)	24250	38800	58200

For example, when internal sampling frequency is 12.8kHz at level 1.0, high-frequency content of signal in 3.2 kHz ∼6.4 kHz range is reconstructed by BWE and the maximal frequency is 6.4 kHz. At the same time, the signal sent to modules to extract and code MFCCs is sampled in 16 kHz, so its MFCCs extracted from high-frequency band can be used for high-frequency reconstruction. In this case, the maximal frequency reconstructed is 8 kHz higher than 6.4 kHz in the previous coder. So our optimization scheme is that the original BWE module can be removed completely when internal sampling frequency is lower than or equal to 16 kHz at low coding level because high-frequency reconstruction can be achieved from MFCCs.

Figure 2 is the framework of improved audio encoder for SVAC based on the above-mentioned idea. In this framework, sample frequency converter converts audio signal to a fixed 16 kHz sampling frequency. The converted signal is output to a low-pass filter (LP) and a high-pass filter (HP). The low-frequency signal is encoded by audio coding module and MFCCs are extracted and coded from full band signal. At last, SVAC bit streams are multiplexed.

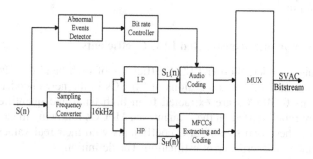

Fig. 2. New block diagram for SVAC audio encoder

In decoder, the MFCCs of high-frequency signal decoded are converted to LPC coefficients to design the synthesis filter which is used to reconstruct the high-frequency content of original signal.

In this new SVAC audio encoder, BWE module is removed completely and the structure of encoder is simplified.

6   Simulation Results

We compare our new encoder with SVAC audio encoder in simulation. 30 mono audio sequences are selected as experimental materials, including 18 male and female speech samples, 5 music samples and 7 abnormal events samples standardized in SVAC. Coding level is 1.0 with 3 different bit rates which are 10.8kbps, 17.2kbps and 24.4kbps respectively. Simulation concerns complexity and bit consumption, comparison of reconstructed LPC Coefficients, contrast of accuracy of reconstructed high-frequency band, and quality evaluation.

6.1   Analysis of Complexity and Bit Consumption

Our encoder removes BWE module in SVAC and avoids calculation and quantization for gain factors and LPC coefficients, without introducing additional operations. So the computational complexity is reduced. The original encoder rebuilds LPC coefficients gained from inverse quantization and interpolation, and also decodes gain factors. However, the new decoder rebuilds LPC coefficients based on MFCCs. Their computational complexities are approximately equal. From the point of view of bit consumption, our encoder saves 16 bits consumption of BWE module in original SVAC, and the saved bits can be used for MFCCs coding or audio coding, which can increase the corresponding quantization accuracy and coding quality. The new encoder does not have to encode energy gain factors due to the energy information contained in MFCCs which namely are log-energies described above, so the new encoder saves bits again.

Here comes to the conclusion that our encoder achieves gains both on complexity and bit consumption.

6.2   Comparison of Reconstructed LPC Coefficients

In SVAC audio encoder, 8-order LPC coefficients of high band (4 kHz~8 kHz) are obtained and quantified. The dimensions of MFCCs in the new encoder is DIM (10, 6) which means 6 MFCCs are extracted from high band and quantified. The LPC coefficients are rebuilt based on these quantified MFCCs following the above method.

We compare these two sets of LPC coefficients with their real values and choose standard variance to describe error estimation. The definition is:

$$\sigma[i] = \sqrt{\frac{\sum_{n=1}^{M}\left(lpc'(i,n) - lpc(i,n)\right)^2}{M}} \tag{5}$$

In the formula above, $lpc'(i,n)$ is distorted value of LPC coefficient and $lpc(i,n)$ is its real value. Also, i ($1 \le i \le 8$) is the index of LPC coefficient, and n represents the frame index. M is the number of frames. 2000 frames of different samples are chosen in our simulation. Table 2 presents average $\sigma[i]$ ($1 \le i \le 8$) in comparison.

Table 2. Error estimation of LPC coefficients

	$\sigma[1]$	$\sigma[3]$	$\sigma[5]$	$\sigma[7]$
	$\sigma[2]$	$\sigma[4]$	$\sigma[6]$	$\sigma[8]$
original encoder	0.166	0.266	0.154	0.136
	0.318	0.200	0.133	0.071
new encoder	0.144	0.191	0.144	0.112
	0.205	0.173	0.119	0.058

It can be seen from the table that compared to the original LPC coefficients, the error of reconstructed LPC coefficients in the new encoder is less than that in SVAC audio encoder. The main reason is that bit rates consumed by LPC coefficients in SVAC audio encoder are relatively less.

Here comes to the conclusion that reconstruction of LPC coefficients has higher accuracy in the new encoder.

6.3   Comparison of Accuracy of Reconstructed High-Frequency Spectrum

Fig.3 is an English female speech signal which is coded by both encoders. One frame is picked up to do comparison on high-frequency spectrum. In Figure 3, picture above is the frequency spectrum of original signal. The intermediate picture is spectrum of signals coded at level 1.0 when the internal sampling frequency is 12.8 kHz. In order to compare effectively, low-frequency part less than 4 kHz uses ACELP to encode, and frequency spectrum in 4 kHz~6.4 kHz range is reconstructed by BWE. Picture below is the reconstructed spectrum in our encoder in which low-frequency part under 4 KHz uses ACELP to encode and frequency spectrum in 4 kHz~8 kHz range is rebuilt based on MFCCs. We can see that two encoders have same performances at low-frequency band, but the new encoder broadens spectrum from 6.4 kHz to 8 kHz, especially this broadening is realized without increasing coding bit rates.

Here comes to the conclusion that the new encoder has higher precision in high-frequency reconstruction.

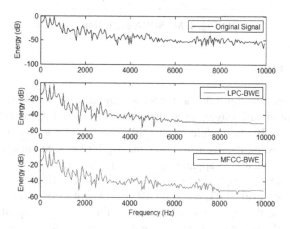

Fig. 3. Reconstructed high-frequency spectrum in two encoders

6.4   Quality Evaluation

We performed MUSHRA experiments for the two encoders. Considering when the spectrum of music signals are below 8 kHz, human's hearing is affected significantly more than speech signals, we only select speech samples and some abnormal events samples in our subjective evaluation.

There are 7 listeners who have rich listening experiences. The experiment follows the MUSHRA method. In order to ensure the score at reasonable range, the original audio samples are hidden in all samples, and 3.5 kHz and 7.0 kHz low-pass filtered signals are added as hidden anchors.

Figure 4 presents score curves of two encoders at different bit rates. The original encoder reconstructs high-frequency below 6.4 kHz, and the new encoder below 8 kHz. Their curves are near the curve of 7.0 kHz anchor. But we can see that coding quality of the new encoder significantly exceeds the original one, and the average gain is about 3.63.

We can get the following conclusion that coding quality of the SVAC audio encoder is improved by the new scheme at same bit rates.

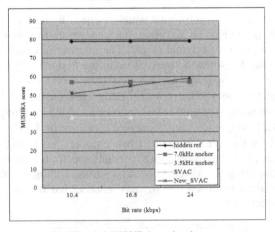

Fig. 4. MUSHRA evaluation

7   Conclusion

We propose an optimization scheme for SVAC audio encoder. We reconstruct high-frequency content of audio signal only dependent on those MFCCs extracted in front-end, and eliminate BWE module to simplify the encoder's structure, along with improvement on coding quality without increasing bit rates. Simulation results show that the new encoder achieves apparent gains in various aspects.

On the other hand, the optimization scheme only works effectively at coding level 1.0. When at higher level, BWE based on MFCCs cannot reconstruct spectrum higher than 8 kHz because the signal for MFCCs extracting is only sampled at 16 kHz. In fact, BWE module in original SVAC audio encoder cannot reconstruct high-frequency content with good quality also. For example, when audio signal is sampled at 19.2 kHz, high band from 9.6 kHz to 19.2 kHz is estimated dependent on 16 bits side information only, and the reconstruction quality is not satisfying. How to design BWE module more effectively at different coding levels for SAVC audio encoder is our future research direction.

Acknowledgment. The work is supported by 'the Fundamental Research Funds for the Central universities' of China (No.12CX04078A).

The authors would like to thank the anonymous reviewers for their constructive comments and corrections and the MUSHRA listening test participants for their effort.

References

GB/T 25724-2010, Technical Specification of Surveillance Video and Audio Coding, Beijing: Standardization Administration of the People's Republic of China, SAC (2010)

Pulakka, H., Remes, U., Yrttiaho, S., Palomaki, K., Kurimo, M., Alku, P.: Bandwidth Extension of Telephone Speech to Low Frequencies Using Sinusoidal Synthesis and a Gaussian Mixture Model. IEEE Trans. Audio, Speech, Lang. Process 20(8), 2219–2231 (2012)

Pulakka, H., Laaksonen, L., Myllyla, V., Yrttiaho, S., Alku, P.: Conversational evaluation of artificial bandwidth extension of telephone speech using a mobile handset. In: Proc. IEEE Int. Conf. Acoust., Speech, Signal Process (ICASSP), pp. 4069–4072 (2012)

Chivukula, R.K., Reznik, Y.A., Devarajan, V., Jayendra-Lakshman, M.: Fast Algorithms for Low-Delay SBR Filter banks in MPEG-4 AAC-ELD. IEEE Trans. Audio, Speech, Lang. Process 20(3), 1022–1031 (2012)

Nour-Eldin, A.H., Kabal, P.: Objective analysis of the effect of memory inclusion on bandwidth extension of narrowband speech. In: Proc. InterSpeech, pp. 2489–2492 (2007)

Nour-Eldin, A.H., Kabal, P.: Mel-frequency cepstral coefficient-based bandwidth extension of narrowband speech. In: Proc. InterSpeech, pp. 53–56 (2008)

Nour-Eldin, A.H., Kabal, P.: Combining frontend-based memory with MFCC features for Bandwidth Extension of narrowband speech. In: Proc. ICASSP, pp. 4001–4004 (2009)

Ramabadran, T., Meunier, J., Jasiuk, M., Kushner, B.: Enhancing distributed speech recognition with back-end speech reconstruction. In: Proc. EuroSpeech, pp. 1859–1862 (2001)

Borgstrom, B.J., Alwan, A.: A Unified Framework for Designing Optimal STSA Estimators Assuming Maximum Likelihood Phase Equivalence of Speech and Noise. IEEE Trans. Audio, Speech, Lang. Process 19(8), 2579–2590 (2011)

Thoma, H.: A system for subjective evaluation of audio, video and audiovisual quality using MUSHRA and SAMVIQ methods. In: 2012 Fourth International Workshop on Quality of Multimedia Experience (QoMEX), pp. 31–32 (2012)

A Traffic Camera Calibration Method
Based on Multi-rectangle

Calibrating a Camera Using Multi-rectangle Constructed by Mark Lines in Traffic Road

Liying Lu, Xiaobo Lu[*], Saiping Ji, and Chen Tong

Key Laboratory of Measurement and Control of CSE
Ministry of Education
School of Automation, Southeast University
Nanjing, 210096, China
8618994100907
xblu2013@126.com

Abstract. As distance detection is frequently needed in modern traffic management, camera calibration has a significant influence on the accuracy of distance detection. The traditional calibration methods usually need a calibration target, and the exact distance of calibration points has to be measured. In this paper, according to the camera imaging model, we put forward a calibration method based on multi-rectangle, which constructs several rectangles with the mark lines of the traffic road. Without measuring the internal or external parameters of the camera or the exact distance of calibration point, we only need the side lengths of rectangles and the image coordinates of the rectangular vertexes to establish the video image distance conversion model. For improving the accuracy of conversion, we use multiple rectangles to determine the coordinates of the vanishing points and modify the coordinates of the rectangular vertexes. Experimental results show that our method is more accurate when compared with the mainstream methods.

Keywords: distance detection, camera calibration, multi-rectangle, parameter modification.

1 Introduction

In modern traffic management, traffic video image processing is frequently used in the field of distance detection, speed detection and visibility detection. As an important part of traffic image processing, camera calibration has a significant influence on the accuracy of detections which related to distance. Camera calibration means establishing the conversion model between the position of image pixel point and the position of actual point. According to the camera model, we get the camera

---

[*] Corresponding author.

Z. Shi et al. (Eds.): IIP 2014, IFIP AICT 432, pp. 230–238, 2014.

parameters using some feature points of a real scene and these points' coordinates in the image. Finally, we build the conversion model between the image coordinates and the actual distance using the camera parameters (Zheng, 1998; Hartley, 2003).

Traditional camera calibration methods usually need some standard calibration targets, whose actual size has been accurately measured before calibrating. These methods use the corresponding relationship of image points and actual points to establish the constraint conditions of camera parameters, then calculate the parameters of camera by some nonlinear optimization algorithm(Zhang, 2000; Zhang, 2004; Ye). It is been proven that the calibration method based on vanishing point is effective and reliable, the position of vanishing point could be determined by one-dimensional target of calibration targets(Guillou, 2000; Yan, 2013). Wei Geng and Yang Liu put forward a new algorithm based on single rectangle, which uses the mark lines in traffic road to calibrate the parameters of camera, then establish mutual conversion model between image coordinate system and the actual distance(Geng, 2012). The method uses the actual lengths of rectangular patterns and the corresponding information of coordinates to establish detection model, without taking manual measurement to get camera internal and external parameters and the absolute distance of standard fixed points.

But under actual situation, as the mark lines may have been worn and the feature points are manual marked, that method may bring serious error. To solve this problem, a calibration algorithm based on multi-rectangle is introduced in this paper, which constructs several rectangles with mark lines of the traffic road, then modifies the Y-axis direction vanishing point by fitting a straight line with multiple points. In the X-axis direction, n straight lines will generate   intersection points, we choose the point with highest accuracy as the X-axis direction vanishing point. To get more accurate calibration parameters, we use the position information of vanishing points and the accuracy of each rectangular vertex to modify the position of every rectangular vertex, then establish distance conversion model with these modified parameters. Through a lot of experiment, this method is more accurate than the single-rectangle algorithm.

2   Calibration Method Based on Multi-rectangle

As the white mark lines and the gray road surface make the contrast ratio of the image not strong enough, and the mark lines may have been worn, it's difficult to localize the position of feature points accurately and the calibration method based on single-rectangle is inclined to bring error. We study a parameter modification method in this paper to reduce the error. As shown in Figure 1, we can construct multiple rectangles in a road surface image.

2.1   Modifying the Coordinate of the Vanishing Point in Y-Axis Direction

The projection line of mark lines in image plane will intersect at point Q_2' , which is the vanishing point in Y-axis direction. Assume that we construct single rectangle

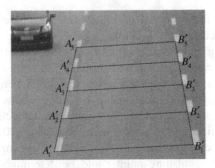

Fig. 1. Construct multiple rectangles in a road surface image

with point $A_1'A_2'B_1'B_2'$, straight line $A_1'A_2'$ and $B_1'B_2'$ can be viewed as the projection lines of mark lines, but error will be brought when any point is localized inaccurately. To reduce the error, we fit straight line L_A' with point $A_1'A_2'A_3'...$ using the least squares method, hence, L_A' can be viewed as the projection line in image plane of the left mark line. Then we fit the projection line in image plane of the right mark line L_B' in the same way. Find the intersection point Q_2' of straight line L_A' and L_B', so Q_2' is the vanishing point in Y-axis direction.

2.2   Modifying the Coordinate of the Vanishing Point in X-Axis Direction

Straight line A_1B_1 , $A_2B_2...A_nB_n$ in road surface plane all parallel to X-axis, so straight line $A_1'B_1'$, $A_2'B_2'...A_n'B_n'$ in image plane should intersect at vanishing point Q_1', but there may be multiple intersection points in actual image, generally, n straight lines will generate C_n^2 intersection points. Assume that straight line $A_i'B_i'$ and $A_j'B_j'$ intersect at point P_{ij}', the closer P_{ij}' near to other straight lines, the closer P_{ij}' to the actual position of vanishing point Q_1'. Define D_{ij} as the accuracy of P_{ij}':

$$D_{ij} = \begin{cases} \sum_{k=1}^{n} d_{ij-k}, i \neq j \\ 0, i = j \end{cases} \tag{1}$$

2.3   Modifying the Coordinates of the Rectangular Vertex

Some feature points may have been inaccurately localized as we use mark lines to construct calibration target, so these coordinates must be modified.

To characterize the accuracy of point A_i' and B_i', we sum all $D_{ik}(1 \leq k \leq n)$ value from point P_{i1}' to point P_{in}' on straight line $A_i'B_i'$, then calculate the normalized summation. Define the accuracy of any rectangular vertex in the Q_1' direction:

$$E_{skq1} = \frac{\sum_{t=1}^{n} D_{kt}}{\max(\sum_{t=1}^{n} D_{kt})} \tag{2}$$

In the formula, s could be a or b, $1 \le k \le n$, $1 \le t \le n$, so sk represents a rectangular vertex, $\max(\sum_{t=1}^{n} D_{kt})$ is the largest value of accuracies on straight line $A_i'B_i'$, $q1$ means the current value is the accuracy in Q_1' direction. But E_{akq1} equals E_{bkq1} in Q_1' direction, which means it's unable to distinguish the accuracy of A_i' and B_i'. So we have to find other straight lines in other direction, make A_i' and B_i' belongs to different straight lines, thus to separate the accuracy of A_i' and B_i'.

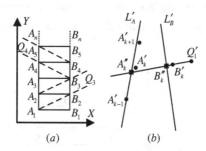

(a)   (b)

Fig. 2. Feature point coordinate modification

As Figure 2(a) shows, every diagonal in the same direction parallels to each other, as an example, we choose a group of straight lines $A_i'B_{i+1}'(1 \le i \le n-1)$, these straight lines will intersect at a vanishing point Q_3'. The coordinate of Q_3' could be calculated in the same way as how we calculate the coordinate of Q_1', then the accuracy of A_i' and B_{i+1}' in Q_3' direction E_{skq3} can be calculated. As an exception, neither A_1' nor B_n' belongs to a straight line in Q_3' direction, so their E_{skq3} value is 0. Similarly, a group of straight lines $A_{i+1}'B_i'(1 \le i \le n-1)$ will intersect at vanishing point Q_4', we get calculate the coordinate of Q_4' and the E_{skq4} value of every feature point. Define the accuracy of a feature point as:

$$E_{sk} = \frac{E_{skq1} + E_{skq3} + E_{skq4}}{1 + efc(E_{skq3}) + efc(E_{skq4})} \tag{3}$$

In the formula, k ranges from 1 to n, $efc(E_{skq3})=0$ when $s=a, k=n$ or $s=b, k=1$; $efc(E_{skq4})=0$ when $s=a, k=1$ or $s=b, k=n$, otherwise $efc(E_{skq3})=1$, $efc(E_{skq4})=1$.

After getting the accuracy of every feature point, we can modify the coordinate of feature points. We take point A'_k and B'_k as an example, the exact feature point A''_k, B''_k should be the intersection point of the straight lines which pass through Q'_1 and straight line L'_A or L'_B, we mark them as square points in Figure 2(b) . The round dots in Figure 2(b) represents the manual marked feature points A'_k, B'_k. As the accuracy of A'_k and B'_k is known, we choose the point which has a smaller accuracy value as reference point, then connect it and Q'_1 to construct the modified straight line $A''_k B''_k$, so the intersection point of $A''_k B''_k$ and L'_A is the modified point A'_k, the intersection point of $A''_k B''_k$ and L'_B is the modified point B'_k.

2.4   Distance Conversion Model

Compared to the single-rectangle method, the multi-rectangle method make Q'_1, Q'_2 and every rectangular vertexes modified, so we can get more accurate calibration parameters.

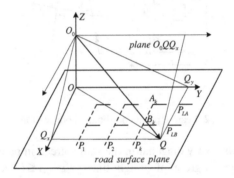

Fig. 3. Calculating calibration parameters

According to camera imaging model, we can calculate scale factor k_{11} and k_{12} in X-axis direction. As Figure 3 shows, P_{LA} is the intersect point of mark line L_A and plane $O_0 Q Q_y$, P_{LB} is the intersection point of mark line L_B and plane $O_0 Q Q_y$. As $Q Q_y$ parallels to X-axis, its projection line in image plane passes through Q'_1, so $Q'_1 Q'$ is the projection line of $Q Q_y$, P'_{LA} is the projection point of P_{LA}, and it is the intersection point of $Q'_1 Q'$ and L'_A, P'_{LB} is the projection point of P_{LB}, and it is the intersection point of $Q'_1 Q'$ and L'_B, let l_{xa} represents the X-axis coordinate of P_{LA}, l_{xb} represents the X-axis coordinate of P_{LB}, according to the one-dimensional distance conversion model:

$$\begin{cases} l_{xa} = k_{11} + k_{12}p_1 \\ l_{xb} = k_{11} + k_{12}p_2 \end{cases} \tag{4}$$

where p_1 is the reciprocal of the pixel-value differencing between point P'_{LA} and Q'_1, p_2 is the reciprocal of the pixel-value differencing between P'_{LB} and Q'_1.

Let s_{AB} represents the distance between L_A and L_B, $s_{AB} = l_{xb} - l_{xa} = k_{12}(p_1 - p_2)$, according to highway construction specifications[6], $s_{AB} = 3.75m$, so:

$$k_{12} = \frac{s_{AB}}{p_1 - p_2} = \frac{3.75}{p_1 - p_2} \tag{5}$$

According to the relationship between k_{11} and k_{12}, we can get k_{11}:

$$k_{11} = -\frac{k_{12}\sin\alpha_1\cos\alpha_1}{f_d} \tag{6}$$

Then we calculate scale factor k_{21} and k_{22} in Y-axis direction. As rectangular vertexes have been modified, every straight line $A'_k B'_k$ passes through Q'_1, so straight line $A_k B_k$ parallels to X-axis, assume the intersection point of $A_k B_k$ and plane $O_0 Q Q_x$ is $P_k (1 \le k \le n)$, the actual distance between every two points is easily calculated according to highway construction specifications. P'_k is the projection point of P_k, and it is also the intersection point of straight line $A'_k B'_k$ and $Q'_2 Q'$. According to the one-dimensional distance conversion model in Y-axis direction, for point P'_1: $l_{y1} = k_{21} + k_{22}p_{21}$, where l_{y1} is the Y-axis coordinate of P'_1, p_{21} is the reciprocal of the pixel-value differencing between point P'_1 and Q'_2, which could be easily calculated in image. Similarly, for P'_k: $l_{y1} + d_k = k_{21} + k_{22}p_{2k}$, where d_k represents the distance between P_k and P_1, p_{2k} represents the reciprocal of the pixel-value differencing between point P'_k and Q'_2. As k ranges from 1 to n, we can get n equations, then we can establish a coordinate system with $(l_{y1} + d_k)$-axis as vertical axis and p_{2k}-axis as horizontal axis, fit a straight line using the points we got in the equations, the slope of the straight line is k_{22}. According to the relationship between k_{21} and k_{22}: $k_{21} = -k_2 \sin\alpha_1 \cos\alpha_1 / f_d$. In conclusion, we can establish the two-dimensional distance conversion model as:

$$\begin{cases} l_x = k_{12}p_1 - k_{12}\sin\alpha_1\cos\alpha_1 / f_d \\ p_1 = 1 / ((u_1 - u_x)^2 + (v_1 - v_x)^2)^{1/2} \\ l_y = k_{22}p_2 - k_{22}\sin\alpha_2\cos\alpha_2 / f_d \\ p_2 = 1 / ((u_2 - u_y)^2 + (v_2 - v_y)^2)^{1/2} \end{cases} \tag{7}$$

3   Results and Analysis of Experiment

3.1   Accuracy Testing Experiment

To evaluate the accuracy of the proposed method, we compare the effect of multi-rectangle method and the single-rectangle method through experiment. We mark six points in image that are corresponding to 21.00m to 214.13m in actual road plane, then calculate the distance of these points using the video image distance conversion model, and compare the results with the actual distance. As Table 1 shows, compared to the single-rectangle method, the proposed method has an improved measuring accuracy on each distance, and as the distance increases, the error of single-rectangle method increases fast, but the error of the proposed method remains in the range of 5%. In the study we found that the accuracy of vanishing point Q_1' and Q_2' have a major impact on the model parameters, for the proposed method, we use multiple rectangles to determine the position of vanishing point, so we could get more accurate vanishing point coordinate than the single-rectangle method.

Table 1. Results for distance detection experiment

Actual distance (m)	Single-rectangle method		Multi-rectangle method	
	Test value (m)	relative error	Test value (m)	relative error
21.00	20.25	3.55%	20.52	2.30%
37.69	36.94	2.00%	36.97	1.91%
62.41	59.59	4.52%	60.91	2.40%
104.08	97.60	6.23%	100.82	3.13%
160.47	147.83	7.88%	154.79	3.54%
214.13	193.56	9.61%	205.52	4.02%

3.2   Stability Testing Experiment

To evaluate the stability of the proposed method, we measure the maximum error of the single-rectangle method and the multi-rectangle method in 10 different traffic images, the distance of each test point is less than 200 meters.

As Table 2 shows, the maximum error of the multi-rectangle method floats around 5%, while the maximum error of the single-rectangle method floats around 10%. In addition, in the 4th and 7th image, the maximum error of the single-rectangle method rise to 16.24% and 21.72% due to the mark line has been worn, while the maximum error of the multi-rectangle is only 6.93% and 7.44%. Through this experiment, we can see the proposed method is more stable than the single-rectangle method.

Table 2. Results of stability testing experiment

serial number	Maximum error (single-rectangle method)	Maximum error (multi-rectangle method)
1	9.61%	4.02%
2	12.37%	4.72%
3	9.88%	5.14%
4	16.24%	6.93%
5	8.75%	4.12%
6	10.03%	5.62%
7	21.72%	7.44%
8	10.94%	5.92%
9	11.41%	5.37%
10	9.18%	4.58%

4   Conclusion

This paper put forward a camera calibration algorithm based on multi-rectangle, which constructs multiple rectangles with mark lines in actual traffic road, then modifies the calibration parameters and establish video image distance conversion model. We modify the coordinate of vanishing point in Y-axis direction by fitting a straight line with multiple rectangular vertexes, modify the coordinate of vanishing point in X-axis direction by choosing the point with highest accuracy among all intersection points, then we modify the position of every rectangular vertex, and establish the distance conversion model using the modified parameters. Compared to single-rectangle method, the proposed method provides more accurate calibration parameters, and the distance conversion model we got has an improve accuracy. Experiment result shows that as distance increases, the detecting error increases too, but the effect of the proposed method is still better than the single-rectangle method.

However, this method still belong to the category of manual calibration, the initial feature points still needs to be manual marked, that would bring some unknown error and reduce the calibration accuracy. How to automatically localize the feature points, that's what we need to improve in further study.

Acknowledgments. This work was supported by the National Natural Science Foundation of China (No.61374194) and China Postdoctoral Science Foundation Funded Project (No.2013M540405).

References

China Ministry of Transportation, Road traffic signs and markings set norms, JTG D82-2009
Geng, W.: Detection Algorithm of Video Image Distance Based on Rectangular Pattern. In: 2012 5th International Congress on Image and Signal Processing, pp. 1013–1017 (2012)

Gonzales: Digital Image Processing. Publishing House of Electronics Industry, Beijing (2007)

Guillou: Using vanishing points for camera calibration and coarse 3D reconstruction from a single image. The Visual Computer 16, 396–410 (2000)

Hartley, et al.: Multiple view geometry in computer vision. Cambridge University Press (2003)

Hautière, et al.: Daytime Visibility Range Monitoring through use of a Roadside Camera. In: IEEE Intelligent Vehicles Symposium, pp. 470–475 (2008)

Yan, H.: Camera Self-Calibration In Highway Dynamic Environment. Computer Aided Design and Computer Graphics 25(7), 892–899 (2013)

Ye, Y.: Study on the Camera Calibration Approach and Algorithm of Edge Detection and Contour Tracking. Dalian University of Technology, Dalian

Yin, W.: Camera calibration based on OpenCV. Computer Engineering and Design 28(1), 197–199 (2007)

Zheng, N.: Computer Vision and Pattern Recognition. Defense Industry Press, Beijing (1998)

Zhang, Z.: A flexible new technique for camera calibration. IEEE Transactions on Pattern Analysis and Machine Intelligence 22, 1330–1334 (2000)

Zhang, Z.: Camera calibration with one-dimensional objects. IEEE Transactions on Pattern Analysis and Machine Intelligence 26, 892–899 (2004)

A Multi-instance Multi-label Learning Framework of Image Retrieval

Chaojun Wang, Zhixin Li[*], and Canlong Zhang

College of Computer Science and Information Technology,
Guangxi Normal University, Guilin 541004, China
wangchaojun009@126.com, {lizx,clzhang}@gxnu.edu.cn

Abstract. Because multi-instance and multi-label learning can effectively deal with the problem of ambiguity when processing images. A multi-instance and multi-label learning method based on Content Based Image Retrieve (CBIR) is proposed in this paper, and the image processing stage we use in image retrieval process is multi-instance and multi-label. We correspond the instances with category labels by using a package which contains the color and texture features of the image area. According to the user to select an image to generate positive sample packs and anti-packages, using multi-instance learning algorithms to learn, using the image retrieval and relevance feedback, the experimental results show that the algorithm is better than the other three algorithms to retrieve results and its retrieval efficiency is higher. According to the user to select an image to generate positive sample packs and anti-packages, using multi-instance learning algorithms to learn, using the image retrieval and relevance feedback. Compared with several algorithms, the experimental results show that the performance of our algorithm is better and its retrieval efficiency is higher.

Keywords: multi-instance multi-Label learning, image retrieval, multi-points diverse density algorithm, image retrieval.

1 Introduction

We live in an era in which multimedia information grows at a speed of geometric index and network technologies are popularized quickly .It is more and more difficult that the images we need are found quickly and effectively from the pictures about the ocean. We can know that text-based image retrieval and content-based image retrieval doesn't solve well the content of the image of the ambiguity through the above introduction, and a complete image includes a lot of small parts. If only a single label does not adequately describe the image, but cannot accurately expressed the users are interested in the parts, which can lead that the part of the image that we are own interested in is not retrieved .The content of this article is to divide the image into several parts, take partial features from several pieces of information as the package of the image generated by examples, and contact the examples of the package with the

---

[*] Corresponding author.

Z. Shi et al. (Eds.): IIP 2014, IFIP AICT 432, pp. 239–248, 2014.

category of the mage. This paper presents a method of the image retrieval based on multi-instance multi-label study. The method takes use of local color of images and texture features to generate multi-sample packages, and give the appropriate subset of classes for objects, and it is no longer the only category tag. We can get the complex high-level concepts derived by the subsets of these categories marked, so that we improve the probability of recall and precision in the process of image retrieval.

2   Multi-instance Multi-Label Learning

In multi-instance learning, every training package is made of more than one sample, and the sample has no concept, but each training package has a concept marked. When we apply this method to image retrieval, we will find that it belongs to a variety of categories for images. If we only use a single category marked words, we cannot fully describe the image, and it will be a problem of semantic gap. Considering the image content with multiple semantics, we should learn that it is actually a map from the sample set to the category tag set on. Considering the image content with multiple semantics, we should learn that it is actually a map from the sample set to the category tag set on. Multi-instance multi-label[1] defined as follows, χ represent example of space γ represent the category space, and represent by mathematical symbols follows:

Multi-instance multi-label learning: given a set of data $\{(X_1, Y_1), (X_2, Y_2),...,(X_m, Y_m)\}$.Objective to learn $f : 2^x \rightarrow 2^y$. $X_i \subseteq \chi$ Is a set of sample $\{ \chi_{i_1}, \chi_{i_2},...,\chi_{in} \}$, $\chi_{ij} \in \chi$ (j=1,2,...,n_i) and $Y_i \in \chi$ is a set of appropriate categories tag of X_i.$\{y_{i1}, y_{i2},...,y_{ili}\}$ $y_{ik} \in \gamma$(k=1,2,...,l_i). n_i is the number of samples among X_i, l_i is the number of tags among Y_i.

There is learning frameworks named traditional supervised learning (single example, single tag) and multi-label learning (single example, multiple tags) and multi-instance learning (multiple example, single tag) besides the above one of sample learning frameworks in this paper. From the above, it is not difficult to find that MIML framework can be converted into other three examples in some special cases. We know that only by using the appropriate expression can capture the important information of image. However ,dealing with the ambiguity of the image by using MIML can better deal with semantic gap problems .MIML also helps to study the high-level semantic concepts, such as elephants, lions on the picture, trees, and some grass. Through the concept of markers we can infer that this is a picture of a description of Africa. This example is the use of MIML and then study the underlying logic of high-level concepts .In order to play the ability of MIML framework, there are generally two kinds of methods, respectively, for example as Bridges and multi-label learning as a bridge.. This paper is to convert Multi-instance multi-label learning algorithm MIMLBOOST [2][3] into Multi-instance learning.

3   The Structure of the Package and Image Retrieval

3.1   The Structure of the Sample Package Process

Image of the sample bag choosing what kind of method determines whether the image semantic information is complete, this will directly affect the retrieval effect. But Maron and Yang's Multiple-Instance learning by maximizing diverse density [4] use the dividing method of fixed generated package. This method will cause originally belong to the same concept divide into different packages or belong to different concepts merged into one package. Sample package need to define in advance. These will only contain color information. Colors of different parts may be similar which can lead to low accuracy of image retrieval. The method [5] of Dai H.Blet each image fixed generates 4 sample packages. What kind of effect will be caused by this method, For a simple image that is bound to introduce noise, and compared with the complex image of some of the information will not wholly include .This paper will use a partition method which is a network clustering of Self-organizing feature Map (Self - Organizing feature Map, SOM)[6].This article uses SOM clustering image segmentation methods[7] to divide images, we use the color features as SOM network input to cluster. After go to the discrete points and region merging after some steps such as each region of the image, we respectively calculate the R, G, B, 3 kinds of color grayscale average to the corresponding color features. If we only use color features cannot describe an area, different images in similar parts of color have some differences and different parts of the color may be similar. In order to improve the accuracy of image retrieval, we have to use textures which are some characteristics of the image .Texture is an important means of visual image .Also it describes gray space distribution of image. Texture images can show some information about the arrangement of surface structure, such as roughness, contrast, direction, the line as the degrees, neat and rough degrees, etc. This paper uses co-occurrence matrix[8] method to deal with the texture feature extraction.

If we break the base of decomposition k, then we can get a $3 \times (k-1)$ co-occurrence matrix. This article will choose 4 most commonly used statistics out of 10. Also it can reflect the scale of statistics as the texture feature. Finally we add color space R, G, B whose gray value is most characteristic's-occurrence matrix has a feature vector of $4 \times (k-1)+3$ dimension. Regard the regions corresponding eigenvectors as an example. So it can be an image in which all areas of its corresponding sample image corresponding to the package.

3.2   Image Retrieval

Figure 1 describes the image retrieval process. There is multi-instance and multi-label learning concept: regard an image as a sample package and the image has a lot of class label. It may be more meaningful in some cases if we understand why an image has some class label, the MIML makes these possible. Using the MIML we can know there is some samples to have such class label. So we can match sample directly with class label images. And we also not only can use image package but also can use the class label between the samples for image retrieval.

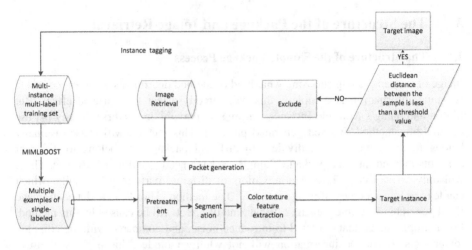

Fig. 1. Flow diagram of image retrieval

In the second chapter we introduce four models of machine learning, the special case of the MIML is the other three kinds of machine learning methods. And in this article the process of solving for multiple tag sample learning problems can be converted the problem into a single tag sample learning problems by using MIMLBOOST algorithm. What is the purpose of doing that? We can find the benefits of doing that through the experiment. This article uses the multi- diversity density algorithm (MPDD) [9] from the image training package to concentrated study some of the concepts for image retrieval. The idea of multi- diversity density algorithm is using density of multiple points of comprehensive information to describe the concept what users are interested in. This algorithm is an improvement compared with the density of the traditional diversity algorithm [10]. When using the density of diversity as a measure, we must select the point of maximum density as a target. If the dimension of the feature subspace is higher, we can't find out the point of maximum density by iterating through the whole feature space. The other is the feature space and vector are both distributions, this will lead to exist multiple local minima points. While using the gradient descent method can't avoid fall into local extremum but get a local optimal solution. We know that the use of one example of a sample point is difficult to contain the contents of the sample, this will cause the description of the contents of the sample packets to have certain content, in order to solve these problems it uses the MPDD algorithm in this article. MPDD algorithm calculation process is as follows.

1. The use of colour and texture extracts a features package, and we enter a positive bag and negative bag, to divided the examples and extract the description of the characteristics of the sample.
2. The use of the diversity of classic learning algorithm to find the divers density points, and output the search of the density of the point.

3. Integrated multiple search the information of the density of the point to calculate the density of the point and the distance between the unknown sample and the point. And we calculate the distance between the unknown samples and the density of multiple diversity point through the following formula.

Set MPDD algorithm output M (M value unknown) a variety of density points, denoted by $F^{*} = \{ F_1, F_2 \ldots, F_M \}$. An unknown sample package has N samples, denoted by $B^M = \{ B_1{}^M, B_2{}^M, \ldots, B_N{}^M \}$ Then the distance between the unknown sample bag B^M and the density of multiple diversity point is

$$\text{Distance}(B^M, F^*) = \frac{1}{M} \min_{i \in N} \text{dis}(B^{M_i}, F_j) \tag{1}$$

In this paper we choose more density of points is selected in the sample package. In the process of calculation is to calculate the density of each variety point and test set are package sample the Euclidean distance between, then calculated the distance between the minimum value, then sum and average. Finding out and testing set minimum Euclidean distance are examples of the package contained. We know when described different images, the importance of the various features of feature space is different, if some image has obvious characteristic of image color and texture feature. Then we can give the characteristics of each assigned a weight to represent different characteristics of different degree. The distance between the instances with the target is measured weighted Euclidean distance. Expressed in mathematical formula is as follows:

$$\left\| B_{ij} - m \right\|^2 = \sum_{n} \omega_i (B_{ijn} - m_n)^2 \tag{2}$$

Image retrieval process is as follows:

— With the method in the article, we can constitute the image of the sample bag by using the image texture and color. For different image, we can merge sample package with the same texture segmentation area and the color characteristics into sample package with new characteristics by using formula 2 in the article.

— We can use the K Nearest neighbour (k-Nearest Neighbour, KNN)[11] classification algorithms to classify the training focused sample packages and test focused sample packages.

— Find the feature vector of the target which is appropriate to users by using formula 1 of the article. Then we can calculate the Euclidean distance between the density point of diversity and all of the samples in the unlabeled sample package. And we can average them as the similarity measure, Order the distance from small to large. After that, present the corresponding image to user.

— According to the rank of similarity measure, we can choice the interested and uninterested images from the retrieval results and label the corresponding classification of package. After that, put it into training set to get relevant feedback.

4   Experimental Result

We do the experiment based on this method, and compare the experiment results with other methods including the Multiple-instance image retrieval[12] proposed by Dai Hongbin, Maron and Ratan and Lozano-Perez's multiple-instance learning method known as the diverse density algorithm[13]. The image dataset this experiment uses has a medium scale. It divides into 10 classes: elephants, flowers, birds, horses, dinosaurs, buildings, and cars, mountains with snow, beaches and plants. The total number of images is up to 1500 and each class has 150 images. We pick 30 images from each class to form the training set as examples. The rest 1200 images form a test set. During each experiment, one class is selected as the target class, and then 5 images which we are care about are selected to generate the positive packet. We select 5 classes randomly from the rest classes and then select an image we are not care about from each of 5 classes randomly to form the negative packet. Now a training set is formed and it includes a positive and negative packet which has 5 images respectively. Here we use more diversity, MPDD density algorithm to learn the training set and retrieve the image in the test set. Then the results are returned by sorting the Euclidean distance of similarity. From the search results, we select the first 5 images which are sorted in target and non-target category, and add the corresponding packet and examples to the training set. We compare the experiment results using the method of Dai Hongbin, Maron and Ratan's Single blob with neighbours method and the method of Yang and Lozano-Perez.

4.1   Experimental Comparison of Four Methods

HongbinDai and ZhihuaZhou use the method of extracting colour features together with the 24 texture features extracted by 24 filters of Gabor wavelet transformation. The feature vector has 27 dimensions: 3 colour features R, G, B and 24 texture features. The feature vector of every region is considered as examples.

Maron and Ratan's Single blob with neighbours' method constitutes the matrix by colour blobs after sampling the image. The feature vector with 15 dimensions is calculated from the R, G and B colour features in each colour blob and four blobs connected with it. There are 9 image packages and each contains a 15-dimensional feature vector.

However, Yang and Lozano - Perez's method is to convert the images to grey ones. After dividing into twenty overlapping regions, every region of the image is conducted mirror filtering, transforming and sampling. There are 40 image packages with 64- dimensional feature vector. The results of the experimental are as follows:

This article is for each category of images 10 times experiment, we used precision recall such a way to measure the efficiency of various methods of retrieval. Precision is when system is the process, the system returns results in the target image and all returned images of a ratio. In this experiment, 5 interested in image and not interested in the image of an initial search results are shown in figure2. And in the process of experiment increases 5 interested in image and not interested in image relevance feedback result is shown in figure 3. For the above all kinds of methods for initial retrieval and related feedback average retrieval time are shown in table1.

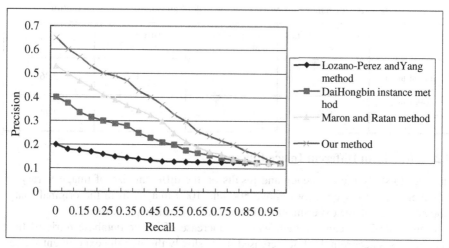

Fig. 2. Initial retrieval results of relevant and 5 irrelevant images

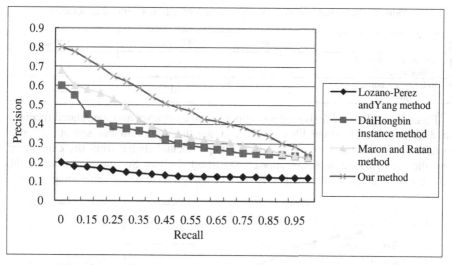

Fig. 3. Relevant feedback results of 10 relevant and 10 irrelevant images

As can be seen from the figure 2 and figure 3, the proposed method takes precedence over these three methods. In this paper, the average time for the initial retrieval and feedback time is prior to the other two methods but longer than another method. The main reason is that MIML learning methods is adopted in this paper. Linking category tag with the sample, the time we use more diversity density algorithm to find the point is less than the former two methods but is more than the density of diversity algorithm. The time this algorithm takes is less than the former two but more than Hongbin Dai multi-instance algorithm. Adopting multi-sample and multi-label algorithm, this article can well describe the image. So through next section in this paper, we learned that the algorithm has better precision and recall radio.

Table 1. Performance comparison on the task of automatic image annotation

	Maron and Ratan method	Lozano-Perez andYang method	DaiHongbin instance method	Our method
The average time of Initial retrieves (s)	3.2	7.63	1.47	2.36
Relevance feedback time (s)	9.56	8.52	2.92	4.24

4.2   Results in Different Image Library

In order to study the efficiency and results on the different size of image library by this method, we respectively remove 500 and 1000 images from the original image library .I repeated the experiment for the method the article put forward.

From table 2 we can see that, with the increase of image database retrieval time also increases accordingly, but increased more slowly through the experiment we can find the time complexity of this article was close to linear.

Table 2. Performance comparison on the task of automatic image annotation

	500 images	1000 images	1500 images
The average time of initial retrieves (s)	1.21	1.56	1.72
Relevance feedback time (s)	2.86	3.12	3.36

Figures 4 and figure 5 show the results of image retrieval and the related feedback in different size of image library. We can see the algorithm has smooth retrieval performance with the increase number of experimental image from the experience, and the performance does not appear significant decline. So, the algorithm of this article can be an applicable method to retrieve a large scale image scale image library.

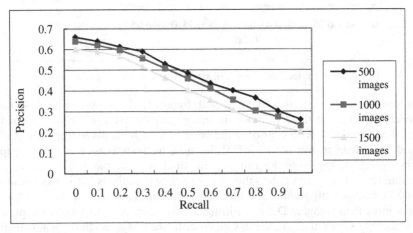

Fig. 4. Relevant feedback results of 10 relevant and 10 irrelevant images

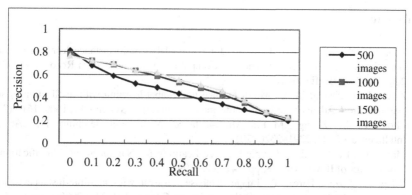

Fig. 5. Relevant feedback results of 10 relevant and 10 irrelevant images

5   Conclusion

This article puts forward a kind of image retrieval based on multiple sample multi-label learning technology. The benefits of this technology are to reduce the semantic gap and solve the problem of ambiguity of the image. It enlighten us use the underlying semantic concept of high-level semantics. Through the above experiment we find this method is helpful to improve the image retrieval precision rate and recall rate. This article implementation process use the method of MIML as the first step to process image, utilize MIMIL Boost algorithm transforms the problem of multiple sample tags into multiple sample single tag questions, then the image segmentation algorithm based on SOM clustering segment image into multiple regions. We combine each region's color characteristics and co-occurrence matrix as a sample package, then use multi-points diverse algorithm for image retrieval and related feedback. It is suitable for medium-sized image library and compare with proposed method in our experiments, the results show that the algorithm of this article retrieval effect is better than that of Maron, Ratan, Yang and Lozano-Perez etc.This paper also introduced to other characteristics to improve the image of package technology. In the future research we learn how to put the MIML into the image retrieval studying, reduce the time of recall and precision of image retrieval ,and design a better algorithm to find out the target class image in the shortest time.

Acknowledgement. This work is supported by the National Natural Science Foundation of China (Nos. 61165009, 61262005, 61363035, 61365009), the National Basic Research Program of China (No. 2012CB326403) and the Guangxi Natural Science Foundation (Nos. 2012GXNSFAA053219, 2013GXNSFAA019345) and the "Bagui Scholar" Project Special Funds.

References

1. Zha, Z.J., Hua, X.S., Mei, T., et al.: Joint multi-label multi-instance learning for image classification. In: IEEE Conference on Computer Vision and Pattern Recognition, CVPR 2008, pp. 1–8. IEEE (2008)
2. Zhou, Z.-H., Zhang, M.-L.: Multi-instance multi-label learning with application to scene classi-fication. Advances in Neural Information Processing Systems 19, 1609–1616 (2007)
3. Zhou, Z.-H., Zhang, M.-L.: Multi-instance Multi-label learning [J]. Artificial Intelligence 176(1), 2291–2320 (2012)
4. Maron, O., Ratan, A.L.: Multiple-Instance Learning for Natural Scene Classification. In: Proceedings of ICML, vol. 98, pp. 341–349 (1998)
5. Yang, C., Lozano-Perez, T.: Image database retrieval with multiple-instance Learning tech-niques. In: Proceedings of ICDE, San Diego, CA, pp. 233–243 (2000)
6. Kohonen, T.: Self-organizing maps. Springer, Berlin (2001)
7. Jiang, Y., Chen, K.J., Zhou, Z.H.: SOM based image segmentation. In: Wang, G., Liu, Q., Yao, Y., Skowron, A. (eds.) RSFDGrC 2003. LNCS (LNAI), vol. 2639, pp. 640–643. Springer, Heidelberg (2003)
8. Marceau, D.J., Howarth, P.J., Dubois, J.M.M., et al.: Evaluation of the grey-level co-occurrence matrix method for land-cover classification using SPOT imagery. IEEE Transactions on Geoscience and Remote Sensing 28(4), 513–519 (1990)
9. Böhm, C., Krebs, F.: The k-nearest neighbour join: Turbo charging the KDD process. Knowledge and Information Systems 6(6), 728–749 (2004)
10. Mian-Shu, C., Shu-Yuan, Y., Zhi-Jie, Z., et al.: Multi-points diverse density learning algorithm and its application in image retrieval. Journal of Jilin University: Engineering Science 41(5), 1456–1460 (2011)
11. Maron, O., Lozano-Pérez, T.: A framework for multiple-instance learning. In: Advances in Neural Information Processing Systems, pp. 570–576 (1998)
12. Hongbin, D., Ling, Z., Zhihua, Z.: A Multi-Instance based approach to image retrieval. Journal of Pattern Recognition and Artificial Intelligence 19(2), 179–185 (2006)
13. ChengYang, L.-P.: Image database retrieval with multiple-instance learning techniques. In: Proceedings of the 16th International Conference on Data Engineering (2000)

The Retrieval of Shoeprint Images Based on the Integral Histogram of the Gabor Transform Domain

Xiangyang Li, Minhua Wu, and Zhiping Shi*

Capital Normal University, College of Information Engineering, Beijing, China
{lixiangyang806,shizhiping}@gmail.com, wumhxxxy@126.com

Abstract. Shoeprint images which are extracted at the scene of cases are a kind of important modern forensic clue and evidence. Retrieving the images of the same or the similar shoeprint images from the database quickly and accurately is very important to criminal investigation. To deal with the fragmental shoeprint images, we propose a shoeprint images matching and retrieval algorithm which computing the integral histogram in the Gabor transform domain. First, through the integral histogram find out the most similar position of the fragmental image in the intact image. Then, extract the features of the region found in the first step. At last, compute the similarity of the two components. Experiment results prove that this algorithm leads an increase of 4.82% in the retrieval precision, compared with computing the global features of two images directly.

Keywords: fragmental images, Gabor filtering, integral histogram, image retrieval.

1 Introduction

Shoeprint images are common found at the scene of the crime, which are easy-to-be-left, multiple, specific, relatively stable, continuous. They play a very important role in the criminal investigation and forensic evidence. The traditional methods of managing and retrieving shoeprint images are mainly based on manual coding. For all kinds of shoeprint images, we need to design a coding standard given different coding for different images. Then according to the standard, implement shoeprint images recognition [1]. The manual coding method relies on the operator too much, so the operator's subjective factors have much influence on the identification result. And at the same time, with the rapid increase of shoeprint images, it's very difficult to achieve a satisfactory result. Retrieving the images according to user's requirements quickly and accurately is related to the efficiency of cases in the department of police. Therefore, it is very important to develop a shoeprint images retrieval algorithm with high retrieval accuracy and high speed.

In general, image features mainly include color, texture, shape and spatial location which are extracted from the entire image or from one region [2-3]. How to extract

---

* Corresponding author.

Z. Shi et al. (Eds.): IIP 2014, IFIP AICT 432, pp. 249–258, 2014.

the features efficiently is the key of content based image retrieval (CBIR). Shoeprint is imprinting of the sole contact with the ground, so color features is meaningless, only features such as texture, shape and space layout are useful. Guan [4] had proposed a method of recognition for shoeprint images based on outline features, which find out the outline points at the longest distance according to the truth that the length of shoe sole was longer than any other parts in a shoe sole, and then calculated the angle shaped by the two points and horizontal axes. It proofreaded the shoe soles in terms of calculated angle , extracted 9 features of shoes sole automatically, and then measured the similarity of shoe soles by Euclidean distance. This method uses the global contour features of the image, so it is not suitable when there are fragmental images among the shoeprint images. Jia [5] had proposed a method which classified the shoeprint images by using the characteristic quantity got from the Laws template convolution of the image. It can identify the type of grain and point pattern. But to complex type such as mixed type, edge block type and circle type, it does not have good recognition efficiency. Xiao [1] had proposed a method based on PSD (power spectral density) and Zernike moment. But if the segmentation of the shoeprint image is inaccurate, the Zernike moment features do not performs well. Zhang [6] had proposed a method for retrieving multi-texture images based on counter and texture segmentation. Gabor wavelet transform is applied to extract the features of images. It has a higher retrieval precision, but it is not suitable to process incomplete images alignment, because it uses the global features.

To deal with the fragmental shoeprint images, we propose a shoeprint images matching and retrieval algorithm based on the integral histogram in the Gabor transform domain. Filter the image with Gabor filter cluster, use the index of the filter which has the maximum response to replace the gray value of the image in every pixel, and then, compute the integral histogram of the transformed image. Through the integral histogram, find out the most similar position of the fragmental image in the intact image. Then, extract the features of the region found in the former step. At last, compute the similarity of the two components. Experiment results prove that this method provides a higher retrieval precision, and it is adaptive to fragmental image processing.

2   Retrieval Process of Shoeprint Images

The retrieval process of shoeprint images is shown in Figure 1. Firstly, pre-process the images obtained on the scene, and then convert them to binary images. Secondly, index these images based on features extracted from them. Finally, we can implicate image retrieval. Due to the complex environment on the scene of the crime, the incomplete contact between the sole and the ground, and the limit of image segmentation technology, a lot of shoeprint images are incomplete. Therefore, after pre-processing, the images have inconsistent size and integrity. As shown in Figure 2, the fragmental image is only part of the full image.

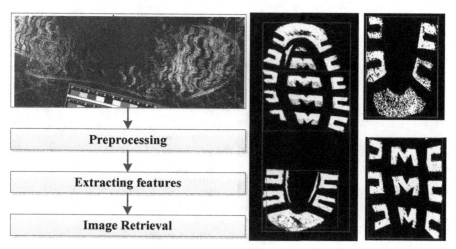

Fig. 1. Shoeprint image retrieval process　　　**Fig. 2.** Intact image and fragmental images

2.1　Shoeprint Preprocessing

Useful information of shoeprint images is included in the binary description of the images, so it is important to determine whether the pixel is shoeprint grains or background. Binary image not only reduces the storage capacity, but also can make big contributions to image identification in the later process. There are four main steps in the binarization preprocessing.

— Smooth the original image with a low-pass filter to reduce the noise.
— Divide the images into three parts: the soles, the arch and the heel, the portion rate are 0.45, 0.25 and 0.3. And then every part is divided into 3*3 blocks. The division of shoeprint is shown in Figure 3.
— To each small block, use the cvAdaptiveThreshold function provided by Opencv 2.4.3 [7] for adaptive threshold binarization., where the first two parameters src and dst are the input and output images, parameter max_val is set to 255, this means the pixel value which is greater than the threshold is set to 255. Parameter adaptive_method is set to ADAPTIVE_THRESH_MEAN_C, block_size is set to the maximum odd value which is not bigger than the width and weight of the small block, the value of the parameter param1 is set to 0. In this way, every pixel is processed by the threshold of the average weighted sum of size b*b from its neighbours.
— Use morphological operations to process the images to eliminate the noise of the small points, and connect the area broken by binarization.

The binarization results of the images are shown in Figure 4.

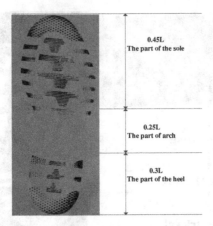

Fig. 3. The division of the shoeprint

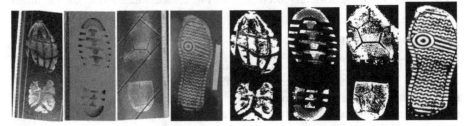

Fig. 4. The original shoeprint images and the images after binarization

2.2 Feature Extraction

Shoeprint images are binary images after pre-processing, so texture feature and shape feature are mainly considered. Texture is the basic structure of the surface or attribute of an object. It is generally acknowledged that texture is the grayscale change or color change of the image on the space. Comparing the Hu invariant moments [8], co-occurrence matrix [9] and Gabor texture feature, the experimental results show that the last one is the best, so we choose the Gabor texture feature as texture descriptor.

Gabor filter is widely used in image representation, recovery and segmentation, as well as other fields, which is mainly based on the following two points: First, Gabor filter can simulate mammalian visual cortex simple cells receptive field, which is conformed to the characteristics of visual physiology. Second, Gabor filter is able to obtain the best joint resolution in frequency domain and space domain [10]. Texture varies in direction, frequency and extent, and the wavelet transform window can be adaptively adjusted to the change of the window center frequency. Therefore, to extract the texture features of an image, we usually use a cluster of self-similar Gabor wavelets obtained from a mother Gabor wavelet instead of a single one.

The Gabor function is a sine wave modulated by Gaussian function. A two dimensional Gabor function h(x, y) and its Fourier transform H(u, v) can be written as [11]:

$$g(x, y) = \frac{1}{2\pi\sigma_x\sigma_y} \exp\left(-\frac{1}{2}(\frac{x^2}{\sigma_x} + \frac{y^2}{\sigma_y})\right) \tag{1}$$

$$h(x, y) = g(x, y)\exp\left(2\pi jWx\right) \tag{2}$$

$$H(x, y) = \exp\left(-\frac{1}{2}(\frac{(u-W)^2}{\sigma_u^2} + \frac{v^2}{\sigma_v})\right) \tag{3}$$

where g(x, y) is the Gaussian function, σ_x and σ_y are   mean square errors in x axis and y axis, which determines the shape of the Gaussian envelope, W is the frequency of the sinusoidal function on the horizontal axis. A cluster of self-similar functions, referred to as Gabor wavelets in the following discussion, is now considered. Let h(x, y) be the mother Gabor wavelet, then   self-similar filter dictionary can be obtained by appropriate dilations and rotations of h(x, y) through the generating function [12]:

$$h_{mn}(x, y) = a^{-m}h(x', y'), a > 1, m, n \in Z \tag{4}$$

$$x' = a^{-m}(x\cos\theta + y\sin\theta),\ y' = (-x\sin\theta + y\cos\theta), \theta = n\pi / K \tag{5}$$

where α^{-m} is the scale factor, S is the number of scales and K is the number of orientations, m=0, 1, …, S-1 and n=0, 1, …, K-1. Assuming that the class of Gabor wavelets contains S scales and K directions, and the frequency ranges from U_l to U_h, the parameter selection method is as follow:

$$a = (U_h/U_1)^{1/(S-1)}, U_{(n)} = a^n U_1$$
$$\sigma_{u(n)} = (a-1)U_{(n)}/[(a+1)\sqrt{2\ln2}] \tag{6}$$
$$\sigma_{v(n)} = \tan(m\pi/(2k))[U(n)-2\ln2(^2/U_{(n)})]/\sqrt{2\ln2-(2\sigma_{u(n)}\ln2/U_{(n)})^2}$$

Gabor filter has the characteristics of multi-channel and multi-resolution analysis, through the Gabor filter to extract the features of an image can achieve high resolution   in both spatial domain and frequency.

2.3   Integral Histogram

Histogram is one of the most widely used features in image recognition and retrieval with high recognition efficiency as it is simple to be calculated. But it cannot be applied to the local alignment of the image because it is the global feature. An improved method is to compute the histogram of the local area we are interested in. Integral Histogram is a method to compute the histograms of all possible target regions in a Cartesian data space with high efficiency [13-14] For the image, H(x, y) is defined as: H(x, y)={Hu, u=1, 2, …, d}, where d is the scope scale of the gray image. H(x, y) is the histogram of the rectangle formed by the origin point and the point (x, y) of the image.

$$H(x, y) = K(R(0, 0, x, y)) \tag{7}$$

where K is the operation of computing the histogram. The histogram of any rectangular in the image can be quickly computed from its Integral histogram.

In Figure 5, there are four vertices. The histogram of the rectangular shaped by the points A, B, C and D is:

$$K(R(x1, y1, x2, y2)) = H(x2, y2) - H(x2, y1) - H(x1, y2) + H(x1, y1) \tag{8}$$

When the integral histogram of an image is computed, the histogram of target region can be computed easily using simple arithmetic operations.

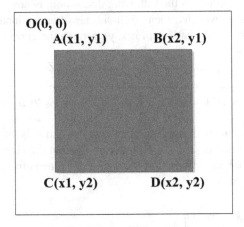

Fig. 5. Integral Histogram

2.4   The Process of the Algorithm

To deal with the fragmental shoeprint images, we propose a shoeprint images matching and retrieval algorithm based on the integral histogram in the Gabor transform domain. First, filter the image with the Gabor wavelets. Because the Gabor wavelets are a cluster of self-similar wavelets, so each pixel can get a response value from each filter. In order to transform the image into the characteristics space, use the index of the filter which has the maximum response to replace the gray value of the image in every pixel. Let the original grayscale value of the image is f(x, y), after the transformation, the value $f'(x, y)$ is defined as:

$$f'(x, y) = k, H_k(f(x, y)) = Max(H_j(f(x, y))(j = 0, 1, 2, \ldots\ldots, N - 1) \tag{9}$$

where N is the total number of the filters, $H_k(f(x, y))$ is the result of the point f(x, y) filtered by the filter numbered k. By this, the image is transformed from spatial domain into Gabor transform domain. We can calculate the histogram of one image, so we get the GTH (Gabor Transform Histogram) features. Meanwhile, we can get the features noted as GTIH (Gabor Transform Integral Histogram) by computing the integral histogram of the image.

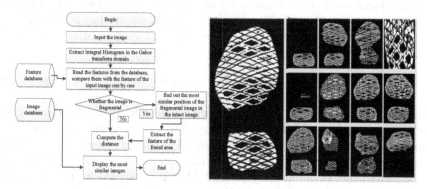

Fig. 6. Flow chart of the algorithm   **Fig. 7.** The retrieval result

The shoeprint images retrieval algorithm is described as shown in Figure 6.The retrieval process is: Filter the image with Gabor filters cluster, use the index of the filter which has the maximum response to replace the gray value of the image in every pixel, and then, compute the integral histogram of the transformed image. To compute the similarity of the fragmental image and the complete image, first, we need to find out the most similar position of the fragmental image in the intact image through the integral histogram. Then, extract the features of the region found in the first step. At last, compute the similarity of the two components. For the image of similar size, compute the similarity directly according to the histogram using the metric method of histogram intersection.

3   Experimental Results and Comparisons

In this paper, the shoeprint images retrieval algorithm based on integral histogram is implemented in Visual Studio 2010 and Opencv 2.4.3 platform with c++ language. The experiment environment is: Intel Core i5-3470 3.20GHz, Windows 7, 4.00 GB. There are 2000 binary images which are the processing results of the shoeprint images captured on the scene of the crime. To evaluate the performance of the algorithm, we choose 195 images as a test set. There are 13 categories in the test set, and each category has 15 images. The images are binary images with black background after a series of processing, such as denoising, segmentation, etc. Sometimes, there are large black blocks at the edge of an image. The black edge not only doesn't carry characteristics of the image, but also increases the amount of calculation when filtering it. Hence, to each image, we compute its vertical histogram and horizontal histogram. After this, we can get the effective area of an image. Then, we extract the features of this area. The effective areas of images are shown on Figure 7. As shown in Figure 2, the outside large rectangle is the effective area of the image. The inner small rectangles of the intact image are the most similar position of the fragmental images computed through integral histogram.

Usually the retrieval performance of a CBIR system is measured by precision and recall [3]. Precision (Pr) is defined as the ratio of the number of relevant images retrieved (N_r) to the number of total retrieved images K. Recall (Re) is defined as the number of retrieved relevant images N_r over the total number of relevant images available in the database Nt.

$$Pr = N_r / K, \quad Re = N_r / N_t \tag{10}$$

Retrieve each image in the test set, and then compute the average precision. Every time, the system returns 15 images which are most similar to the input images, the former, the more similar, and the most similar 13 images are shown. Figure 7 is the shoeprint retrieval system interface showing the retrieval results. We use Pr to measure the performance of different algorithm. Table 1 is a comparison of several different algorithms. They are described as follow.

— Hu invariant moments features [8]. Calculate the Hu invariant moments features of the image, and use Euclidean distance to measure the similarity (Hu).
— Co-occurrence matrix features [9]. Compute the gray level co-occurrence matrix of the image. The matrix is generated by the parameters: step length is 1 pixel and the orientations are 0, 45, 90, 135 degrees. And then, compute the deficit, energy and entropy of these matrixes and use Euclidean distance to measure the similarity (CoMatrix).
— Extract the GTH feature from the image, and use Euclidean distance to measure the similarity (GTH-E).
— Extract the GTH feature from the image, use histogram intersection to measure the similarity (GTH-I).
— Filter the image with Gabor wavelets. For each filter, we can get a middle image after the filtering, compute the mean and variance of the middle image. Use them as the features of the image (GMV), and use histogram intersection to measure the similarity. GMV are multi-channel features get from the filtered image, so it is not suitable to use Euclidean distance to measure the similarity (GMV-E).
— Extract GTIH features from the image and use histogram intersection to measure the similarity. If the image is fragmental image, find out its most similar position in the intact image, through the integral histogram, compute the feature of the found region (GTIH-I).
— Use the method of GTIH-I to find out the most similar position. Use Gabor cluster filters to filtering the found region to extract the features. Then use histogram intersection to measure the similarity (GTIH-I-R).

In Table 1, Time is the average retrieve time of one image. The experimental results show that the features extracted by Gabor filters are effective to describe the characteristics of the images. Compared with GTH-I,   GTH-E proved that the distance measurement histogram intersection is a little better than Euclidean distance, but the difference is small. GTIH-I search for the local features, so its accuracy is better than GTH-I which uses the global features. GTIH-I-R re-extracts the features of the found region, so it increases the retrieval accuracy obviously. GTIH-I-R improves the precision by 4.82% on average compared with the GMV-E.

Table 1. The Performance of Different Algorithm

Algorithm	Percision	Time (second)
Hu	0.228718	0.002523
CoMatrix	0.233504	0.011149
GTH-E	0.396923	4.688564
GTH-I	0.403077	4.277297
GMV-E	0.419487	4.580621
GTIH-I	0.419145	208.085226
GTIH-I-R	0.467692	317.529205

As shown in Table 2, to measure the similarity of fragmental image and intact image, the result of the method using integral histogram to compute the features of the most similar region is 0.902385. The result of the method using the cluster of Gabor filters to re-extract the features of the most similar region is 0.938849. The results show that the re-extract features has higher retrieval efficiency. Because the re-extract features of the most similar region have no characteristics of other regions, so they are more localized and more accurate.

Table 2. Compute the features of the most similar region

Method	Features (12-demensional)		
The feature of the original fragmental image	0.205336	0.083475	0.139575
	0.107085	0.063219	0.031357
	0.054153	0.038420	0.075421
	0.006755	0.059016	0.007900
Use the integral histogram to compute the features directly	0.183289	0.093500	0.194598
	0.122024	0.073280	0.038685
	0.069721	0.049539	0.127830
	0.008835	0.025142	0.013555
Use the cluster of Gabor filters to re-extract the features	0.194752	0.076310	0.151491
	0.096853	0.063219	0.033249
	0.092727	0.041732	0.062361
	0.007493	0.042909	0.008626

4   Conclusion

The paper has presented a shoeprint images retrieval algorithm based on Gabor transform domain. The algorithm can provide a higher retrieval precision, especially when the images contain fragmental ones. The innovation of the algorithm is: use the index of the filter which has the maximum response to replace the gray value of the image in every pixel, and then, compute the integral histogram of the transformed image. Use the integral histogram to efficiently find the region in the intact image which the fragmental one is most similar to; avoid comparing the global features directly. Measure the similarity of the features re-extracted from the found region and the features of the fragmental image, to improve the shoeprint images retrieval precision.

Acknowledgment. This work was supported in part by National Nature Science Found Grants No. 60903141 and No. 61165009. The authors are grateful for the many specific and valuable comments made by the reviewer.

References

1. Xiao, R., Lu, N., Shi, P.: Methods on shoeprint matching. In: The 13th National Conference on Image and Graphics, Nanjing, pp. 256–360 (2006)
2. Zhang, D., Islam, M., Lu, G.: A review on automatic image annotation techniques. Pattern Recognition 45(1), 146–162 (2012)
3. Chadha, A., Mallik, S., Johar, R.: Comparative study and optimization of feature extraction techniques for content based image retrieval. International Journal of Computer Application 52(20), 25–42 (2012)
4. Guan, Y., Li, C., Zhong, M.: Research and realization of recognition for shoe soles based on outline feature. Application Research of Computers 25(8), 2413–2415 (2008)
5. Jia, S., Shi, W., Zeng, J., Chen, S.: Shoe prints identification and classification methods based on texture characteristics. Journal of Da Lian Jiao Tong University 19(1), 59–62 (2008)
6. Zhang, Z., Shi, Z., Shi, Z., Shi, Z.: Image retrieval based on contour. Journal of Software 19(9), 2461–2470 (2008)
7. Bradski, G., Kaehler, A.: Learning Opencv, pp. 155–161. Tsinghua University Press, Beijing (2009)
8. Hu, M.: Visual pattern recognition by moment invariant. IRE Transaction on Information Theory 8(2), 179–187 (1962)
9. Gao, C., Hui, X.: GLCM-Based texture feature extraction. Computer System Application 19(6), 195–198 (2010)
10. Sing, S., Hemachandran, K.: Content-Based image retrieval using Color Moment and Gabor texture feature. International Journal of Computer Science Issues 9(1), 229–309 (2012)
11. Roslan, R., Jamil, N.: Texture feature extraction using 2-D Gabor filter. In: Proceedings of IEEE Symposium on Computer Applications and Industrial Electronics, Kota Kinabalu, pp. 173–178 (2012)
12. Manjunath, B., Ma, W.: Texture feature for browsing and retrieval if image data. IEEE Transactions on PAMI 18(8), 837–842 (1996)
13. Porikli, F.: Integral histogram: a fast way to extract histograms in Cartesian spaces. In: Proceedings of IEEE Computer Society Conference on Computer Vision and Pattern Recognition, San Diego, pp. 829–836 (2005)
14. Park, J., Park, J., Kim, T.: Block-based fast integral histogram, Xi'an. Spring on Engineering and Technology, pp. 1–4 (2012)

Scene Classification Using Spatial and Color Features

Peilong Zeng, Zhixin Li[*], and Canlong Zhang

College of Computer Science and Information Technology
Guangxi Normal University, Guilin, 541004, China
zplhang@126.com, {lizx,Zhangcl}@gxnu.edu.cn

Abstract. With the increment of images in modern time, scene classification becomes more significant and harder to be settled. Many models have been proposed to classify scene images such as Latent Semantic Analysis (LSA), Probabilistic Latent Semantic Analysis (PLSA) and Latent Dirichlet Allocation (LDA). In this paper, we propose an improved method, which combines spatial and color features and bases on PLSA model. When calculating and quantizing spatial features, chain code is used in the process of feature extraction. At the same time, color features are extracted in every block region. The PLSA model is applied in the scene classification. Finally, the experiment results between PLSA and other models are compared. The results show that our method is better than many other state-of-the-art scene classification methods.

Keywords: Probabilistic Latent Semantic Analysis, scene classification, spatial feature, chain code, KNN-SVM classifier.

1   Introduction

In recent years, image understanding and classification has been frequently researched and widely applied to all kinds of practical systems [9]. As an important issue of image classification, scene classification aims to label an image among a set of semantic categories (such as mountains and tall buildings) automatically [3]. For example, images in Fig. 1 can be classified into the category of "coast".

Fig. 1. The coast scene images with different illumination

It differs from the conventional object classification. Scene classification is an extremely challenging and difficult task because of the ambiguity, rotation, variability and the wide change of illumination and scale conditions of the scene images even if for

---

[*] Corresponding author.

Z. Shi et al. (Eds.): IIP 2014, IFIP AICT 432, pp. 259–268, 2014.

the same scene category [5]. The images in Fig. 1 also show that a category may include many coast images with different illumination. What's more, a scene is generally composed of several entities and the entities are often organized in an unpredictable layout [12]. It is difficult to define a set of properties that would include all its possible manifestations and extract effective common features to classify the images to the same category.

As was mentioned in [4], there were two basic strategies in the literature about scene classification. The first strategy uses low-level features such as global color, texture histograms and the power spectrum. It is normally used to classify small number of scene categories (city versus open country, etc.) [9]. The second one uses an intermediate representation before classifying scenes [10,11] and it has been applied to cases where there are a larger number of scene categories [4]. Biederman [1] showed that humans can recognize scenes by considering them as a whole one, without recognizing individual objects. Oliva and Torralba [11] proposed a low dimensional representation of scenes, based on global properties such as naturalness and openness.

A lot of efforts have been made to solve the problem in greater generality, through design of techniques capable of classifying relatively large number of scene categories in the last few years [2, 13]. In this paper, we calculate spatial features via chain code method instead of Hough transform in [9] first and then combine them with color features, and then use PLSA to classify scene images.

The main contributions of our paper lie in two aspects. One is the application of chain code method to the spatial features' extraction. We use the chain code to describe the shape and spatial information of a scene image. The combination of spatial and color features improve the performance of scene classification. The other is that we use PLSA for learning and SVM and KNN classifiers for classifying. An improved classifier KNN-SVM [17], the hybrid of KNN and SVM classifier, is used. The experiment results prove the good effect of the classifier.

The next section briefly describes the PLSA model. Then, in Section 3, we describe the classification method by applying PLSA model and KNN-SVM classifier to images. Section 4 describes the spatial and color features used to form the visual vocabulary as well as the feature extraction. The details of experiments and results are provided in section 5. Conclusion is drawn in the 6th section.

2   PLSA Model

Probabilistic Latent Semantic Analysis (PLSA) was proposed by Hofmann in 1999 to solve the problem of ambiguity between words [6]. It is a generative model from the statistical literature [7] which is researched a lot and generates many variations.

In text analysis, it is used to discover topics in a document. Here in scene classification, we treat images as documents and discover categories as topics (such as mountain and road). The model is applied to images by using visual words which is formed by vector quantizing color, shape, texture and SIFT features [4].

A collection of scene images $D=d_1, ..., d_N$ with words from a visual vocabulary $W=w_1, ..., w_V$ are given.   The data in a $V \times N$ co-occurrence table are defined as $N_{ij}=n$

(w_i, d_j), where $n (w_i, d_j)$ denotes how often the word w_i occurred in an image d_j. In PLSA a latent variable model for co-occurrence data associates an unobserved class variable $z \in Z = z_1, ..., z_Z$ with each observation.

The algorithm of PLSA is as follows:

1.   Select a document d_i with probability $P(d_i)$;
2.   Pick a latent class z_k with probability $P(z_k/d_i)$;
3.   Generate a word w_j with probability $P(w_j/z_k)$;

As a result, an observation pair (d_i, w_j) is obtained, and z_k is discarded. (d, w) is a joint probability model over $N \times V$ which is defined by the mixture:

$$P(d,w) = \sum_{z \in Z} P(d,w,z) = P(d) \sum_{z \in Z} P(w \mid z) P(z \mid d) \tag{1}$$

And then from $P(d,w) = P(d)P(w/d)$, we get

$$P(w \mid d) = \sum_{z \in Z} P(w \mid z) P(z \mid d) \tag{2}$$

$P(w/z)$ is the topic specific distribution and each image is modeled as a mixture of topics $P(z/d)$. The graphical model is represented in Fig. 2.

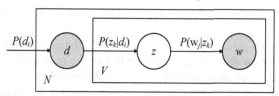

Fig. 1. Graphical model of PLSA

3   KNN-SVM Classification

The process of the training mainly includes two steps: In the first step, the probabilistic distributions of topics $P(w/z)$ are learned from the training images. $P(w/z)$ and $P(z/d_{train})$ are determined by applying the PLSA model to the whole set of training images. A Z-vector $P(z/d_{train})$ represents each training images where Z is the number of learned topics. In the second step, KNN-SVM classifier is trained by using the vector $P(z/d_{train})$ of each training image and the class label. In this stage, we propose a hybrid classifier of KNN and SVM to improve the classifier performance.

When classifying the unseen test images, the specific coefficients $P(z/d_{test})$ are computed and then they are used to classify the test image using the KNN-SVM classifier. The unseen image is projected onto the simplex using the $P(w/z)$ learned during the training process. The mixing coefficients $P(z/d_{test})$ are computed such that the divergence between the distribution and $P(w/d_{test})$ is minimized. The EM algorithm is run in similar manner to achieve the result. Now only the coefficients $P(z/d_{test})$ are updated in each M-step with the learned $P(w/z)$ kept unchanged. The test image is then classified by the KNN-SVM classifier. Fig. 3 shows graphically the process for both training and testing.

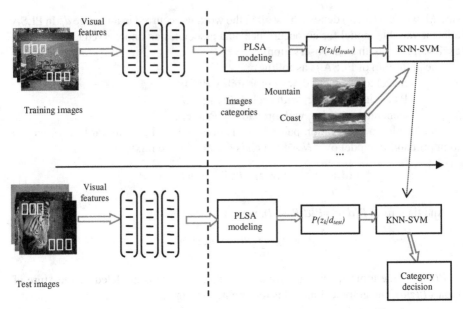

Fig. 3. Overview of visual vocabulary formation, modeling and classification

In detail, the KNN selects K nearest neighbors of the new images in the training dataset using Euclidean distance, and then the test image is classified according to the category label which is fit most in the K nearest neighbors; while for SVM classifier, an exponential kernel *exp-αd* is used, where d is the Euclidean distance between the vectors and α is the learned weight of the training image example [15]. The perspectives of the KNN-SVM classifier are: The algorithm behaves as KNN classifier and it can easily deal with huge multiclass of the scene images when K is small, while it becomes a SVM model when K is large [17]. No matter the image dataset is huge or small, the classifier can perform very well.

The algorithm is as follows:

— Use a crude and simple distance function to find a collection of Ko neighbors.

— Compute the accurate distance on the Ko sample images and select the K nearest neighbors.

— Compute (or read from cache if the query repeat) the pairwise distance of the K neighbors and the query.

— Convert the pairwise distance matrix into a kernel matrix.

— Apply SVM to the kernel matrix to classify the test image and return the result.

4   Spatial and color features extraction

4.1   Spatial Feature Extraction by Chain Code

Chain code is a compact way to represent the contour of an object for shape coding and recognition. The chain code histogram (CCH) [8] and minimize sum statistical direction code (MSSDC) [16] are two methods of all and their advantages are translation and scale invariant. However, they don't take the direction and spatial information into consideration. In the paper, we adopt an advanced method, chain code spatial distribution entropy (CCSDE) [14], to calculate the spatial features because it takes full advantage of the statistical feature, the distribution and the relativity of the chain code sequence [14].

The chain code is defined as follows: for one pixel on the boundary of an image or object, it has n neighbors, numbered from 0 to $n-1$, which are called direction codes. There are 4-direction and 8-direction chain code illustrated in Fig. 4.

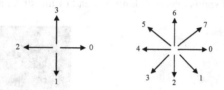

Fig. 2. 4-direction and 8-direction chain code

CCH defines $h_i=n_i/N$, where n_i is the number of chain code with i-direction, and N is the number of code links. CCSDE defines I as a contour, $I(x, y)$ as the direction at the point (x, y). $A_i = \{(x, y)/(x, y) \in I, I(x, y) = i, 0 \le i \le n-1\}$ denotes the chain codes set with i direction. $|A_i|$ is the number of chain codes in set A_i and $C_i (x_i, y_i)$ as the centroid with

$$x_i = \sum_{(x,y)\in A_i} x / |A_i| \tag{3}$$

$$y_i = \sum_{(x,y)\in A_i} y / |A_i| \tag{4}$$

Let r_i be the maximum distance from C_i to i-direction chain code:

$$r_i = \max_{(x,y)\in A_i} (\sqrt{(x-x_i)^2 + (y-y_i)^2}) \tag{5}$$

With C_i as center and jr_i/M as radius, we draw M concentric circles. Let $|A_{ij}|$ be the count of the chain codes with i direction inside jth circle. After normalizing, CCSDE is denoted as

$$SE_i(m_i) = -\sum_{j=1}^{M} m_{ij} \log_2(m_{ij}) \tag{6}$$

Combined with CCH, the new feature vector is given as

$$\langle (h_0, SE_0), \cdots, (h_i, SE_i), \cdots, (h_{n-1}, SE_{n-1}) \rangle \tag{7}$$

If we compare the similarity of two contours c_1 and c_2, the similarity can defined as

$$S(c_1, c_2) = \sum_{i=0}^{n-1} \min(h_i^{(c_1)}, h_i^{(c_2)}) \times \frac{\min(SE_i^{(c_1)}, SE_i^{(c_2)})}{\max(SE_i^{(c_1)}, SE_i^{(c_2)})} \tag{8}$$

4.2   Color Feature Extraction

The color features are vital to a scene image but is not sufficient to distinguish similar scenes. To present semantic description better, we combine color feature with the spatial feature. There are a large number of different scene images which have similar structural and spatial features and they are easily classified into the same category using the spatial features only. For instance, Fig. 5 shows two distinctly different scene images of coast and open country but they have similar edge detection images. They both have 3 levels from top to bottom: the former has sky, sea and sands, while the latter has sky, flowers and grassland. The main difference between the two scene images is the color of sands and grassland, instead of the structure and spatial features.

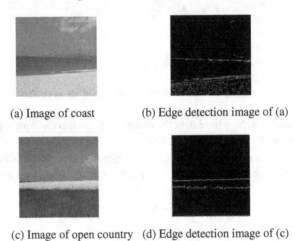

(a) Image of coast   (b) Edge detection image of (a)

(c) Image of open country   (d) Edge detection image of (c)

Fig. 3. Similar edge detection images of different scene images

Computer system can display over 2^{24} colors with 24 bits. We take advantage of this capability to define color data directly as RGB values and eliminate the step of mapping numerical values to locations in a colormap. We define an m-by-n-by-3 matrix to specify truecolor for the scene image. To reduce the number of colors in an image, some functions can be used to approximate the colors in the original image.

4.3   Features Combination

In our model, the visual words contain two types: RGB color and spatial description. If the image can be easily classified in a category using only color or spatial feature when training the images, the classification is easy done. On the contrary, we take advantage of both color and spatial feature to classify.

When using color features, the image is split into block regions. The RGB color features are extracted in every block of each image first and then global RGB color features are formed as visual words. They are vectors of RGB values. After the spatial and color features are extracted, the K-means algorithm is applied to cluster these features. Each cluster center corresponds to a visual word and a visual vocabulary includes K visual words. A set of visual words make up an image, which is similar to that the text is consist of words.

5   Experiments and Results

According to the previous description, in our model, the scene image dataset divides into two parts. One of them is used to extract spatial and color features and train the parameters for the PLSA model, and another is used to test the model we proposed and do experiments to compare with other models. The experiments and the results display as follows.

5.1   Datasets and Parameters Setting

In this paper, we choose Oliva and Torralba (OT) and Fei-Fei and Perona (FP) as the image datasets to conduct our experiment.

The OT dataset has 2688 images and is divided into 8 categories: four nature scenes (328 forest, 374 mountain, 360 coasts and 410 open country) and four artifact ones (308 inside cities, 260 highway, 292 streets and 356 tall buildings).The FP dataset has more than 2688 images but it is only available in gray scale. It includes 13 categories: 8 OT categories plus 174 bedrooms, 151 kitchens, 241 residences, 216 offices and 289 living rooms.

When comparing with [9], we use the OT dataset. What's more, we also compare our experiment results with other models based on both datasets.

The values of the latent semantic variables (Z), the number of visual vocabulary (V) in PLSA models and the number of neighbors (K) in KNN classifier are especially important. So, firstly, we choose 100 random scene images from the training set to find the optimal parameters, and the rest of the training images are used to calculate the visual vocabulary and topics of PLSA. However, for the other parameters (the number of block regions etc.), we assign them by referring to the experience or other similar experiments.

5.2   Experiments and Classification Results

As to the classification result, the performance is measured by the average of precision and recall. The precision is defined as the number of images correctly classified in the topic category divides the total number of images in the same category. While the recall is defined as the number of images correctly classified in their categories divides the number of images which are relative to the topic.

We use a group of different values of K, V and Z to experiment and analysis the effect and the optimal value. To achieve objective variations, we repeat the experiments 5 times with selecting different random training images and test sets. The mean variation curves are displayed below in Fig.6.

According to the figures below, we can observe that we get good performance above 75 percent when the number of visual words *V=700*, the number of latent topics *Z=30* and the number of neighbors *K=10*.

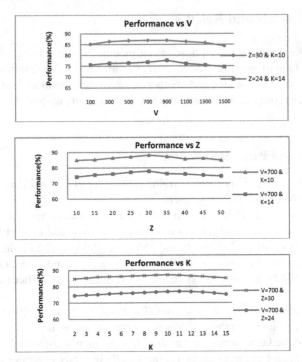

Fig. 4. Performance under variation of different V, Z and K

We apply our methods to the experiment and use the OT dataset, the same one as [9]. The experiment results are comparing in Table 1. The categories of forest, highway, street and tall building have the best result in Ken Shimazaki's experiment. The average is 0.549. However, our proposed model has a better result and almost all 8 categories have a higher mean recall and precision rate.

Table 1. Performance comparison between our model and [9]

Categories	Coast	Forest	Mountain	Open country
Mean Recall and Precision	0.516	0.697	0.354	0.409
Our model	0.632	0.799	0.532	0.527

Categories	Highway	Inside city	Street	Tall building
Mean Recall and Precision	0.670	0.380	0.766	0.656
Our model	0.783	0.502	0.860	0.837

The experiments using color and spatial features respectively and a hybrid of both are done when extracting image features. We use the KNN-SVM classifier we propose. Table 2 displays the performance for both OT and FP dataset by using color and spatial features. After that, the experiments of different classifiers (KNN, SVM and KNN-SVM) are conducted using the method of extracting both color and spatial features we propose when classifying the scene images. Table 3 shows that our classifier is improved the classification performance on OT dataset as well as FP dataset.

Table 2. Comparison among color, spatial and both features

# Categories	Color	Spatial	Both (Our model)
8 OT dataset	0.742	0.683	0.830
13 FP dataset	0.736	0.622	0.815

Table 3. Comparison among KNN, SVM and KNN-SVM classifiers

# Categories	KNN	SVM	KNN-SVM (Our model)
8 OT dataset	0.853	0.864	0.871
13 FP dataset	0.724	0.738	0.743

Other researches on scene classification such as LDA [13] and modified PLSA (based on SIFT features) model [2] were proposed. We compare the average performance of our model with them and the results are showed in Table 4.

Table 4. Comparison among PLSA (SIFT), LDA and our model

# training images	128	256	512	1024	1344
PLSA (SIFT)	0.682	0.753	0.795	0.846	0.859
LDA	0.772	0.783	0.794	0.810	0.816
Our model	0.690	0.764	0.798	0.853	0.859

6   Conclusion

We have proposed an improved method to classify scene images using PLSA model. When extracting the features of the scene image, the color and spatial features are combined to obtain full data of the image. The spatial features are calculated using the chain code and color features are represented in RGB space. Then we use the PLSA to model and train some images to get the proper parameters of the model. At last, the other images in the dataset are used to test our proposed model.

According to our experiments, the results certify that our approach is excellent for scene classification. After the comparison with other models, higher mean precision and recall of our method is achieved, although they are not the best. What's more, the scalability and robustness of our model are also excellent. Our future works are improving the method we proposed to enhance the efficiency and conducting more experiments with different parameters to increase our method's performance.

Acknowledgement. This work is supported by the National Natural Science Foundation of China (Nos. 61165009, 61262005, 61363035, 61365009), the National Basic Research Program of China (No. 2012CB326403), the Guangxi Natural Science Foundation (Nos. 2012GXNSFAA053219, 2013GXNSFAA019345) and the "Bagui Scholar" Project Special Funds.

References

1. Biederman, I.: Aspects and extension of a theory of human image understanding. In: Computational Processes in Human Vision: an Inter-Disciplinary Perspective, New (1988)
2. Bosch, A., Zisserman, A., Muñoz, X.: Scene classification via pLSA. In: Leonardis, A., Bischof, H., Pinz, A. (eds.) ECCV 2006. LNCS, vol. 3954, pp. 517–530. Springer, Heidelberg (2006)
3. Bosch, A., Muñoz, X., Martí, R.: Which is the best way to organize/classify images by content? Image and Vision Computing 25(6), 778–791 (2007)
4. Bosch, A., Zisserman, A., Muoz, X.: Scene classification using a hybrid generative/ discriminative approach. IEEE Trans. Pattern Analysis and Machine Intelligence 30(4), 712–727 (2008)
5. Chen, S., Tian, Y.: Evaluating effectiveness of latent dirichlet allocation model for scene classification. In. In: 2011 20th Annual Wireless and Optical Communications Conference (WOCC), pp. 1–6. IEEE (2011)
6. Hofmann, T.: Probabilistic latent semantic indexing. In: Proc. 22nd Annual Int'l. ACM SIGIR Conf. on Research and Development in Information Retrieval, pp. 50–57 (1999)
7. Hofmann, T.: Unsupervised learning by probabilistic latent semantic analysis. Machine Learning 42(1-2), 177–196 (2001)
8. Iivarinen, J., Visa, A.: Shape recognition of irregular objects. . *In: Intelligent Robots and Computer Vision XV: Algorithms, Techniques, Active Vision, and Materials Handling, pp. 25–32. SPIE (1996)*
9. Shimazaki, K., Nagao, T.: Scene classification using color and structure-based features. In: Sixth International Workshop on Computational Intelligence and Applications (IWCIA), pp. 211–216. IEEE (2013)
10. Fei-Fei Li, P.: Perona. 2005. A bayesian hierarchical model for learning natural scene categories. In: IEEE Computer Society Conf. on Computer Vision and Pattern Recognition, vol. 2, pp. 524–531 (2005)
11. Oliva, A., Torralba, A.: Modeling the shape of the scene: A holistic representation of the spatial envelope. International Journal of Computer Vision 42(3), 145–175 (2001)
12. Quelhas, P., Monay, F., Odobez, J.M., et al.: Modeling scenes with local descriptors and latent aspects. In: ICCV, vol. 1, pp. 883–890 (2005)
13. Rasiwasia, N., Vasconcelos, N.: Latent Dirichlet Allocation Models for Image Classification. IEEE Trans. PAMI 35(11), 2665–2679 (2013)
14. Sun, J., Xu, H.: Contour-Shape recognition and retrival based on chain code. In: Proc. Computational Intelligence and Security, CIS 2009, vol. 1, pp. 349–352 (2009)
15. Schölkopf, B., Smola, A.J.: Learning with kernels: Support vector machines, regularization, optimization, and beyond. MIT Press (2002)
16. Wang, X.L., Xie, K.L.: A novel direction chain code-based image retrieval. In: Proceedings of the Fourth International Conference on Computer and Information Technology, CIT 2004, pp. 190–193 (2004)
17. Zhang, H., Berg, A.C., Maire, M., Malik, J.: SVM-KNN: Discriminative nearest neighbor classification for visual category recognition. In: 2006 IEEE Computer Society Conference on Computer Vision and Pattern Recognition, vol. 2, pp. 2126–2136 (2006)

Multipath Convolutional-Recursive Neural Networks
for Object Recognition

Xiangyang Li[1,2], Shuqiang Jiang[2], Xinhang Song[2], Luis Herranz[2], and Zhiping Shi[1]

[1] Capital Normal University, College of Information Engineering, Beijing, China
{lixiangyang806,shizhiping}@gmail.com
[2] Key Lab of Intelligent Information Processing, Institute of Computing Tech., Beijing, China
sqjiang@ict.ac.cn, {xinhang.son,luis.herranz}@vipl.ict.ac.cn

Abstract. Extracting good representations from images is essential for many computer vision tasks. While progress in deep learning shows the importance of learning hierarchical features, it is also important to learn features through multiple paths. This paper presents Multipath Convolutional-Recursive Neural Networks(M-CRNNs), a novel scheme which aims to learn image features from multiple paths using models based on combination of convolutional and recursive neural networks (CNNs and RNNs). CNNs learn low-level features, and RNNs, whose inputs are the outputs of the CNNs, learn the efficient high-level features. The final features of an image are the combination of the features from all the paths. The result shows that the features learned from M-CRNNs are a highly discriminative image representation that increases the precision in object recognition.

Keywords: Multiple paths, convolutional neural networks, recursive neural networks, classification.

1 Introduction

Visual recognition is a major focus of research in computer vision and machine learning. In the past few years, many works have been done on visual descriptors [1-4]. These methods have shown robustness against visual complexities, such as changes in scale, illumination, affine distortions, and pose variations [5]. For example the SIFT operator [1], which can be understood and generalized as a way to go from pixels to patch descriptors, applies oriented edge filters to small paths and determines the dominant orientation through a winner-take-all operation. Finally, the resulting sparse vectors are added over large patches to form local orientation histograms. Designing algorithms to get meaningful features is important but requires deep domain knowledge, and it is often difficult to expand the scope and ease of its applicability.

To overcome this weakness of feature engineering, feature learning learns transformations of the data that make it easier to extract useful information when building classifiers or other predictors [6]. There are a variety of works about feature learning, such as hierarchical sparse coding [7], deep autoencoders [8], convolutional deep

Z. Shi et al. (Eds.): IIP 2014, IFIP AICT 432, pp. 269–277, 2014.
© IFIP International Federation for Information Processing 2014

belief networks [9], and convolutional neural networks [10]. These approaches learn image features from raw pixels through multiple feature transforms layers.

Recent work on convolutional-recursive deep learning [11] is very efficient. Combining CNNs to extract low-level features from RGB and depth image, and RNNs to map the learned features into a lower dimensional feature space, the result of the model outperforms a lot of designed features and algorithms. Compared to standard feed forward neural networks with layers of similar size, CNNs have much fewer connections and parameters, thus being easier to train, while their theoretically-best performance is likely to be the same [12]. A single convolutional neural network layer provides useful translational invariance of low level features such as edges and allows part of an object to be deformable to some extent. Recursive neural networks, combine convolution and pooling into one efficient and hierarchical operation, project inputs into a lower dimensional space through multiple layers.

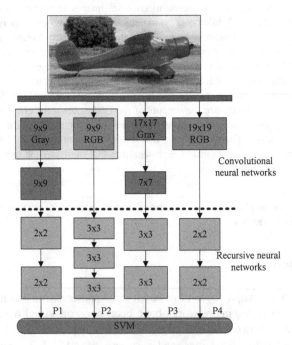

Fig. 1. Architecture of M-CRNNs

Concerned about the multi-facet nature of visual structures, Bo et al. [13] proposed multipath hierarchical matching pursuit. Discriminative structures, which we want to extract in feature learning procedures, may appear at varying scales with varying amounts of spatial and appearance invariance. Multipath hierarchical matching pursuits capture the heterogeneity, and build it into the learning architecture.

Motivated by this, and using the simplicity and high efficiency of convolution-recursive neural networks (CRNNs), we propose multipath convolutional-recursive neural networks (M-CRNNs). The overview of this framework is illustrated in Fig. 1.

A single path M-CRNNs is built upon a single-path CRNNs and learns features through many pathways on multiple bags of patches of varying size, by encoding each patch through multiple paths with a varying number of layers. The M-CRNNs architecture is important as it significantly and efficiently expands the richness of the image representation and contributes to improvement to image classification.

The remainder of this paper is organized as follows. In section 2, we propose M-CRNNs model and present learning procedures in detail. In section 3, experiment results are presented and discussed. Finally we conclude the paper in section 4.

2   Multipath Convolutional-Recursive Neural Networks

In this section, we describe our multipath convolutional-recursive neural networks. We first learn CNNs filters and then feed image patches into CNNs layers. The invariant features extracted from CNNs are given to recursive neural networks. RNNs learn higher order features that represent the image. All features from different paths are concatenated to form the final features that can then be used to classify images.

2.1   Convolutional Neural Networks

The main idea of CNNs is to convolve filters over the input image to extract distinctive features. We convolve an image of size (height and width) d_L with N square filters of size d_P, resulting in N filter responses, each of dimensionality $d_L - d_P + 1$. We then pool them with square regions of size d_l and a stride size of s, to obtain a pooled response with width and height equal to $r = (d_L - d_P + 1 - d_l)/s + 1$. So the output Y of the CNNs layer applied to one image is a $N \times r \times r$ dimensional 3D matrix. We apply this same procedure to both RGB images and grayscale images separately.

The structure of our layer of CNNs is similar to the one proposed by Jarrett et al. [14]. Different systems use different strategies to construct filters to extract features from images. For single layer models, filters sometimes are hard-wired, such as Gabor Wavelets [15]. For multi-stage vision systems, there is no prior knowledge that would conduct us to design sensitive filters for the second or higher layers. Hence filters are learned using supervised or unsupervised methods in many models. We use Predictive Sparse Decomposition (PSD) [16] and k-means to learn the filters.

A single stage of CNNs model includes four layers: filter bank layer, rectification layer, local contrast normalization layer and average pooling and subsampling layer. In the filter bank layer, the input is a 3D array with n_1 2D feature maps of size $n_2 \times n_3$. Each element is denoted as x_{ijt}, and each map is denoted as x_i. The filter bank has N filters. Each filter k_{ij} has a size of $l_1 \times l_2$, and transforms the input feature map x_i into the output feature map y_i. The module computes:

$$y_j = g_j \cdot f(\sum_i k_{ij} \otimes x_i) \qquad (1)$$

where f is a nonlinear function such as tanh, $\otimes$ is the 2D discrete convolutional operator and g_j is a trainable scalar coefficient. The output of the filter bank layer is a 3D array y composed of N feature maps of size $m_1 \times m_2$. We have $m_1 = n_2 - l_1 + 1, m_2 = n_3 - l_2 + 1$. The rectification layer simply applies the function: $y_{ijk} = | x_{ijk} |$. The third layer performs local subtractive and divisive normalizations, enforcing a sort of local competition between adjacent features in a feature map, and between features at the same spatial location in different maps. The objective of average pooling and subsampling layer is to build robustness to small distortions, which plays the same role as the complex cells. Output value is $y_{ijk} = \sum_{pq} w_{pq} \cdot x_{i,j+p,k+q}$, where w_{pq} is uniform weight window ("boxcar filter"). Each output is then subsampled spatially by size S horizontally and vertically.

For multi-layer CNNs models, the higher layer feature extractor is fed with the results of its prior layer. The feature extraction procedure is the same as in the single-layer model.

2.2   Recursive Neural Networks

Recursive neural networks have been used in natural language parsing [18]. The main idea of RNNs is to learn hierarchical feature representations by repeatedly applying the same neural network recursively in a tree structure.

Unlike previous works, the structure of the RNNs we use allows each layer to merge blocks of adjacent vectors instead of only pairs of vectors. Each image is denoted as a 3D matrix $X \in \mathbb{R}^{K \times r \times r}$, and the columns are K-dimensional. We define a block to be a list of adjacent column vectors which are merged into a parent vector $p \in \mathbb{R}^K$. In our work, we only consider square blocks. Blocks are of size $K \times w \times w$. For example, if we merge vectors in a block with $w = 5$, the input of one merging procedure is a structure of total size $K \times 5 \times 5$. In general, we have w^2 vectors in each block. The neural network for computing the parent vector is:

$$p = f\left(W \begin{bmatrix} x_1 \\ \cdot \\ \cdot \\ \cdot \\ x_{w^2} \end{bmatrix} \right) \tag{2}$$

where the parameter matrix $W \in \mathbb{R}^{K \times b^2 K}$, f is a nonlinear function such as tanh. Eq. (2) will be applied to all blocks of vectors in X with the same weights W. The structure of recursive neural networks is illustrated in Fig. 2. In our work N RNNs are used, each of which output a k-dimensional vector. So the output of RNNs is a NK-dimensional vector.

2.3 Multipath Convolutional-Recursive Neural Networks

In many visual recognition models, images were regarded as unordered collections of small patches, such as in the bag-of-features model. These traditional models do not consider the spatial positions of patches and relationships between patches, which are useful for visual recognition. The spatial pyramid pooling strategy overcomes this drawback by organizing patches into spatial cells at multiple levels and then concatenating them to a long feature vector [19]. The compelling advantages of spatial pyramid model are:

— The pyramid model generates a larger number of features and decreases the chance of overfitting.
— Adds local invariance and stability of the learned features using spatial pooling.
— It can capture invariant structures at multiple levels, because of different pooling sizes in the pyramid.

A single path convolutional-recursive neural network has the first two advantages. To combine the advantage of multiple patch sizes and the strength of hierarchical architectures, our multipath convolutional-recursive neural networks learn image features in multiple pathways, varying patch sizes and number of layers. Different paths with different receptive fields can capture different features. This combination of layers contains more information.

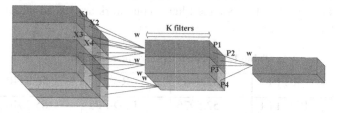

Fig. 2. Recursive neural network

The overview framework can be seen in Figure 1. Images of different size are put into CNNs to get low-level features. Each path corresponds to a specific patch size and number of layers. The final layer of each path is a feature vector for the whole image. All the features are then concatenated and used by a svm classifier for object recognition. The inputs of P1 and P3 are grayscale images, and the inputs of P2 and P4 are RGB images, so the features we finally get not only contain gray information, but also contain color information.

3 Experimental Results

In this section, we report the result of M-CRNNs for object category recognition. All our experiments are carried out on Caltech-101. The dataset contains 9,144 images of 101 object categories and one background category. Following the standard experiment setting, we trained models on 30 images and tested on no more than 50 images per category.

For the grayscale path, we first pre-process images with a procedure similar to [20]. First, we convert the image to grayscale and resize it with its largest dimension set to 151 pixels, while preserving its aspect ratio. Second, we subtract its mean deviation. Third, we apply divisive normalization. Finally, the image is zero-padded to 143×143 pixel. For the RGB path, the image was resized to 143×143.

3.1   Random Filters and Unsupervised Learned Filters

While convolutional pooling architectures can be inherently frequency selective and translation invariant, even with random weights [17], the filters in our model can be set to random values and kept fixed. In our framework, the filters convoluted with the RGB images are learned by k-means, and the ones convoluted with the grayscale images are learned by PSD. We compared the performance of learned filters and random-fixed filters in Table 1.

We estimate the accuracies of random filters and filters learned by PSD on the model. PP1 denotes: one-layer CNNs with 256 filters of size 9×9, 10×10 boxcar, 5 ×5 down-sampling (ds), and three-layer RNNs. PP2 denotes two-layer CNNs, the same configuration as P3 (as described in 3.2). Prefix R denotes the filters are randomly initialized, and PSD means the filters are learned in an unsupervised way using of PSD. As we can see in Table 1, for one-layer CNNs, the performance of the leaned filters and the randomly initialed one are nearly the same, but in two-layer CNNs, the learned ones are much better. So, the filters in our work are leaned through unsupervised training.

Table 1. Random filters and filters learned

R-PP1	52.6995	R-PP2	56.9440
PSD-PP1	52.5976	PSD-PP2	65.7046

3.2   Architecture of M-RCNNs

We outline the four paths used in our experiment with multipath convolutional-recursive neural networks in our experiments.

Path one (P1) is done on gray images. First images are pre-processed, two stages of CNNs and two stages of RNNs are performed to extract the features. The first CNN stage has 64 filters of size 9×9, 10×10 boxcar filter, and 5×5 ds. The second CNN stage is composed of 256 output feature maps, each of which combines a random subset of 16 feature maps from the first one. So the total number of convolutional filters is 256×16=4096. The average pooling model uses 6x6 boxcars filter with a 4×4 downsampling step. The output of a 3D matrix of size 256×4×4 is fed to a two- layer CNN with 128 RNNs. The final output of P1 is a 256×128 vector.

Path two (P2) is done on RGB images. First, 400000 random patches of size 9×9 ×3 are extracted. Then the patches are normalized and whitened. Pre-processed patches are clustered into 256 centers by running k-means. The boxcar filter and ds of the one layer CNNs are both 5×5. The output feature of size 256×27×27 is fed to a three layers RNNs. The size of the filters is 9×9×3 pixels (see Fig. 3).

Path three (P3) has two layers CNNs and three layers RNNs. The first stage of CNNs is with 64 filters of size 17×17, 7×7 boxcar and 5×5 ds. The second stage of CNNs is with 256×16=4096 filters of size 7×7, 3×3 boxcar and 2×2 ds.

Path four (P4) has one stage CNNs and two stages of RNNs. The size of filters is 19×19×3.

The filters in our architecture are shown in Figure 3. Left are the 64 filters of P1 in the first CNNs layers. In the middle, there are 64 filters in path P3 of the first CNNs layers. Right are the 256 filters in P2.

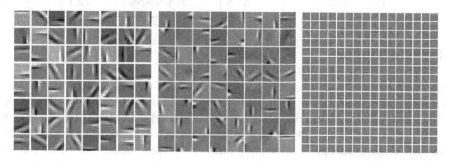

Fig. 3. Visualization of filters

The final features of an image are the concatenated features extracted from all the paths. The accuracy of different paths is shown in Table 2. Generally speaking, P1 works well for object classification due to its hierarchical architecture. Compared with the accuracy 65.5 of [14], the architecture of P1 illustrates that CRNNs works slightly better than single CNNs. Different paths can capture visual invariant features at different scales. By concatenating the entire paths, the combined features have higher accuracy than each single path.

Table 2. Accuracy of each path

P1	66.9610	P4	56.9779
P2	56.1290	P2+P3	71.4092
P3	65.7046	P1+P2+P3+P4	72.1902

3.3   Comparison to Other Methods

In this section we compare our model to related models in literature in Table 3. SVM-KNN [19] is a hybrid of SVM and NN which focus on shape and texture. Soft threshold coding (SIFT+T) [21] and spatial pyramid matching [22] approaches based on SIFT features. Invariant predictive sparse decomposition (ISPD) [23], convolutional deep belief networks (CDBN) [9], convolutional neural networks (CNNs) [14], and deconvolutional networks (DN) [24] are approaches of hierarchical feature learning. Our method outperforms many SIFT-based approaches as well as hierarchical feature learning methods.

Table 3. Classification accuracy on Caltech-101

SIFT+T	67.7	CDBN	65.4
SPM	64.4	CNNs	65.5
SVM-KNN	66.2	DN	66.9
IPSD	65.5	M-CRNNs	72.1

4   Conclusion

We have proposed multiple path convolutional-recursive neural networks to learn meaning full representations from raw images. The proposed method combines convolutional neural networks and recursive neural networks, and learns the filters in an unsupervised way, which allows for parallelization and high speeds. Learning features through several pathways with varying number of layers, makes the visual features capture invariances at different scales. The architecture feeds grayscale patches and RGB patches to the CNNs, so it contains gray information and color information. Our future work will aim at the combination of multiple paths and deep convolutional neural networks.

Acknowledgements. This work was supported in part by National Natural Science Foundation of China: 61322212 and 61350110237, in part by the Chinese Academy of Sciences Fellowships for Young International Scientists: 2011Y1GB05, in part by the Key Technologies R&D Program of China under Grant no. 2012BAH18B02.

References

1. Lowe, D.: Distinctive image features from scale-invariant keypoints. International Journal of Computer Vision 60(2), 91–110 (2004)
2. Dalal, N., Triggs, B.: Histograms of oriented gradients for human detection. In: CVPR, San Diego (2005)
3. Wang, J., Yang, J., Yu, K., Lv, F., Huang, T., Gong, Y.: Locality-constrained linear coding for image classification. In: CVPR, San Francisco (2010)
4. Bo, L., Ren, X., Fox, D.: Kernel descriptors for visual recognition. In: NIPS, Vancouver (2010)
5. Lobel, H., Vidal, R., Soto, A.: Hierarchical joint Max-Margin learning of mid and top level representations for visual recognition. In: ICCV, Sydney (2013)
6. Bengio, Y., Courville, A., Vincent, P.: Representation learning: a review and new perspectives. IEEE Transaction on Pattern Analysis and Machine Intelligence 35(8), 1798–1828 (2013)

7. Yu, K., Lin, Y., Lafferty, J.: Learning image representations from the pixel level via hierarchical sparse coding. In: CVPR, Colorado Springs (2011)
8. Le, Q., Ranzato, M., Monga, R., Devin, M., Chan, K., Gorrado, G., Dean, J., Ng, A.: Building high-level features using large scale unsupervised learning. In: ICML, Scotland (2012)
9. Lee, H., Grosse, R., Ranganath, R., Ng, A.: Convolutional deep belief networks for scalable unsupervised learning of hierarchical representations. In: ICML, Montreal (2009)
10. Lawrence, S., Giles, C., Tsoi, A., Back, D.: Face recognition: a convolutional neural-network approach. IEEE Transaction on Neural Networks 8(1), 98–113 (1997)
11. Socher, R., Huval, B., Bhat, B., Manning, D., Ng, A.: Convolutional-recursive deep learning for 3D object classification. In: NIPS, Nevada (2012)
12. Krizhevsky, A., Sutskever, I., Hinton, G.: ImageNet classification with deep convolutional neural networks. In: NIPS, Nevada (2012)
13. Bo, L., Ren, X., Fox, D.: Multipath sparse coding using hierarchical matching pursuit. In: CVPR, Portland (2013)
14. Jarrett, K., Kavukcuoglu, K., Ranzato, M., LeCun, Y.: What is the best multi-stage architecture for object recognition. In: ICCV, Xi'an (2009)
15. Serre, T., Wolf, L., Poggio, T.: Object recognition with features Inspired by visual cortex. In: CVPR, San Diego (2005)
16. Kavukcuoglu, K., Ranzato, M., LeCun, Y.: Fast inference in sparse coding algorithm with applications to object recognition. Technical report, Computational and Biological Learning Lab, Courant Institute, NYU (2008)
17. Saxe, A., Koh, P., Chen, Z., Bhand, M., Suresh, B., Ng, A.: On random weights and unsupervised feature learning. In: ICML, Washington (2011)
18. Socher, R., Maning, C., Ng, A.: Learning continuous phrase representation and syntactic parsing with recursive neural networks. In: NIPS, Vancouver (2010)
19. Lazebnik, S., Schmid, C., Ponce, J.: Beyond bags of features: spatial pyramid matching for recognizing natural scene categories. In: CVPR, New York (2006)
20. Pinto, N., Cox, D., DiCarlo, J.: Why is real-world visual object recognition hard. PLOS Computational Biology 4(1) (2008)
21. Coates, A., Ng, A.: The importance of encoding versus training with sparse coding and vector quantization. In: ICML, Washington (2011)
22. Zhang, H., Berg, A., MaireM.,Malik, J.: SVM-KNN: discriminative nearest classification for visual category recognition. In: CVPR, New York (2006)
23. Kavukcuoglu, K., Ranzato, M., Fergus, R., LeCun, Y.: Learning invariant features through topographic filter maps. In: CVPR, Florida (2009)
24. Zeiler, M., Krishnan, D., Taylor, G., Fergus, R.: Deconvolutional networks. In: CVPR, San Francisco (2010)

Identification of Co-regulated Gene Network by Using Path Consistency Algorithm Based on Gene Ontology

Hanshi Wang[1], Chenxiao Wang[1], Lizhen Liu[1], Chao Du[1], and Jingli Lu[2]

[1] Information and Engineering College, Capital Normal University, Beijing, China
necrostone@sina.com, {wangchenxiao0329,liz_liu,dc0213}@126.com
[2] Agresearch Ltd, Hamilton, New Zealand
Janny.jingll@gmail.com

Abstract. Recently, the reconstruction of co-regulated gene network has become increasingly popular in the area of bioinformatics. It tries to find the associated information and network topology among genes through large numbers of biological data. In this paper, we proposed a novel method PC-GO by using path consistency (PC) algorithm based on gene ontology (GO). GO provides a basis for measuring the conditional semantic similarity between genes, and then PC algorithm is applied to remove links between genes with less correlation in the network. We successfully applied our algorithm to yeast data. Experimental results show that the accuracy and integrity of the co-regulated network acquired by our method outperforms previous methods.

Keywords: Gene ontology, conditional semantic similarity, path consistency algorithm, co-regulated gene network.

1   Introduction

High-throughput techniques have produced vast amounts of sequence, expression and structure data [1]. As the data source of bioinformatics, gene expression data is now in an increasingly important position. Increasing evidences suggest that interactions between genes have an impact on the regulation of gene expression [2]. Currently, many scholars are committed to find the association of genes [3-5] since during getting the associated information between genes, one of the major challenges is the identification of co-regulated gene network. Gene co-regulated network, which aims to find the associated information among genes through large numbers of biological data expressed by genes and visualize the network topology representing gene interactions, as well as reveals the complex reaction mechanism among genes, is regarded as one of the most important objectives in the field of bioinformatics.

Numerous analytical methods have been developed to identify gene co-regulated network from gene expression profiles [6]. Genes that are part of the same operon in prokaryotes, or have the same expression pattern in eukaryotes, are co-regulated transcriptionally [7]. Generally speaking, we considered a network of genes co-regulated if the percentage that has one or more common TFs is above 80%. Researches on the regulation relationship and regulatory mechanism provide the opportunity for

Z. Shi et al. (Eds.): IIP 2014, IFIP AICT 432, pp. 278–283, 2014.

understanding the underlying and predicting genes function, which can help to systematically characterize the process of life activities.

Gene Ontology (GO) [8] is a standard vocabulary of functional terms and allows for coherent annotation of gene products. These annotations provide a basis for new methods to compare the similarity of genes and gene products regarding their molecular function and biological role. The semantically similarity of annotation information on genes can be an evidence for functionally similarity of genes. Meanwhile, gene products that participate in the same biochemical reaction, have similar biological functions [9]. Constructing a gene co-regulated network automatically based on GO is still a big challenge at present.

In this paper, the Yeast dataset was used as the test data. We present a new PC algorithm using the CSS based on GO. It consists of two parts: the first part is to calculate the CSS between genes based on GO, and the second is to remove links between genes based on pair-wise similarity by using PC algorithm. Experimental results show that our method has improved the accuracy and integrity of the co-regulated network.

2   Method

In this section, we will introduce the CSS calculation method, as well as the PC-GO method for identifying co-regulated gene network.

2.1   Conditional Semantic Similarity (CSS) of Genes

The function similarity between genes can be determined by comparing the semantic similarity. To improve the accuracy of the semantic similarity of genes, we must consider the semantic similarity of GOs annotating genes, and the key is to measure the semantic similarity of GO terms. This is achieved by considering the Wang's method [10] of computing semantic similarity between gene pairs. Based on Wang's method, we proposed a new concept – conditional semantic similarity (CSS), which can be applied to the PC algorithm.

Since the semantic of a GO term are determined by its location in the GO graph and semantic relations with all of its ancestor terms, the directed acyclic graph (DAG) starting from the specific GO term and ending at the root GO term (a sub-DAG of an ontology) is used to show all the relationship of this specific GO term in the ontology. Formally, a GO term A can be represented as $DAG_A = (A, T_A, E_A)$, where T_A is the GO terms set in DAG_A, including term A and all of its ancestor terms in sub-DAG, and E_A is the set of edges (semantic relations) connecting the GO terms in sub-DAG. The semantic value of GO term A is defined as the aggregate contribution of all terms in sub-DAG. The contribution of a GO term (including A) to the semantic meaning of A is defined as S-value. S-value is calculated by:

$$\begin{cases} S_A(A) = 1 \\ S_A(t) = \max\{\omega_e * S_A(t') \mid t' \in childrenof(t)\} \qquad t \neq A \end{cases} \qquad (1)$$

where ω_e is the semantic contribution factor of two edge types of semantic relationship – "is a" and "part of". Through large numbers of repeated experiments, Wang obtained that the semantic contribution factors for "is - a" and "part - of" are 0.8 and 0.6, respectively.

Given $DAG_A = (A, T_A, E_A)$ and $DAG_B = (B, T_B, E_B)$ for GO terms A and B respectively, the semantic similarity, $S_{GO}(A,B)$, is defined as

$$S_{GO}(A,B) = \frac{\sum_{t \in T_A \cap T_B} S_A(t) + S_B(t)}{SV(A) + SV(B)} \qquad (2)$$

where t is the intersection GO term of sub-DAG of A and B, $SV(A)$ and $SV(B)$ are the semantic value of GO term A and B. The semantic value of GO term M is calculated by $SV(M) = \sum_{t \in T_M} S_M(t)$.

As each gene is annotated by GO terms, the semantic similarity between two genes A and B can be represented by the semantic similarity of two GO terms sets, which annotated the corresponding gene. Assuming $GO_1 = \{go_{11}, go_{12}, \cdots, go_{1m}\}$ and $GO_2 = \{go_{21}, go_{22}, \cdots, go_{2n}\}$ are two GO terms sets that annotate genes G_1 and G_2 respectively, their semantic similarity is as follows:

$$Sim(G_1, G_2) = \frac{\sum_{1 \le i \le m} Sim(go_{1i}, GO_2) + \sum_{1 \le j \le n} Sim(go_{2j}, GO_1)}{m + n} \qquad (3)$$

In this formula, $Sim(go, GO) = \max_{1 \le i \le k} S_{GO}(go, go_i)$, that is, $Sim(go, GO)$ is defined as the maximum semantic similarity between term go and any of the k terms in set GO.

In this work, we proposed a new concept - conditional semantic similarity (CSS), which indicates the similarity of two genes given the state of another gene. CSS between three genes can be written as:

$$Sim(G_1, G_2 \mid G_3) = \frac{\mid Sim(G_1, G_2) - Sim(G_1, G_3) * Sim(G_2, G_3) \mid}{\sqrt{1 - Sim^2(G_1, G_3)} * \sqrt{1 - Sim^2(G_2, G_3)}} \qquad (4)$$

2.2   The PC-GO Method

After we obtain the semantic similarity and CSS through formulation (3) and (4), the PC algorithm is used to remove the edges, which the semantic similarity or CSS value is smaller than the threshold θ given in advance. The θ value is experimentally set 0.8 for the reason that genes in a co-regulated network are highly functionally related.

The process of PC-GO begins with a complete graph and attempts to remove as many links as possible. First, for adjacent gene pair i and j, compute the semantic similarity $Sim(i, j)$ (0-order). If the gene pair i and j has a lower semantic similarity, it presents functionally irrelevant, then we delete the edge between genes i and j. Second, for adjacent gene pair i and j, select the adjacent gene k of them and compute

CSS value $Sim(i, j \mid k)$ (1-order). If the gene pair i and j has a low CSS, delete the edge between them. The next step is to compute higher order CSS until there are no more adjacent edges. At each round, the number of neighbors in the conditional set increases one by one. A conceptual representation of this approach is presented in Figure 1. In this figure, $Sim(\cdot, \cdot)$ is the semantic similarity and $Sim(\cdot, \cdot \mid \cdot)$ is the CSS. The semantic similarity and CSS lower than the given threshold represent functionally irrelevant between genes.

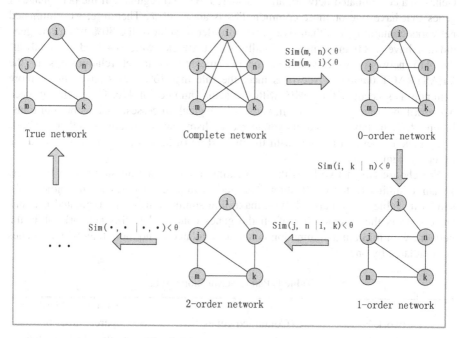

Fig. 1. Diagram of PC-GO method

3   Evaluation

Yeast Biochemical Pathway Database (YeastCyc) was used as the test data set. The YeastCyc Biochemical is a collection of manually curated metabolic pathway and enzymes of Saccharomyces cerevisiae. There are 217 metabolic pathways in the YeastCyc database. A total of 6381 yeast genes got from 217 pathways were selected for our investigation and GO terms search for all 6381 genes were performed based on the April 2014 releases of GO terms and gene annotations for Saccharomyces cerevisiae from the Gene Ontology Consortium[8]. We calculated the semantic similarities and CSS of gene pairs, and then partitioned genes into functionally related co-regulated networks.

Genes that share the same TF(s) are co-regulated, which can be a criterion to estimate whether genes are in a co-regulated network. The YEASTRACT database [11] (Yeast Search for Transcription Regulators And Consensus Tracking) is a curated repository of more than 106000 regulatory associations between TFs and target genes in Saccharomyces cerevisiae, which can be used to search for common TFs of genes in each network. In the YEASTRACT database, we used the "Rank by TF" function and considered only expression evidence in the "documented" regulation to obtain the percentage of the genes in each network that are commonly regulated by one or more known TFs.

For each gene network, the summation of all the semantic similarity was calculated. Genes in a co-regulated network are considered to be co-regulated if the percentage of genes that have one or more common TFs is above 80%. Therefore, we eliminated networks containing less than five genes in order to achieve the 80% when one gene did not have a TF in common. Finally, 623 networks were obtained for analysis. All 415 networks were searched for documented TF-target relationships in the YEASTRACT database. Networks meet the rule that 80% genes have one or more common TFs account for 76.4% with the actual number of 476. Genes that in each network functionally participate in various biological processes, such as protein translation, RNA metabolism, carbohydrate metabolism, etc. Compared to other methods using gene expression profiles with the proportion of 52%, our method improved the accuracy significantly.

We choose two networks with the maximum and minimum summation if all the semantic similarity for verification. The verification result is shown in Table 1. As shown in Table 1, the network with maximum summation is a co-regulated network since Cst6 is the common TF of all the genes contained in this network while the network with minimum summation is not co-regulated as no more than 63.64% genes are regulated by one TF.

Table 1. Part of verification results

Network	Contained genes	Common TFs and percentages of commonly regulated	
Network with maximum summation	TFC3, EFB1, YAL004W, SPO7, CYS3, YAL018C, MAK16, POP5, YAL037W, ERV46, YAR023C, YAR066W, YAR075W, YBL077W	Cst6	100%
		Ace2	78.57%
		Spt20	71.43%
		Msn2	71.43%
		Sfp1	71.43%
		Gcr1	57.14%
		Snf2	50%
Network with minimum summation	ECM4, CSN9, ARI1, YAR1, ATG15, SAW1, GAL7, RPB9, PIIM7, PCA1, COA4	Msn2	63.64%
		Msn4	63.64%
		Yap1	54.55%
		Bas1	54.55%
		Hsf1	54.55%
		Yhp1	45.45%
		Snf2	45.45%

4   Conclusions

We early proposed a novel PC-GO method by using path consistency algorithm based on gene ontology to identify gene co-regulated network, which provides a new way to the bioinformatics research. The effectiveness of the proposed approach has been experimentally demonstrated through its application to a well-researched yeast dataset. One of the strengths of our approach is that the prediction of co-regulated gene group does not require the availability of gene expression profiles. However, the result generated by the gene conditional semantic similarity calculation method may have an impact on the accuracy of PC algorithm, which require us to improve the accuracy of each procedure in the future research work.

Acknowledgements. This work was supported in part by National Science Foundation of China under Grants No. 61303105; the Humanity & Social Science general project of Ministry of Education under Grants No.14YJAZH046; the Beijing Educational Committee Science and Technology Development Planned under Grants No.KM201410028017; Academic Degree Graduate Courses group projects and the Beijing Key Disciplines of Computer Application Technology.

References

1. Obayashi, T., et al.: ATTED-II: a database of co-expressed genes and cis elements for identifying co-regulated gene groups in Arabidopsis. Nucleic Acids Research 35(suppl. 1), D863–D869 (2007)
2. Lanctôt, C., et al.: Dynamic genome architecture in the nuclear space: regulation of gene expression in three dimensions. Nature Reviews Genetics 8(2), 104–115 (2007)
3. Korbel, J.O., et al.: Systematic association of genes to phenotypes by genome and literature mining. PLoS Biology 3(5), e134 (2005)
4. Lee, I., et al.: Rational association of genes with traits using a genome-scale gene network for Arabidopsis thaliana. Nature Biotechnology 28(2), 149–156 (2010)
5. Perez-Iratxeta, C., Bork, P., Andrade, M.A.: Association of genes to genetically inherited diseases using data mining. Nature Genetics 31(3), 316–319 (2002)
6. Berri, S., et al.: Characterization of WRKY co-regulatory networks in rice and Arabidopsis. BMC Plant Biology 9(1), 120 (2009)
7. Teichmann, S.A., Babu, M.M.: Conservation of gene co-regulation in prokaryotes and eukaryotes. Trends in Biotechnology 20(10), 407–410 (2002)
8. Ashburner, M., et al.: Gene Ontology: tool for the unification of biology. Nature Genetics 25(1), 25–29 (2000)
9. Wei, H., et al.: Transcriptional coordination of the metabolic network in Arabidopsis. Plant Physiology 142(2), 762–774 (2006)
10. Wang, J.Z., et al.: A new method to measure the semantic similarity of GO terms. Bioinformatics 23(10), 1274–1281 (2007)
11. Teixeira, M.C., et al.: The YEASTRACT database: a tool for the analysis of transcription regulatory associations in Saccharomyces cerevisiae. Nucleic Acids Research 34(suppl. 1), D446–D451 (2006)

Case Retrieval for Network Security Emergency Response Based on Description Logic

Fei Jiang, Tianlong Gu, Liang Chang, and Zhoubo Xu

Guangxi Key Laboratory of Trusted Software, Guilin University of Electronic Technology,
Guilin 541004, China
jiangfei0128@sina.com, {cctlgu,changl,xzbli_11}@guet.edu.cn

Abstract. Network security emergency response (NSER) is an important topic in information security. Nowadays, a large number of NSER systems and tools are developed, which can effectively detect part of security incidents and provide general best-practice guidelines for handling some type of security incidents, but not give a reasonable, fast, effective processing method for every security incidents in actual environment. An intelligent method based on case-based reasoning (CBR) and description logic (DL) is proposed for NSER. Firstly, a case base for NSER is organized in such a way that domain knowledge of NSER is described by the DL ALCO(**D**). Secondly, based on refinement operator and refinement graph in DLs, an algorithm for measuring the similarity of ALCO(**D**) concepts is designed and used for retrieving cases from the case base. It is demonstrated that our method can reuse past experiences on security incidents to generate response automatically.

Keywords: Emergency response, Network security incident, Case based reasoning, Description logic, Case retrieval.

1 Introduction

Network security emergency response (NSER) is a kind of service that helps to mitigate further damage when network security incident occurred, which has a positive role in protecting the security of enterprise and terminal, and it is also the centre of future information security policy. Since Cliff Stoll's the first book, the Cuckoo's Egg, introduces methods to hack the computer, as well as a large number of Computer Emergency Response Team Coordination Centers (CERT/CC) are established in the world, the ideas about NSER began to get attention. In recent years, Mitropoulos[1] et al. has theoretically investigated the research and application of NSER before 2006, and gave a reasonable security incident processing system framework. To provide data model for CERT and share the information of incidents and vulnerabilities related to the information exchange standard—IODEF[2] (Incident Object Description and Exchange Format) has been developed. In order to help to effectively handle security incidents, the literature [3] provides best-practice guidelines to detect,

Z. Shi et al. (Eds.): IIP 2014, IFIP AICT 432, pp. 284–293, 2014.

analyze and handle part of security incidents, and the brief steps to handle some different types of security incidents.

However, most literature are focused on "high impact" incidents (which have high impact on society and network security techniques) on the research of security incidents, and does not help to improve the overall level of NSER and to handle specific security incidents. So some scholars proposed using CBR[4] method for NSER. Considering new various types of security incident are ongoing, these incidents always have the similarity, and the similar incidents have similar incident response. By using the past similar cases to help to solve the current security incidents is found. Capuzzi et al. [5] combined with CBR develops a complete security tool based on ID/PS and which integrates the log Association, attack classification and response plan generation, but it ignores the fact that IDS with high false alarm rate, almost 98%, and it can be said that is unpractical. Considering the high false-positive rate of IDS, Kim et al. [6] proposes using RFM (Recency, Frequency, Monetary) method combined with the analysis of log file to directly detect abnormal incidents and to reduce the false-positive rate of IDS, and then combine with CBR to find the most similar case, but its primary intention is only to detect security incidents, rather than respond to incidents. In order to effectively respond to security incidents, the literature [7] proposes by using some meta knowledge in the NSER domain to help organize case base, and then implementing CBR reasoning to provide NSER, but its meta knowledge representation and retrieval algorithm for solving security incidents too rough to solve the incidents.

Considering the knowledge of case base with good structure, automatic classification concept, and the implementation of corresponding CBR reasoning for NSER, this paper introduces ALCO(**D**) logic (a form of DL) to describe the knowledge of NSER. DL is a form of knowledge representation, which has good semantic, expression and inference capability, and allow for an automatic classification of concepts. DL has been systematically researched for several decades, its application can be used very effective and fast. Making full use of the advantages of CBR and DL, this paper develops an intelligent method to help to solve the problem of NSER, and to prevent and handle network security incidents.

This paper is organized as follows: Section 2 using ALCO(**D**) logic represents the knowledge of NSER. Section 3 and 4, design a case retrieval method for retrieving the most similar incidents, and illustrates the given method and its validity. Finally, discuss the existing problems and future work.

2   Knowledge Representation of NSER

A case CA can be described in three tuples, i.e. $CA = (P, S, O)$, where P, S, and O are used to describe the problem or situation of network security incident, its solution or method of NSER, and the outcome obtained by the solution S to the given problem P, respectively. For a given case base CB, with P_i, S_i, and O_i respectively denoting the problem description, solution and outcome of a case CA_i, so that $CA_i \in CB$, $0 \leq i \leq n$, n is the number of cases in the CB.

A large variety of case representation formalisms have been proposed. Depending on different applications, case representation will be different, mainly includes feature vector representations, structured representations, and textual representations. NSER is a knowledge intensive domain, and this paper adopts the structured approach—DL to represent the problem(situation) of security incidents and its outcome for classifying and integrating the knowledge, and the text formalism to present its solution.

Problems (P_i): An incident is generally related to the time, location, executors, recipient, state and effects, and network security incident is also not exceptional. For describing security incidents, we also need to describe the information about the type, time, organization, information about potential attackers, affected resources and its information, effects, state and information about measures has been taken to solve incidents in a security incident. All of this information will help to describe security incident quite clearly.

Type: According to the literature [3], network security incidents can be divided into the following categories: malicious code incidents, denial of service(DoS) incidents, unauthorized access incidents, inappropriate usage incidents or a single incident that encompasses two or more incidents above. When security incidents occurred, the type of security incidents should be first considered in order to determine the most appropriate response strategies. A security incident is a DoS incident, inappropriate usage incident or other incidents, only by determining the type of security incident, its response strategy will become quite clear. DoS attack incident does not involve in actual invasion, so it is the easiest to respond and the most difficult to prevent. Inappropriate usage of resources is usually the insider using others computer in inappropriate ways, it usually needs to consider more the internal factors. Fig.1 shows a hierarchical relationship for different types of security incidents, including the hierarchical relationship between the virus, worm and DDoS incident. It also expresses an inclusion relationship between some concepts about security incidents by using ALCO(**D**) logic (e.g. *DoS_incident⊒DDoS_incident*). Besides the hierarchical relationship mentioned above，some incidents have their own unique characteristics, for example, virus incidents has the characteristics of propagation ways. These characteristics are also considered to better distinguish different incidents, and to provide more information for retrieving and proposing emergency repose strategy.

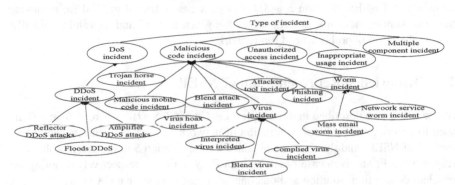

Fig. 1. Hierarchy relationship diagram of type of security incident

Time: With the progress of computer science and technology, security incidents is designed with more and more powerful function, its destructive force and influence are also increasing. After some new security incidents arise, security defense level for the security incidents will be improved. The old methods of attack in security incidents have no effect on new environment in the future. Considering the time factor, we can make it associated to the recent security incidents, facilitate to find more and better similar incident for the current incident. In addition, in the process of NSER, the longer a security incident lasts, the more potential there is for damage and loss. So we describe the time of a incident for NSER as:

$$Time \sqsupseteq \exists hasOccurrenceTime. = h \text{ year} \sqcap \exists hasDuration. = f \text{ hour}$$

Where h is a natural number, and f is a positive real number.

Organization: Security incidents often occurred in different personal hosts and organizations, whose response and focus will be different. For example, the response is different between commercial organizations and governmental agencies. The response to security incidents in financial institutions emphasizes the continuity of business and the pecuniary loss, and government emphasizes the publicity and the confidential data loss. In order to handle security incidents better, this paper considers the organization, i.e. governmental agencies, commercial organizations, military institutions, medical institutions, scientific and educational institutions, network service providers. By using ALCO(**D**) logic, it can be represented as:

$$Organization \sqsupseteq (Commerce \sqcup Education \sqcup ISP \sqcup Military \sqcup Govement)$$

Attacker's information in the incident: Information such as IP address, communication protocol or port which the attackers used, will help to NSER to track the attacker or close the channels which being attacked and further to block the attack. So the attacker's information can be described as:

$$Attacker\_Info \sqsupseteq \exists hasAttackerInfo. (IPAddress \sqcup Port \sqcup Protocol)$$

Affected resource: Different resources (firewalls, web servers, network connection, user workstations and applications, etc.) to organizations and individuals have different significance. The different resources which have different effects on the organizations and individuals, the priority to NSER will be different. The incidents which refer to criticality resources or resources which have great potential influence need to handle first. These resources includes: the compromised host, network and its services, network equipment, etc. In order to NSER, we need to know the situation (information) about the compromised host. The information includes the function of host, the number of hosts, the type of operating system, the anomalous phenomena of system, security tools, applications, services, and hardware in affected host. The anomalous phenomenon such as the host always pops up the suspicious content of message or window and suddenly slows down, the security tools (antivirus software, anti-spam software, etc.) warn that the system has viruses or abnormal attacks, the audit logs from the operating system, service or application was found with intrusion will help to analyze the security incident. Consider network and its service with different state such as network can't be connected, network traffic anomaly, NSER will be different.

By using ALCO(**D**) logic, some knowledge about the affected resource can be represented as follow:

Affected-Resource⊒(Affected-Host⊔Affected-Network⊔Affected-NetworkDevice).
Affected-Host⊒(∃hasOS.OS⊓∃hasHostFunction.Host_Function⊓∃hasNumber.=
m tai⊓∃hasAabnormal. (Host_System ⊔ Host_Application⊔Host_Hardware)).
OS⊒Windows⊔Linux⊔Unix⊔Android⊔iOS.   Host_Function⊒(Client⊔Server).
Server ⊒(FTP_server⊔Web_server⊔Data_server⊔Mail_server).
Host_Application⊒Security_Tool⊔IE.
Host_Hardware⊒Keyboard⊔Loudspeaker.   Host_System⊒Host_SystemService.

Effect: The security incidents may impact on victims and lead to some disruption and loss, including the loss of money, damage of reputation and data loss, leakage, and destruction. We also consider the factor for NSER and describe it as:

Effect⊒{Money_Loss, Publicity_Loss, Data_Loss}

State: Reponses-taken will be use to different security incidents under different conditions. Some attacks will have symptoms, before the damage occurs, the response we performed is to prevent, while some attack has occurred and destroyed the service, we must quickly mitigate the damage which cause by incident, deal with the security incidents, and restore the system. It can be represented as follows:

State⊒ {finished, ongoing, unhappened, unknown}.

Response-taken:When the security incidents occurred, organizations or individuals will handle it by disconnecting from the network or closing the infected host. It can be represented as follows:

Response-taken ⊒{Close_Host, Disconnect_Network}.

In order to handle security incidents better, all of the information about security incident should be retrieved or revised, and can transform into the data that computer can be identified. After there are all of the needful elements to depict the network security incidents mentioned above, this paper using ALCO(**D**) logic represents the problem (*P*) of a case.

Solution (*S*): Describe the whole process of NSER to deal with specific security incidents. The textual representation is mainly used to describe solution to the given problem (security incident).

Outcome (*O*): The results of NSER may be good or bad, how to measure it? The user's satisfaction is used to evaluate the result. The satisfaction is high, the solution is good, and can be adopted; the satisfaction is very low, the solution is not appropriate, and used to learn the lessons. The value of user's satisfaction we can evaluation by the mean value of acceptability, feasibility, flexibility, operability, integrity and consistency of the solution about incident after user use the solution and assess. The Outcome is also described by using the ALCO(**D**) logic.

The structure of case CA_i is shown as follow.

$CA_i = (P_i, \; S_i, \; O_i)$

$P_i = \exists hasType.Type \sqcap \exists hasOccurredTime.=n$ year$\sqcap \exists hasDuration.=f$ hour$\sqcap \exists hasOrg.$ Organization

  $\sqcap \exists hasAttackerInfo.$ Attacker_Info$\sqcap \exists hasAffectedResource.Affected$-Resource

  $\sqcap \exists hasEffect.Effect \sqcap \exists hasState.State \sqcap \exists hasResponse$-taken. Response-taken.

S_i : Omission

$O_i : \exists\ hasSatisfaction.Satisfaction$

Fig. 2. Structure of cases

As can be seen, ALCO(**D**) logic with the strong ability of description, have clear semantics for describing network security incidents, which can describe the internal structure of cases, depict more comprehensive knowledge, close to man's mind-set and expression powerfully. It will also benefit case retrieval and revise.

3   Case Retrieval for Network Security Incident

When new network security incidents occur, the system needs to retrieve the similar security incidents from case base. Case retrieval is a key stage in CBR design. In case retrieval, similarity is usually used. The more close to 1 it is, the higher degree of similarity between the two cases are. Case retrieval directly affected the relevancy of cases, and affected whether to generate the appropriate solution to problem or not. Depending on different application, about similarity, different case representation has different measuring methods. Cunningham[8] investigated the mainstream method of similarity measure for different applications in CBR areas. Sánchez-Ruizet[9,10] et al. put forward the similarity measure in the space of concepts and in the space of conjunctive queries between concepts and individuals about εL logic. According to the type of different attributes about disaster events such as the numerical, interval, character type in the field of disaster emergency, Amailef [11]et al. use the attributes of disaster emergency based ontology to define case structure, and give different similarity metrics.

Based on the previous research, this paper uses the ALCO (**D**) logic represent the case, this section will further give a similarity strategy based on refinement operator and refinement graph for network security incidents. Now the section briefly summarizes the notation for refinement operator and the relevant concepts for this paper. Refinement operators are defined over quasi-ordered sets. A quasi-ordered set is a pair $(S, \; \leq)$,where S is set, and $\leq$ is a binary relation among elements of S. If $a \leq b$ and $b \leq a$, we say that $a \approx b$. Refinement operator are defined as follows: A down refinement operator ρ over a quasi-ordered set $(S, \; \leq)$ is a function such that $\forall a \in S : \rho(a) \subseteq \{ b \in S | b \leq a\}$; A up refinement operator γ over a quasi-ordered set $(S, \; \leq)$ is a function such that $\forall a \in S : \gamma(a) \subseteq \{ b \in S | a \leq b\}$.Down operator refinement operators generate elements of which are smaller (which in this paper will mean "more specific"), in contrast, up operator refinement operators generate elements of which are bigger (which in this paper will mean "more general").

The Least common subsumer (LCS) of a set of given concepts, $C_1,...,C_n$ is another concept $C=LCS$ $(C_1,...,C_n)$ such that $\forall_{i=1...n} C_i \sqsubseteq C$, and for any other concept C' such that $\forall_{i=1...n} C_i \sqsubseteq C'$, $C \sqsubseteq C'$ holds.

If given two concepts C and D such that $C \sqsubseteq D$, it is possible to reach C from D by applying a downward refinement operator ρ to D a finite number of times, i.e. $C \in \rho^*(\underset{\rho}{D})$. The length of the refinement chain to reach C from D, which we will note by $\lambda(D \xrightarrow{\rho} C)$, is an indication of how much more information C has that was not contained in D. Given any two concepts, their LCS is the most specific concept which subsumes both. The LCS of two concepts contains all that is shared between two concepts, and the more they share the more similar they are. So, we can now define similarity between two concepts C and D. i.e. the similarity between two concepts C and D is assessed as the amount of information contained in their LCS divided by the total amount of information in C and D.

To measure similarity of two cases of network security incident, we suppose the problem of cases CA_1, CA_2 is $P_1 \equiv C_1 \sqcap C_2 \sqcap ... \sqcap C_n$ and $P_2 \equiv D_1 \sqcap D_2 \sqcap ... \sqcap D_m$ respectively. Where $C_i, D_j (i=1,..., n; j=1,..., m)$ are the ALCO(**D**) formula, then we can define the similarity between CA_1 and CA_2.

$$Sim(CA_1, CA_2) = \alpha \cdot Sim_\rho (P_1, P_2) + (1 - \alpha) \cdot sim_c \left(con_F \left(C_i, D_j \right) \right)$$

$$(0 \le \alpha \le 1) \tag{1}$$

Where
$$Sim_\rho (P_1, P_2) = \frac{\lambda_1}{\lambda_1 + \lambda_2 + \lambda_3} \tag{2}$$

$$\lambda_1 = \lambda(\top \xrightarrow{\rho} LCS(P1, P2)) \tag{3}$$

$$\lambda_2 = \lambda(LCS(P1, P2) \xrightarrow{\rho} C) \tag{4}$$

$$\lambda_3 = \lambda(LCS(P1, P2) \xrightarrow{\rho} D) \tag{5}$$

$$sim_c \left(con_F \left(C_i, D_j \right) \right) = \sum_{i=0}^{k} \omega_i \cdot \frac{|d_1 - d_2|}{|max - min|}$$

$$(k \text{ is a natural number}, 0 \le \omega_i \le 1, \Sigma_{i=0}^k \omega_i = 1) \tag{6}$$

Where α and ω_i are weighted factors, k is the number of concept with the same type of role and numerical value in concrete domain. If $F. = d_1$, $F. = d_2$ and max not equal min, then max and min are the maximum and minimum value of concrete role with data type d_1 and d_2, respectively. If $F.=d_1$, $F.=d_2$ and max equal min, then the right side of equation equals to ω_i, i.e.| d_1- d_2|/|max-min| equal 1.

Due to the description of P (problem) in the case of security incident in essence is a concept, formula (2) defines the overall similarity Sim_ρ between two cases.

The calculation of sim_c is the correction similarity between two cases, which used to assess similarity of two different concept with the same type role and different numerical value in concrete domain of case representation, such as the similarity between $\exists hasOccurredTime.= 2010$ year and $\exists hasOccurredTime.= 2009$ year. According to the need of knowledge representation, the concrete domain D only used one feature and a predicate formula, i.e. the situation of $\exists F. d$ and $\forall F. d$.

4   Case Study

Suppose there are two cases CA_1 and CA_2 which are described by two concepts *Tiger_virus_incident* and *Dummycom_virus_incident* respectively as follow

Tiger_virus_incident ≡
  ∃*hasType*.(*Virus_Incident* ⊓ *Worm_Incident* ⊓
        ∃*hasTransMethod*.(*Webshell* ⊔ *Mobile_memory_media* ⊔ *LAN_WeakPW*
                ⊔{*IE_day_vulnerability, Affected-exefile*}))
  ⊓ ∃*hasOccurenceTime*. = 2010 year ⊓ ∃*hasOrg. Commerce*
  ⊓ ∃*hasAffectedResource*.(∃*hasHostFunction.Client* ⊓ ∃*hasOS*.(*Win7*⊔*Win-xp*) ⊓
            ∃*hasNumber*.=20 tai ⊓
            ∃*hasAbnormal*.(*Antivirus*⊓*hasSign*.{*unavailable*}⊓
                CPU⊓*hasSign*.{*usage_high*}⊓*IE*⊓*hasSign*.{*abnormal*}⊓
                System⊓*hasSign*.{*slowdown*}⊓*Exe-file*⊓*hasSign*.{*infected*}⊓
                Hard-disk ⊓*hasSign*.{*Read_fast*}))
  ⊓∃*hasState*.{*on-going*} ⊓ ∃*hasEffect*.{*Money_Loss*}.
Dummycom_virus_incident ≡
  ∃*hasType*.(*Virus_Incident*
        ⊓∃*hasTransMethod*.(*LAN_ARP*⊔*Mobile_Memory_media*⊔{*Affected-exefile*}))
  ⊓∃*hasOccurenceTime*.=2009 year⊓∃*hasOrg. Education*
  ⊓∃*hasAffectedResource*.(∃*hasHostFunction.Client* ∃ *hasOS.Win-xp*⊓
            ∃*hasNumber*,=100 tai⊓
            ∃*hasAbnormal*.((*Exe-file*⊓∃*hasSign*.{*infected*})⊓
            Antivirus⊓*hasSign*.{*unavailable*})⊓
            Hidden_file⊓*hasSign*.{*Not_display*})⊓
            System⊓*hasSign*.{*slowdown, time_distorted, blue_screen*}⊓
            IE⊓∃*hasSign*.{*Sec_tool_web_no_access*}))
  ⊓∃*hasState*.{*ongoing*}⊓∃*hasEffect*.{*Data_lost*}.
Then ,their least common subsumer (LCS) is:
  LCS(*Tiger_virus_incident,Dummycom_virus_incident*) ≡
    ∃*hasType*.(*Virus_Incident*⊓
        ∃*hasTransMethod*.(*LAN*⊔*Mobile-memory-media*⊔{*Affected-exefile*}))
    ⊓∃*hasOccurrenceTime*.=*n* year⊓∃*hasOrg.Organization*
    ⊓∃*hasAffectedResource*.((∃*hasOS.Win-xp*)⊓ ∃*hasHostFunction.Client*⊓∃*hasNumber*.=*m* tai
            ⊓∃*hasAbnormal*.((*Antivirus*⊓∃*hasSign*.{*unavailable*})
                ⊓*System*⊓*hasSign*.{*slowdown*}⊓ *Exe-file*⊓*hasSign*.{*infected*}
                ⊓ (*IE*⊓*hasSign*.{*abnormal*}))
    ⊓∃*hasState*. {*ongoing*}⊓∃*hasEffect*. *Effect*.

In this paper, the correlation coefficient α we take 0.98, ω_i take 0.5, then 24/(24+12+7)=0.558, the similarity of CA_1, CA_2 is 0.554.

In order to verify the validity of the proposed algorithm, this paper collected more than 20 typical cases for nearly 3 years from the CNCERT and calculated their similarity measure. As shown in Fig. 3, the case 1-7 are virus or worm incidents, case 8-11 are mobile malware incidents, case 12-14 are DDoS incidents, case 15-17 are phishing site or Trojan incidents, case 18-20 are webpage tamper incidents. Their similarity with three different types of incident is calculated. As can be seen from the graph, The conficker worm incident is concentrated in case 7, 8 with higher similarity, and DDoS incident is concentrated in case 12-14 with its high similarity. This can be explained that the retrieval algorithm has a certain degree of differentiation, which can effectively distinguish the different form of the virus, worm and DDoS incidents, retrieve previous incidents to match the target case, and obtain a method for appropriately handling security incidents in actual environment.

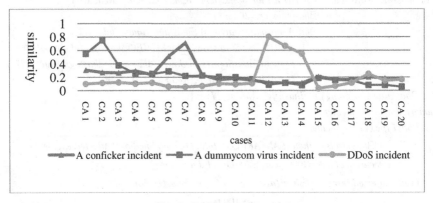

Fig. 3. Similarity of cases

5   Conclusion

This paper presents a method for appropriately handling security incidents under specific environment by exploring the past similar cases. Consider the characteristics of DL with clear semantics and good inference capability, this paper mainly use the formalization DL to represent NSER cases, thus providing an effective way to classify and search the knowledge of case base. In order to enhance the overall effect of retrieval, it design a good matching algorithm of similarity based on refinement operator and refinement graph to distinguish different cases in case base and retrieve the most similar cases. Finally, this paper proves the similarity metric with good effects by experiments. The further work as following: design the case reuse and revise process, combine the failing cases to design some more reasonable method of retrieve, reuse and revise.

Acknowledgements. This work is supported by the National Natural Science Foundation of China (Nos. 61262030, 61363030, 61100025), the Natural Science Foundation of Guangxi Province (No.2012GXNSFBA053169) and the Science Foundation of Guangxi Key Laboratory of Trusted Software.

References

1. Mitropoulos, S., Dimitrios, P., Christos, D.: On Incident Handling and Response: A state-of-the-art approach. Computers & Security 25(5), 351–370 (2006)
2. Danyliw, R., Meijer, J., Demchenko, Y.: RFC 5070:The Incident Object Description Exchange Format. Internet Engineering Task Force (IETF) (2007)
3. Scarfone, K., Grance, T., Masone, K.: Computer security incident handling guide. NIST Special Publication 800(61), 38 (2008)
4. Lopez De Mantaras, R., McSherry, D., Bridge, D., et al.: Retrieval, reuse, revision and retention in case-based reasoning. The Knowledge Engineering Review 20(03), 215–240 (2005)
5. Capuzzi, G., Spalazzi, L., Pagliarecci, F.: IRSS: Incident Response Support System. In: International Symposium on Collaborative Technologies and Systems (CTS), pp. 81–88. IEEE (2006)
6. Kim, H.K., Im, K.H., Park, S.C.D.: for computer security incident response applying CBR and collaborative response. Expert Systems with Applications 37(1), 852–870 (2010)
7. Ping, L., Haifeng, Y., Guoqing, M.: An incident response decision support system based on CBR and ontology. In: Proc. of the 2010 Int Conf. on Computer Application and System Modeling (ICCASM), vol. 11, pp. 337–340. IEEE (2010)
8. Cunningham, P., Taxonomy, A.: of Similarity Mechanisms for Case-Based Reasoning. IEEE Trans. on Knowledge and Data Engineering 21(11), 1532–1543 (2009)
9. Sánchez-Ruiz, A.A., Ontañón, S., González-Calero, P.A., Plaza, E.: Measuring similarity in description logics using refinement operators. In: Ram, A., Wiratunga, N., et al. (eds.) ICCBR 2011. LNCS, vol. 6880, pp. 289–303. Springer, Heidelberg (2011)
10. Sánchez-Ruiz, A.A., Ontañón, S., González-Calero, P.A., Plaza, E.: Refinement-Based Similarity Measure over DL Conjunctive Queries. In: Delany, S.J., Ontañón, S. (eds.) ICCBR 2013. LNCS, vol. 7969, pp. 270–284. Springer, Heidelberg (2013)
11. Amailef, K., Lu, J.: Ontology-supported case-based reasoning approach for intelligent m-Government emergency response services. Decision Support Systems 55(1), 79–97 (2013)

On the Prevention of Invalid Route Injection Attack

Meng Li, Quanliang Jing, Zhongjiang Yao, and Jingang Liu

Capital Normal University Haidian District, Beijing City, China
limeng7065@gmail.com, QuanliangJing@ict.ac.cn,
{zhongjiangyao,liujg2003}@163.com

Abstract. In recent years, the Internet is suffering from severe attacks from the global routing system, such as prefix hijack, digital cannon. These attacks break down network services and sabotage infrastructures. In this paper, we present a novel attack on route control plane---invalid route injection, and simulate the attack by adding controlled software routers in the stub and check the resource consumptions of routers before and after the attack. The experimental results demonstrate that invalid route injection can bring severe damage to routers in the Internet. Based on the analysis of the results, we discuss the prevention of the attack and propose some effective protection and countermeasures to the attack.

Keywords: Invalid route, injection, OSPF, BGP, route attack.

1   Introduction

With the rapid development and growth of IP technology, Internet has penetrated into politics, economy, military and daily life and brought it a corresponding increased reliance on the underlying infrastructures. Therefore, it is essential to ensure the security of the Internet. Route security is an important part of Internet security. It is becoming vitally important and facing significant challenges at the same time.

The Internet is composed of tens of thousands of Autonomous Systems (ASes) which operate individual parts of the infrastructures. ASes exchange route information via an external gateway protocol like BGP [1] (Border Gateway Protocol). Within an AS, routers communicate with each other through intra-domain routing protocols [2] such as OSPF (Open Shortest Path First Protocol) which has been widely used currently, and IS-IS (Intermediate System to Intermediate System) which is developing and spreading gradually. These protocols can efficiently distribute dynamic topological information among its participants, facilitate route calculations and make packet forwarding decisions. They form the heart of Internet infrastructures.

However, these routing protocols are not so robust for their disadvantages and loopholes which can cause some incidents or be used by malicious attempts to compromise the availability of the network [3]. For example, since its own route information cannot be validated by BGP itself, it has to fully trust all the other peering routers[5]. Based on that, On 24th February 2008, Pakistan Telecom started an unauthorized announcement of prefix 208.65.153.0/24. PCCW Global, one of Pakistan Telecom's upstream providers, forwarded this announcement to the rest of the

Z. Shi et al. (Eds.): IIP 2014, IFIP AICT 432, pp. 294–302, 2014.
© IFIP International Federation for Information Processing 2014

Internet, which resulted in the hijacking of YouTube traffic on a global scale [6]. For another, Max Schuchard, from the University of Minnesota in 2010 put forward the concept of 'digital cannon'[7], which showed that a digital cannon using a 'botnet' composed of 25000 computers can destroy the entire Internet using BGP protocol. These routing incidents result in irreparable damage to the politics and economics.

In order to detect the routing attacks effectively and provide a theoretical foundation for routing attack detection, in this paper we present a novel network routing attack named invalid route injection. Then we do some simulate experiments to verify the attack. The results show that the invalid route injection attacks do have an effect on the communication among the hosts or routers, causing a serious impact on the stability and security of the network. Finally we discuss some preventive measures.

2   Related Work

Using the control plane to attack the Internet has recently been proposed to the literature. 'Digital cannon', which interferes information exchange by using data plane to affect control plane, is one of such attacks. The attack causes routers frequently exchanging neighbor routing information, resulting in the exhaustion of CPU, memory and other resources of the routers, and eventually crushing control plane. However, this method needs to get the topology of the entire network and critical devices. Once the network devices filter the ICMP packets, the attack will not happen [8]. Currently, there are also many instances of 'prefix hijacking' attacks in the Internet, which are caused by mis-configuration of routing information or malicious attempts. The AS that hijacks a prefix can intercept all the hijacked traffic, result in a denial-of-service attack against the hijacked AS. Also the hijacks can redirect the traffic to an incorrect destination for phishing attack. Although prefix hijack has many cases in theory, it is hard to implement in reality since the Internet has not only very strict access strategy of the Internet, but also the attack need to configure the routes directly [9]. This paper introduces a new method of routing attack—invalid route injection attack. This methodology avoids directly manipulating core routers running BGP, and is easy to be implemented.

There are also many measurement studies on routing attack related problems, for example, analyzing route changes and their causes, measuring how the end-to-end performance is impacted by protocol policy, and studying black-holes in the Internet [10-13]. Here, we make some control measurement to prevent the potential invalid route attack, to narrow down the scope of the attack. Since the attack is carried on by following the protocols, we identify changes when the attack happens, and record some parameter thresholds, helping better understand the characteristics of the attack.

3   Attack Description

3.1   Methodology

Invalid route injection attack aims at routers running intra-domain protocols. The attackers first establish neighbor relationship with routers in the AS, after that it will advertise many invalid routes to impact the border routers, eventually make other routers in the area affected. Since the border routers are core equipment of the network, it is difficult to attack them directly. Instead of acquiring control of BGP, invalid route injection attack is manipulated by using features of OSPF or IS-IS protocols. It will cause depletion of routers' resource and instability even a loss of connectivity in the network.

To validate the effectiveness of the attack presented above, we design a topology similar to the real network, and simulate the attack on the topology. Since the results of this experiment have nothing to do with the specific autonomous system, the real autonomous system number is not necessary. We use Cisco 1800 series routers in the experiment.

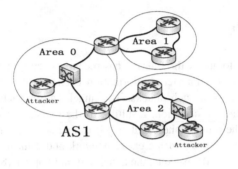

Fig. 1. Invalid Route Injection

The set of intra-domain routers is defined as R_r = {router-id} each element represents a router that is identified by its router-id. $C_{router-id}$ represents the memory size of each router. The maximum memory within the router can be expressed as:

$$C_{max} = \max \{ C_{router - id} \}$$

The minimum memory within the router can be expressed as:

$$C_{min} = \min \{ C_{router - id} \}$$

The amount of memory occupied by the attacker's invalid route is defined as Y. Theoretically, $Y \geq C_{max}$ means that memory resource of the routers within the entire domain is exhausted after the injection; $C_{min} \leq Y \leq C_{max}$ means that part of routers' memory resources has been used up; $Y \leq C_{min}$ means the injection would not affect the area. Since the router not only handles protocol interactions but also does packet forwarding and calculation of routing paths, the impact of the amount of memory above on a router is an upper bound. Figure1 illustrates the invalid route injection attack, the basic principle of the attack is to add controlled router in the stub by establishing OSPF/IS-IS neighborhood with normal routers, meantime declaring a large number of false or invalid transient messages, thus resulting in many routers' resources deplete in the network. By means of DDoS (Distributed Denial of Service) theory, it will impact the routers of the entire network including high-performance boarder routers through deploying distributed controlled routers.

3.2   Implementation of the Attack

1)   Attack position
First we choose a position to attack. Since previous attacks require to know about the core routers or to get the status of the entire network topology in advance, they are difficult to be carried out. However, invalid route injection attack only needs to find routers running OSPF/IS-IS protocol in the stub and to establish OSPF/IS-IS neighbor relationship with normal routers. Hence, it is much easier to implement in practice.

2)   Invalid Route Generation
Next we generate a lot of static routes as invalid injection. There are many ways to introduce static routes. For instance, we can probe the data layer such as traceroute and ping to get the network segment address, and accordingly to generate a lot of static routes. Also it can be generated in random. After that, the invalid routers can be propagated into the network by flooding mechanism of the protocol, and exhaust the CPU, memory and other resources, thereafter affecting the entire network.

3)   Invalid Route Injection
Then, we inject route in the network. The way of injection depends on routing protocol running among routers in the network. If using OSPF, the attacker would establish OSPF neighbor relationship with the routers and generate invalid routes redistributed into OSPF, and affect other local routers and other routers in the network; If using IS-IS, the attacker would establish IS-IS neighbor relationship, and directly inject the invalid entries into the IS-IS network.

4)   Invalid Route Revocation
Finally, in order to advance the attack impact, we revoke or add static route repeatedly to affect the entire network. It will cause the control plane unstable and make the network in a turbulent environment. Thus routers would consume routers' resources, and packets forwarding and neighbor relationship establishment would be affected among routers.

4   Simulation and Experimental Analysis

4.1   Definition of Evaluation

In order to demonstrate the effect of invalid route implantation to the network, we define three evaluation indicators for routers: average CPU utilization, average memory utilization and average packet loss rate. Routers' average CPU utilization is calculated as follows:

$$\text{AVE\_CPU} = \frac{\sum_{i=1}^{p}\sum_{j=1}^{q_p} D_{ij}}{M} \tag{1}$$

Let p be the number of the AS, q_p be routers' number in AS p, we define D as CPU utilization, M as total number of routers in the network.

Memory utilization of a router is expressed as:

$$AVE\_Mem = \frac{\sum_{i=1}^{p}\sum_{j=1}^{q_p}\frac{B_{ij}}{O_{ij}}}{M} \tag{2}$$

As the same, p is the number of the AS, q_p to be router number in AS p, we denote B as memory footprint of a router, O as a router's total amount of memory, M as total number of the routers in the network.

Average packet loss rate is calculated as equation [3]. Here, p represents the number of packets sent, S_i means packets number of the i-th, A means the packets number that have been sent successfully for the i-th.

$$AVE\_Loss = \frac{\sum_{i=1}^{p}\frac{A_i}{S_i}}{p} \tag{3}$$

4.2   Network Topology and Experiment Design

As shown in Figure 2, AS1, AS2 and AS3 represent different autonomous system; connection among them has been marked. The topology of each AS is shown in Figure 3, there are three areas in one AS, each AS has eight routers, and three routers in area 2 can connect multiple hosts or packet tester.

In the experiment, we first make the network reach a steady state and add attacker in area 2 or area 0 respectively according to the methodology described in Section3. Then we inject a large number of static routes and observe changes in CPU utilization, memory utilization and packet loss rate. Since the value of CPU utilization and memory usage have instantaneous effect, we need to wait for the network to be stable again after the injection and calculate the statistics for router's CPU, memory usage.

We wait about 15 minutes after the injection. The router CPU utilization for each router is calculated using the following formula:

$$B = \frac{\sum_{i=1}^{p} R_i}{p} \tag{4}$$

We denote p as times of calculation, Q measures CPU utilization rate for each time. When checking the router CPU utilization, we type in 'show processes' in the command line of the router. Three values of CPU utilization will be showed, which respectively represent the value within nearest 5 seconds, nearest 1 minute and nearest 5 minutes. In order to ensure the accuracy of the data, we adopt the CPU utilization values within five minutes for each time.

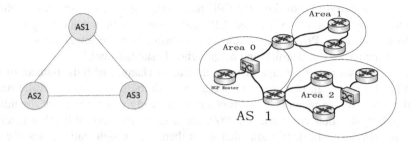

Fig. 2. Topology of Simulation   **Fig. 3.** Topology of AS1

Memory footprint is calculated as equation (5), let p represent times we check, R to be the footprint of each view. We type 'show memory' in the command line to check the memory utilization of the router.

$$D = \frac{\sum_{i=1}^{p} Q_i}{p} \tag{5}$$

4.3   Experimental Analysis

Figure 4 reflects changes about the average CPU utilization of the router. In two attackers' situation, with the invalid route entry increasing from 10000 to 80000, the percent of average CPU utilization rise rapidly from 5% to 90%, followed by it is in smooth change. Comparing two attackers' case with the only one attacker situation, the former one has higher average CPU utilization. Also there is a large gap between them, indicating a plurality of attackers would have a greater impact on the network.

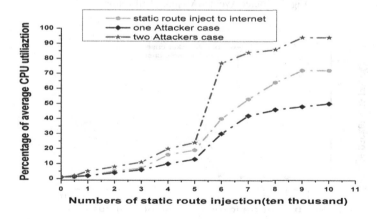

Fig. 4. Average CPU Utilization Trend

Within a certain period of time, the CPU processing capacity of a router is limited. When too many invalid routes injected, CPU utilization is so high that other processes are blocked to take CPU. Thus we draw a conclusion that the attack does have an impact on the packets forwarding and neighborhood establishment.

Figure5 shows router average memory utilization changes with the number of injected invalid routes. In two attackers' situation, with the increasing of invalid route injection, the average memory utilization strikes a steady upward trend at the initial phase. When injection entry reach 80000, the average memory utilization reaches 90%, and then leveled off. It means that when there are enough static routes, the attack will consume almost all the memory of routers. Then we continue to increase the number of static routers, the utilization increasing is no longer in linear trend but stabilized. Since the router's memory is limited, when the invalid routes occupy too much memory, it will affect the normal operation of routers, and have an impact on the establishment of neighbor relations and packets forwarding.

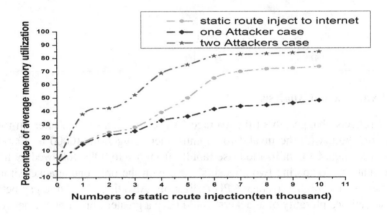

Fig. 5. Router Average Memory Utilization

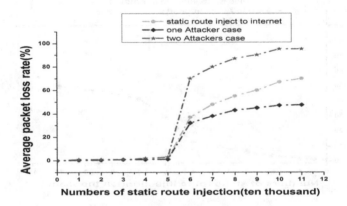

Fig. 6. Packet Loss Rate Trend

With the injection of invalid route, packet loss rate throughout the network is shown in Figure 6. In two attackers' situation, the basic network data packets are rarely lost when the invalid route entry is less than 50000, However, the invalid entry and the packet loss rate are increased significantly with the increasing of invalid entry. While the invalid routes reached 60000, packet loss rate is 70%. While 100000 entries, the packet loss rate could even reach 95%. As shown in the figure, the tendency of the network packet loss rate is in line with the average CPU utilization and average memory utilization. The number of Attackers could result in a big gap in the router packet loss rate, before and after the attack, providing that more Attackers, greater impact on the network.

5   Preventive Measures

5.1   Method for the Time Introducing Route of BGP

According to the attack principle, routers will be affected if an invalid route injects into the network in the corresponding AS. However, if BGP router's reach- ability information in the corresponding AS is brought into the network through the static mode, the attack will not spread to other AS, thereby narrow down the affected scope of invalid route injection attack.

In Figure 4, within two Attackers in the network, the average CPU utilization of the router changes significantly after we set one static route of into the network. When invalid route entry injected reach 60000, there is 35% gap increase, and the gap will remain at this level following the increasing number of injection. Routers' memory utilization has direct relation with the number of injection. Introducing a router in a static way will influence number of invalid route in the network. It can be seen in Figure 5, at the initial stage, the average memory utilization of the router already has about 20% gap, then gradually stabilized. Similarly, the packet loss rate shown in Fig.6, the difference began to become larger, about 25% when invalid route injection entry reach 60000, in line with the trend of average CPU utilization and average memory utilization.

5.2   Warning Threshold for Routers

The invalid routes injected to the network could be learned and propagated by routers. We can deploy route entry detection system in service provider's network. Once the route entry detected is beyond a certain baseline, we will make a warning alarm to remind the provider to take measures and reduce losses caused by the attack.

The above methods and other preventive measures such as: the use of QOS mechanism to control the level of security resources [8], modify the route protocol [14-16], with a variety of detection methods [10-13] to check the attacks. Those are not the fundamental solution for static route injection attacks. To prevent attacks fundamentally, we need redesign the existing router or amend the protocols to strengthen network security and prevent attacks against the protocols radically.

6   Conclusion

Through in-depth study of network route protocols, we investigate the invalid route injection attack which can cause severe damage to the network. The key point of the attack is to declare a large number of invalid route information into the network resulting in massive route updates, eventually causing routers' resources exhausted in the network. Meanwhile we provide the specific steps in the generation and injection of invalid route. We also verify the attack by experiment and analyze the changes in the network after attacks. Finally, we discuss some preventive measures of the attack. These measures can reduce the losses caused by the attack effectively.

Acknowledgment. This is supported by Research Foundation of Education Bureau of Hebei Province (No. Z2013124).

References

1. Kent, S., Lynn, C., Seo, K.: Secure border gateway protocol (S-BGP). IEEE Journal on Selected Areas in Communications 18(4), 582–592 (2000)
2. Ospfversion2, http://www.ietf.org/rfc/rfc-2328.txt
3. Kevin, B., Toni, F., Patrick, M., et al.: A survey of BGP security. In: Proceedings of ACM Internet Measurement Workshop, New Orleans, LA (November 2005)
4. Murphy, S.: BGP security vulnerabilities analysis. IETF Internet RFC, RFC 4272 (2006)
5. RIPE, "Youtube hijacking: A ripe nccris case study" (2008), http://www.ripe.net/news/study-youtube-hijacking.html
6. Schuchard, M., Aohaisen, A., Kune, D., et al.: Losing Control of the Internet: Using the Data Plane to Attack the Control plane. In: Proc. of the 17th ACM Conf. on Computer and Communications Security, pp. 726–728. ACM, New York (2010)
7. Bornhauser, U., Martini, P.: About prefix hijacking in the Internet. In: 2011 IEEE 36th Conference on Local Computer Networks (LCN), pp. 143–146. IEEE, Bonn (2011)
8. Liu, Y., Su, J., Chang, R.K.C.: LDC: Detecting BGP Prefix Hijacking by Load Distribution Change. In: 2012 IEEE 26th International Parallel and Distributed Processing Symposium Workshops & PhD Forum (IPDPSW), pp. 1197–1203. IEEE, Shanghai (2012)
9. Lad, M., Massey, D., Pei, D., et al.: PHAS: A prefix hijack alert system. In: Proc. USENIX Security Symposium, vol. 2, pp. 153–166 (2006)
10. Zheng, C., Ji, L., Pei, D., et al.: A light-weight distributed scheme for detecting IP prefix hijacks in real-time. ACM SIGCOMM Computer Communication Review 37(4), 277–288 (2007)
11. Hu, X., Mao, Z.M.: Accurate Real-time Identification of IP Prefix Hijacking. In: IEEE Symposium on Security and Privacy, SP 2007, pp. 3–17. IEEE, Berkeley (2007)
12. White, R., Securing, B.G.P.: through secure origin BGP (soBGP). The Internet Protocol Journal 6(3), 15–22 (2003)
13. Oorschot, P.C., Wan, T., Kranakis, E.: On interdomain routing security and pretty secure BGP (psBGP). ACM Transactions on Information and System Security (TISSEC) 10(3), 11 (2007)

A Formal Model for Attack Mutation Using Dynamic Description Logics

Zhuxiao Wang[1,*], Jing Guo[2], Jin Shi[2], Hui He[1], Ying Zhang[1],
Hui Peng[3], and Guanhua Tian[4]

[1] School of Control and Computer Engineering,
State Key Laboratory of Alternate Electrical Power System with Renewable Energy Sources,
North China Electric Power University, Beijing 102206, China
{wangzx,huihe,yingzhang}@ncepu.edu.cn
[2] National Computer Network Emergency Response Technical Team/Coordination Center
of China, Beijing 100029, China
guojing.research@gmail.com, shijin@cert.org.cn
[3] Education Technology Center, Beijing International Studies University,
Beijing 100024, China
penghui@bisu.edu.cn
[4] Institute of Automation, Chinese Academy of Sciences, Beijing 100190, China
guanhua.tian@ia.ac.cn

Abstract. All currently available Network-based Intrusion Detection Systems (NIDS) rely upon passive protocol analysis which is fundamentally flawed as an attack can evade detection by exploiting ambiguities in the traffic stream as seen by the NIDS. We observe that different attack variations can be derived from the original attack using simple transformations. This paper proposes a semantic model for attack mutation based on dynamic description logics (DDL(X)), extensions of description logics (DLs) with a dynamic dimension, and explores the possibility of using DDL(X) as a basis for evasion composition. The attack mutation model describes all the possible transformations and how they can be applied to the original attack to generate a large number of attack variations. Furthermore, this paper presents a heuristics planning algorithm for the automation of evasion composition at the functional level based on DDL(X). Our approach employs classical DL-TBoxes to capture the constraints of the domain, DL-ABoxes to present the attack, and DL-formulas to encode the objective sequence of packets respectively. In such a way, the evasion composition problem is solved by a decidable tableau procedure. The preliminary results certify the potential of the approach.

Keywords: Intrusion Detection/Prevention Systems, Multi-protocol Evasions, Advanced Evasion Techniques, Knowledge Representation and Reasoning, Dynamic Description Logics.

---

* Corresponding author.

Z. Shi et al. (Eds.): IIP 2014, IFIP AICT 432, pp. 303–311, 2014.

1   Introduction

A weakness of most currently available Network-based Intrusion Detection Systems (NIDS) that rely upon passive protocol analysis is their inability to recognize an attack that evades detection by exploiting ambiguities in the traffic stream as seen by the NIDS. Exploitable ambiguities may arise in different ways: (1) The NIDS may lack complete analysis for the full range of behavior allowed by a particular protocol. (2) Without detailed knowledge of the victim end-system's protocol implementation, the NIDS may be unable to determine how the victim will treat a given sequence of packets if different implementations interpret the same stream of packets in different ways. (3) Without detailed knowledge of the network topology between the NIDS and the victim end-system, the NIDS may be unable to determine whether a given packet will even be seen by the end-system[1,2]. Advanced Evasion Techniques (AET's) is a lately established term in network security industry referring to a set of non-trivial and expensive means to bypass NIDS in order to deliver an exploit, attack, or other form of malware to a target network or system, without detection. Advanced evasion techniques can be identified according to certain underlying principles: (1) Delivered in a highly liberal way; (2) Employ rarely used protocol properties; (3) Use of unusual combinations; (4) Craft network traffic that disregards strict protocol specifcations; (5) Exploit the technical and inspection limitations of security devices: memory capacity, performance optimization, design flaws[3].

Since we are interested in testing the ability of a NIDS to properly identify real intrusions, we need a way to ensure that executing each of mutants generated by advanced evasion techniques against the vulnerable application, we are going to obtain the same effect as executing the original exploit script. In this article, we define an attack mutation model that describes all the possible transformations and how they can be applied to the original attack to generate a large number of attack variations. The semantic model for attack mutation is based on dynamic description logics (DDL(X)), extensions of description logics (DLs) with a dynamic dimension. The attack mutation model is self explanatory. Given an original attack, the attack mutation model can provide a proof that a sequence of transformations used for obfuscation is a real attack. Developers can use the model to analyze attacks and to identify the exact transformation that their NIDS fails to handle. The attack mutation model is exhaustive, capable of generating all attack variations from a known base attack using a set of rules. All the mutation techniques are individually sound, and also that any possible composition of them is sound. So the model is sound, generating only instances that implement the original attack.

In this article, we present an approach for automatic evasion method plans based on dynamic description logics[4-6] named DDL(X), extensions of DLs [7] with a dynamic dimension[8,9]. Our approach used classical DL-TBoxes to capture the domain constraints, DL-formulas to encode the objective sequence of packets, and DL-ABoxes to be a special representation of the attack that provides to the underlying mutation engine the mechanism to manipulate the attack, respectively. Actions in DDL(X) were used to abstract the functionalities of the available evasion methods. In such a way, the automatic evasion plans can be reduced to formula satisfiability checking in DDL(X) and solved by a decidable tableau procedure.

In the following sections, we firstly illustrate how variants of a real exploit can be derived from the original exploit script in Section 2. In Section 3, we demonstrate the descriptions of evasion techniques can be formalized as actions in DDL(X), and formalize the notion of the attack mutation model using dynamic description logics. Afterwards, in Section 4 we detail the problem of evasion composition can be solved by reasoning about actions in DDL(X) and present a heuristic planning algorithm for automated composition of evasions. Finally, we summarize the paper in Section 5.

2   Approach Overview

First of all, we need an instance of the exploit script that we want to mutate. The base instance is then used to derive another attack instance by repeatedly applying single step transformations. Our example vulnerability is a published buffer overflow in a commonly used Windows XP SP2 host (MSRPC Server Service Vulnerability, CVE-2008-4250 in www.cve.mitre.org.); exploiting the overflow may allow arbitrary code execution. The exploit causes the overflow by providing a crafted RPC request that triggers the overflow during path canonicalization. We call this exploit E_{MSRPC}.

Given an exploit E, Trace(E) denotes a sequence of packets from a NIDS perspective and the function Post(E) to be the post-conditions of the execution of E against a target system. Consider the following twelve transformation rules in Table 1 that can be applied to an existing operational description of how a vulnerability is exploited to generate a new different version of the same exploit.

We call these rules semantics preserving because they do not alter the semantics of E, i.e. the transformation does not affect the results of the execution of the exploit. If E is an instance of the E_{MSRPC} attack, then by using rules in Tab.1 it is possible to derive the conclusion that the E' is also an instance of E_{MSRPC} and the instance E' contains the necessary data for a successful E_{MSRPC} attack. For example, we apply r_1 on E to send extra NetBIOS packets to break the packet flow. Then, we apply r_9 and change the order of TCP segments. In dynamic description logic terminology, starting with an exploit E, we can successively apply a set of transformations $T = \{r_0, r_1, ..., r_n\}$ to create a complex mutant exploit E'.

To formalize the notion of the semantics preserving transformation, we get:

$$\text{Trace(E)} \neq \text{Trace(E')}$$

$$\text{Post(E)} = \text{Post(E')}$$

The first condition requires that the transformation manifests itself as a change in the sequence of packets. The second condition requires that the transformation preserves the attack post-conditions. While the E and the E' might look different from a NIDS perspective, from the attacker point of view the E' generated by evasion technique is still an "effective" attack (i.e., that executing the E' against the vulnerable application, we are going to obtain the same effect as executing the original exploit script E). Intuitively speaking, one can infer E' from E, and vice versa.

Table 1. Atomic evasions

	Name	Description
NetBIOS	NetBIOS chaff(r_1)	Send extra NetBIOS packets to break the packet flow
	NetBIOS initial chaff(r_2)	Send chaff NetBIOS packets when establishing the NetBIOS connection
SMB	SMB filename obfuscation(r_3)	Obfuscate the tree name used in the SMB NT Create AndX method
	SMB WriteAndX padding(r_4)	Insert extra padding between the WriteAndX header and payload
MSRPC	MSRPC request segmentation(r_5)	Set the maximum number of bytes written in a single MSRPC fragment
	MSRPC NDR modifications(r_6)	Set NDR types not related to endianness
TCP	TCP Chaff(r_7)	Send chaff TCP segments to baffle inspection
	TCP timestamp option settings(r_8)	Set initial TCP timestamp option settings
	TCP segment order(r_9)	Change the order of TCP segments
IP	IPv4 chaff(r_{10})	Send chaff IPv4 packets interleaved with normal packets
	IPv4 fragmentation(r_{11})	Fragment IPv4 packets to given size
	IPv4 fragment order(r_{12})	Change the order of IPv4 fragments

In our description frameworks for evasions, functional descriptions are essentially the state-based and use at least pre-state and post-state constraints to characterize intended executions of an evasion. Many evasion techniques can be combined. When evasion techniques are combined at different levels, NIDSs that detect each separate evasion technique often fail at detecting some permutations. First, we apply the transformation rules in a Breadth-First order: we first apply application level rules because they are independent against TCP-level and IP-level rules, then we segment each instance into small pieces, change the order of TCP segments we get, and send chaff IPv4 packets interleaved with normal packets(Figure 1). Second, we prune away some of the derivation branches to decrease the number of instances. In some cases, only a subset of possible factors evades NIDS detection; adding additional evasion measures to an evasive attack may cause the NIDS to detect it.

3   A Formal Model for Attack Mutation

A DDL(X) model is a tuple M = (W, T, Δ, I), where,

W is a set of states;

$T : \mathrm{N_A} \rightarrow 2^{W \times W}$ is a function mapping action names into binary relations on W;

Δ is a non-empty domain;

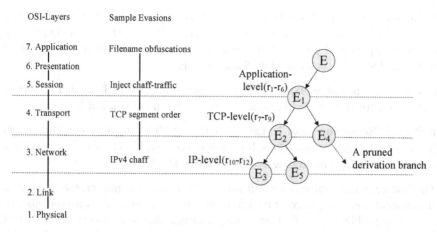

Fig. 1. An Illustration of a Sample Multi-protocol Evasion

I is a function which associates with each state $w \in W$ a description logic interpretation $I(w) =< \Delta, \cdot^{I(w)} >$, where the mapping $\cdot^{I(w)}$ assigns each concept to a subset of Δ, each role to a subset of $\Delta \times \Delta$, and each individual to an element of Δ.

Definition 1 (Atomic evasions). An atomic evasion is a tuple t=<Pre, Effects>, where Pre is a finite set of formulas in DDL(ALCO) specifying the preconditions for the execution of t; and Effects is a finite set of assertions or their negation in ALCO, which is the facts holding in the newly-reached world by the evasion's execution.

Composite evasions are constructed from atomic evasions with the help of classic constructors in dynamic description logics[4].

Below we formally define a mutation model for an exploit as well as some reasoning tasks and the planning problem.

Definition 2 (Mutation Model of an Exploit). Let E be an instance of an exploit and T be a set of sound inference rules with respect to E.

A mutation model of E is a DDL(X) model (W, T, Δ, I).

Such a model enables derivation of new exploits by applying the inference rules (like evasion methods) on already known exploits. For an attack E, we envision the attack mutation model that, with respect to a set of rules, is sound: derives only sequences of packets that implement E; complete: can derive any sequence of packets that implements E; and decidable: given a sequence of packets, there is an algorithm that determines whether or not the sequence is derived from the already known exploit.

Definition 3 (NIDS View). Let N be a NIDS. N's view with respect to an exploit E, denoted S^E_N, is the set of sequences of packets that N recognizes as E.

Given a NIDS and an instance of an exploit E, the basic reasoning task for DDL(X) is to find an instance of E that evades the NIDS.

Definition 4 (Reasoning Task 1). Let (W, T, Δ, I) be an attack mutation model of E, and N be a NIDS. Let S^E_N be the view of N with respect to E. The reasoning problem is to find a sequence of packets S that is derivable from E, but is not in S^E_N. More formally, find $S \notin S^E_N$ such that $M \vDash S$ and $M \vDash E$.

Given an instance of an exploit E and a sequence of packets S, another important reasoning task for DDL(X) is to determine whether S is an instance of E.

Definition 5 (Reasoning Task 2). Let $M = (W, T, \Delta, I)$ be an attack mutation model of E and S be a sequence of packets. S is an instance of E if and only if there exists a model $M = (W, T, \Delta, I)$ and a state $w \in W$ such that $M \vDash E$ and $(M, w) \vDash S$.

The last inference problem we will investigate is the planning problem. Given a goal statement (i.e. a sequence of packets) and a set of actions(i.e. evasion methods), the planning problem is to find an action sequence that will lead from the initial state(i.e. an already known exploit) to states in which the goal will hold. With DDL(X), we define the plans as follows.

Definition 6 (The Planning Problem). An action sequence $r_1, ..., r_n$ is a plan for a goal S w.r.t. $M = (W, T, \Delta, I)$ if and only if (i) the sequence-action $r_1;...; r_n$ is executable on states described by E and (ii) S is a consequence of applying $r_1;...; r_n$ on states described by E.

It is intuitive to model evasions by actions in DDL(ALCO). As demonstrated in this section, the functionalities of evasions can be semantically transformed into actions in DDL(ALCO) by a proper domain ontology (TBox). All kinds of reasoning tasks concerning the functionalities of evasions thus can be reduced to the reasoning about actions in DDL(ALCO), which are the topic of the next section.

4   An Efficient Algorithm for Evasion Composition

In this section we demonstrate that the problem of evasion composition can be reduced to satisfiability checking of formulas in DDL(ALCO) and then be solved by a decidable procedure after the transformation process that transforms evasions to actions in DDL(ALCO). Afterwards, we propose a heuristic planning algorithm for automated composition of evasions. The algorithm achieves good balance between computational performance and accuracy.

When facing a problem of evasion composition, we firstly collect the relevant evasions and transform these preexisting evasions to atomic actions by constructing the specification of the domain.

Let us analyze the following two formulas:

$\neg([(\alpha_1 \cup ... \cup \alpha_n)^*] \Pi \wedge \text{Con j}(E)) \vee < \text{plan} > \text{true}$, where $\text{Conj}(E)$ represents the conjunction of all the elements of the set E, Π the formula $\wedge_{i=1}^{n} (\neg \text{Con}$ $j(P_i) \vee <\alpha_i> \text{true})$ and P_i the precondition of action r_i for each i: $1 \leq i \leq n$.

The whole formula above states that the action sequence plan is executable on states described by E. The above formula is valid if its negation (labeled as Eq. 1) is not satifiable:

$$[(\alpha_1 \cup ... \cup \alpha_n)^*] \ \Pi \wedge \text{Con } j(E) \wedge \neg< \text{plan} > \text{true} \qquad (1)$$

$\neg\text{Con } j(E) \vee [\text{plan}]S$.

This formula indicates that the goal S is a consequence of applying the action sequence plan on world states described by E. Similarly, the above formula is valid if its negation (labeled as Eq. 2) is not satifiable:

$$\text{Con } j(E) \wedge \neg [\text{plan}]S \qquad (2)$$

Given a goal statement S and a set of actions $\sum = \{r_1, r_2, ..., r_m\}$, Algorithm 1 shows how to produce a plan (i.e. an action sequence) that will lead from the initial state to states in which the goal S will hold. The heuristics algorithm ActionPlan() travels the possible world states in a Breadth-First manner and terminates at a successful plan, or failure after the algorithm attempts nearly exhaustively. However, instead of travelling all possible world states, the heuristic selects world states more likely to produce an successful plan than other world states. It is selective at each decision point (Line.11), picking world states that are more likely to produce solutions.

Algorithm 1 ActionPlan(*E, T, $\sum$, S, plan*)
Input: initial ABox *E*; TBox *T*; the set $\sum = \{r_1, r_2, ..., r_m\}$ of available actions; and formula *S*, seen as a goal statement.
Output: an successful plan or *nil* as failure;
Begin
    1. initialize *plan* with an empty list;
    2. Initialize queue *QueOfPlans* with *plan*;
    3. while (*QueOfPlans* is unempty) do
    4.   remove head of *QueOfPlans* and set it to *plan*;
    5.   if (Eq. 2 is unsatifiable) then
    6.   return *plan* as a successful plan;
    7.   else
    8.   for each $r_i \in \sum$ do
    9.   *newplan* $\leftarrow$ < *plan*, r_i >;// appending r_i to the rear of the list *plan*.
    10.   if([$(r_1 \cup ... \cup r_k)^*$] $\Pi \wedge Con \ j(E) \wedge \neg<$ *newplan* $>$ *true* is unsatisfiable) then
    11.   if(EvaluatePlan(*E, T, S, plan, newplan*) $\geq$ 0) then
    12.   queue *QueOfPlans* with *newplan*;
    13.   end if
    14.   end if
    15.   end for
    16.   end if
    17. end while
    18. return *nil*.

End

Algorithm 2 EvaluatePlan(*E*, *T*, *goal*, *srcplan*, *dstplan*)

Input: initial ABox *E*; TBox *T*; formula *goal*, seen as a goal statement; *srcplan*, a plan for left-hand side of comparison; *dstplan*, a plan for right-hand side of comparison.

Output: an integer greater than, equal to, or less than 0, if the number of primitive formula achieved through dstplan is greater than, equal to, or less than the number of primitive formula achieved through srcplan, respectively.

Begin

    1. srccount←0;

    2. dstcount←0;

    3. rewrite formula *goal* to a disjunctive normal form formula $G = \vee_{i=0}^{d} (\wedge_{j=0}^{m} \varphi_{ij})$;

    4. for each primitive formula φ_{ij} do

    5.   if *Con j(E)* $\wedge \neg$ [*srcplan*] φ_{ij} is unsatifiable then

    6.   srccount←srccount+1;

    7.   end if

    8.   if *Con j(E)* $\wedge \neg$ [*dstplan*] φ_{ij} is unsatifiable then

    9.   dstcount←dstcount+1;

    10.   end if

    11. end for

    12. return dstcount − srccount;

End

5   Summary

In this article, we aim at providing an effective framework for the composition of evasion techniques to test the quality of intrusion detection signatures. Our approach supports multiple evasion techniques and allows the developer of the test to compose these techniques to achieve a wide range of attack mutations. We define an attack mutation model that describes all the possible transformations and how they can be applied to the original attack to generate a large number of attack variations. The attack mutation model is based on DDL(X), extensions of description logics (DLs) with a dynamic dimension. In particular, we proposed a heuristic planning algorithm for automated composition of evasion methods by a reduction to the formula satisfiability checking in DDL(X) and a selection of world states more likely to produce a successful plan. The functionalities of the evasion methods are abstracted by actions in DDL(X), while the domain constraints, exploits, and the objective sequence of packets are encoded in TBoxes, ABoxes and DL-formulas, respectively. Then the problem of evasion composition can be reduced to formula satisfiability checking in DDL(X) and solved by a decision procedure. Afterwards, instead of travelling all possible world states in a Breadth-First manner, a heuristic planning algorithm selects world states more likely to produce an successful plan than other world states. Our approach has several important advantages: the attack mutation model permits the application of analytical methods for deriving sound evasion composition; executing each of mutants generated by our approach against the vulnerable application, we are going to obtain the same effect as executing the original exploit script.

Acknowledgements. This work is supported by the National Natural Science Foundation of China (No. 61300132, No.61363058, No.61372182, No.61305056, No. 61303172), the Specialized Research Fund for the Doctoral Program of Higher Education (No.20120036120003), the Fundamental Research Funds for the Central Universities (No.12ZP09, No.13ZP010) and the Scientific Plan Project of Beijing Municipal Commission of Education(No.SQKM201420131001).

References

1. Ptacek, T.H., Newsham, T.N.: Insertion, evasion, and denial of service: Eluding network intrusion detection. Secure Networks INC Calgary Alberta (1998)
2. Handley, M., Paxson, V., Kreibich, C.: Network Intrusion Detection: Evasion, Traffic Normalization, and End-to-End Protocol Semantics. In: USENIX Security Symposium 2001, pp. 115–131 (2001)
3. Niemi, O.P.: Protect Against Advanced Evasion Techniques, McAfee (2014),
   http://www.mcafee.com/us/resources/white-papers/
   wp-protect-against-adv-evasion-techniques.pdf
4. Chang, L., Shi, Z., Gu, T., Zhao, L.: A Family of Dynamic Description Logics for Representing and Reasoning About Action. J. Autom. Reasoning, 1–52 (2010)
5. Wang, Z., Yang, K., Shi, Z.: Failure Diagnosis of Internetware Systems Using Dynamic Description Logic. J. Softw. China 21, 248–260 (2010)
6. Wang, Z., Peng, H., Guo, J., Zhang, Y., Wu, K., Xu, H., Wang, X.: An architecture description language based on dynamic description logics. In: Shi, Z., Leake, D., Vadera, S. (eds.) Intelligent Information Processing VI. IFIP AICT, vol. 385, pp. 157–166. Springer, Heidelberg (2012)
7. Baader, F., Calvanese, D., McGuinness, D., Nardi, D., Patel-Schneider, P.F.: The description logic handbook: theory, implementation, and applications. Cambridge University Press (2003)
8. Artale, A., Franconi, E.: A temporal description logic for reasoning about actions and plans. J. Artif. Intell. Res. USA 9, 463–506 (1998)
9. Baader, F., Lutz, C., Milicic, M., Sattler, U., Wolter, F.: Integrating description logics and action formalisms: First results. Proc. Natl. Conf. Artif. Intell. USA 2, 572–577 (2005)

Efficient Integrity Protection for P2P Streaming

Lingli Deng[1,*], Ziyao Xu[2], Wei Chen[1], Lu Lu[1], and Xiaodong Duan[1]

[1] Network Department of China Mobile Research Institute,
32 Xuanwumenxi Ave, Beijing, 100053, China
{denglingli,chenweiyj,lulu,duanxiaodong}@chinamobile.com
[2] Alibaba Corp.,
12 Xidawang Street, Chaoyang, Beijing, 100022, China
ziyao.xu@alibaba-inc.com

Abstract. This paper proposes an efficient checksum consolidation method, where the content source publishes only a few consolidated checksums (as the direct verification proof) for the media content. A tree-based and a chain-based scheme are proposed to consolidate chunk checksums respectively for VoD and live streaming application scenarios.

Keywords: hash, chunk, integrity, P2P streaming.

1 Introduction

In a typical P2P streaming system[1], there exist several well-known portals for the content sources to publish their local media content to potential audience accessing the network through the other peers. For a published media program, its responsible tracker server as well as other human readable description information, are returned to a querying peer by the portal, as the response to the latter's program selection request. The responsible tracker holds the IP/port address list for correspondent swarm (i.e. peer group uploading/downloading the same program) members. By submitting a swarm joining request to the responsible tracker, a downloading peer receives the peer list (containing sharing peers' IP/Port addresses, etc.) and hands in its own IP address and Port number to the tracker to update the swarm's peer list. From this point, a downloading peer can initiate media downloading request to other intermediate peers in the list.

To exploit the many-to-many sharing pattern of the P2P protocol, a large media file is divided into a group of chunks and each chunk is distributed independently afterwards [2]. In other words, a downloading peer needs only a single chunk, rather than the complete chunk set for the original media file, to become a server for uploading the local chunk to other peers. It is desirable from the respects of both the service provider and the downloading peers, to ensure that the media content is exactly the same as published and not manipulated by any intermediate party, especially when the provider holds certain reputation/authority/responsibility for the information's authenticity/validity it delivers to the public.

---

* This work is supported by Chinese National Grant 2012ZX03002008.

Z. Shi et al. (Eds.): IIP 2014, IFIP AICT 432, pp. 312–321, 2014.

However, previous P2P streaming systems put little concern on integrity protection. Moreover, particular requirements for streaming applications to achieve chunk-level integrity protection make previous solutions for P2P file-sharing networks inapplicable to streaming settings. On the contrary, the consolidated method proposed in this paper is highly applicable to P2P streaming systems, for its cost-efficiency in achieving chunk-level integrity protection for chunk sets, achieved by employing hash consolidation methods (i.e. hash tree based scheme for VoD applications and hash chain based scheme for live streaming applications). It is expected to bring down the storage and verification cost for downloading peers, the publishing cost for content sources, and the storage cost for system portal/trackers.

The rest of this paper is organized as follows: Section 2 gives a brief review on related work. Section 4 analysis integrity protection requirements for P2P Streaming applications. Section 3 introduces the hash tree and hash chain mechanisms to be utilized by our proposals in Section 5. In section 6, a preliminary analysis is conducted. Section 7 concludes.

2 Related Work

Previous P2P file-downloading systems take measures to ensure file integrity from the publisher all the way to the ultimate downloading peer. For example, BitTorrent users can publish a file's hash checksum to the system portal/tracker, to be used later by the downloading peer to verify a file's integrity after downloading the complete chunk set and rebuild the original file. However, in BitTorrent's scheme, publisher only provides the single checksum for the whole file, a downloading peer cannot locate the specific chunk(s) if only part(s) of the chunk set get manipulated, which means a single manipulated chunk would cause the searching and downloading the complete file all over again. Therefore, eMule enhances BitTorrent's scheme allowing content sources to incorporate the hash checksums for each individual chunk into the published content handle, so that chunk-level integrity verification and re-transmission are viable for the downloading peer. Another drawback of BitTorrent's scheme is that the total verification work is performed after the whole set of chunks are downloaded completely, indicating an insufferable delay to the completion of file downloading transaction, indicating devastated user experience. To conquer the problem, [3] proposed to exploite the iterative feature of the hash algorithm, in order to hide the check delay into the file downloading process, where the downloading peer uses the hash algorithm to compute a chunk's checksum and updates the file's hash vector whenever it received a new chunk. After downloading the last chunk, a peer only has to compute a single chunk's hash checksum before building the file's checksum and is able to complete integrity verification rapidly.

3 Background

A single hash tree[4] is able to protect the integrity of a large number of unsafely stored data objects, once the tree's root hash is safely stored. A leaf node is created for each data object, and each internal node is the hash value of the concatenation of its direct children. It is computationally infeasible for an attacker to construct another hash tree out of some modified leaf nodes to produce the same root hash, as the internal nodes (including the root hash) are constructed with collision-free hash algorithms [5]. See Algorithm 1, where $tree(S, H, d)$ calculates the root hash r for the d-degreed tree constructed from an array S using a hash functions from H.

Algorithm 1. $tree(S, H, d)$

$k \leftarrow 1; s \leftarrow S.length$
while $s > 1$ **do**
    $O \leftarrow \emptyset; n \leftarrow \lceil s/d \rceil; temp \leftarrow S[k]$
    for $i = 1$ to n **do**
        for $j = 1$ to $d - 1$ **do**
            $temp \leftarrow temp || S[k+1]; k \leftarrow k + 1$
        end for
        $O[i] \leftarrow H(temp)$
    end for
    $s \leftarrow n; S \leftarrow O$
end while
return $O[1]$

To check for the integrity of a leaf node, one needs to repeatedly compute the hash checksum for the parent node by concatenating the values of the node in question and its siblings and move upwards along the tree until the root is arrived. If the computed value is the same as the securely stored root hash, the leaf's integrity is intact. To update the value of a leaf node, one needs first to check the leaf's integrity (as stated above). If the value is intact, he/she can go further to modify the leaf's value, and update the leaf's parent, its grandparent, until the root hash is updated accordingly.

Intuitively, a hash chain is a hash tree, where "the height of each right-side sub-tree of any internal node is 1". Hash chains can be used to protect the integrity of a sequence.

4   Problem Statement

Definition 1 (P2P Chunk Integrity). *A chunk set has integrity, if each of its element maintains hash-integrity during the transmission chain from the content provider all the way to the downloading peer (including all the intermediate peers who downloaded the media chunk and uploaded to the next peer along the way).*

To ensure chunk-level integrity, previous P2P file-downloading systems allow a content source to provide the hash checksums for each chunk (in the form of a hash vector) as part of the descriptive information of the intended file.

A downloading peer is able to verify a chunk's integrity by locally calculating its hash checksum and check the value against correspondent element in the hash vector published by the portal.

However, as summarized by Table 1, streaming applications differ from the file downloading applications in dictating the following specific requirements for chunk integrity protection.

Table 1. Special Requirements for Media Chunk Integrity Verification

Application	Real-Time Verification	Partial Publication	Off-line Verification
Live	•	•	
VoD	•		•

First of all, in streaming applications, where the downloading peer is always holding a small number of chunks (not necessarily the whole chunk set for the media program) and is required to perform chunk-level integrity verification before playing the media program locally. In other words, they dictate *real-time verification requirement* that the local verification of individual/subset of media chunks cannot rely on the program's global information (e.g. the complete chunk set or the whole vector of hash checksums).

Secondly, the non-stop feature of some 7×24 live programs makes their content sources themselves is incapable of computing and publishing the global checksum for the media beforehand statically. Indicating *partial publication requirement* that the publication of checksum information cannot rely on the program's global information.

Thirdly, it is intuitively that until the portal delete the relevant downloading link(s) from its resource sharing list out of management reasons, the system should support the efficient downloading and correct verification of media chunks of a VoD media file, once there is a live peer group who together own a complete chunk set of the file, even when the content source leaved the system temporarily or permanently. In short, VoD applications also demand *off-line verification requirement* that the local verification for individual/subset of media chunks of a VoD media program cannot rely on the program's source's presence.

5   Consolidated Integrity Verification

The heart of our proposal, is the introduction of a peer-to-peer paradigm for the hash checksum generation and distribution, for the purpose of reducing the publication and storage signaling expenditures for potentially centralized portal/tracker in ensuring chunk-level integrity protection.

The proposed method mainly involves three procedures in a P2P streaming system, including[1]: (1) *a source-portal interaction* for media and direct

---

[1] For systems employing centralized tracker architecture, the tracker may take the responsibility of checksum publication and our proposal also applies by replacing the portal's role with tracker.

checksum publication, when the content source to compute the root hash and its signature locally and publishes them to the system portal; (2) *a peer-peer interaction* for indirect checksum; (3) *generation and distribution*, when the intermediate sharing peer to compute the verification paths for locally verified chunks and share them with other downloading peers; and (4) *a local verification procedure*, when the downloading peer verifies a newly downloaded chunk's integrity employing the indirect checksums (verification paths) from other swarm members as well as the direct checksums (root hashes) from the system portal.

5.1   Chain-Based Scheme for Live Streaming

The root hash of a published live channel $s(t)$, is published to the portal and updated at time t on a periodic basis (e.g. for every u chunks), by the content source using the hash chain algorithm, while the previous value $s(t-1)$ is also stored by the portal to be used by the downloading peer to verify chunk(s)'s integrity within the latest updating period (t-1,t]. The intermediate sharing peer, after downloading a media chunk x within the latest updating period, computes its hash checksum h_x and stores the value into a u-length verification vector D as $D[x]$. The verification vector $D[1..u]$ constitutes the verification path s_x for chunk x, i.e. $s_x \doteq D[1..u]$. The downloading peer needs only to acquire the verification path $D[1..u]$ from any intermediate sharing peer, as well as the root hash pair $s(t)$ and $s(t-1)$ from the portal, to verify chunk x's integrity.

Root Hash's Formulation and Publication. Assume that a content source $Peer - S$ wishes to publish a live channel X to the portal. The correspondent process of the channel's root hash's formulation and publication is as follows.

- **Step1:** $Peer - S$ locally produces the descriptive information $m(X)$ for the live channel X, and submits it, as well as the first set of u chunks x_{1u}'s root hash $s(0)$ (computed through the hash chain algorithm), to the portal for publication.

$$s(0) = chain(X[1..u], H, d) \tag{1}$$

- **Step2:** For each updating period (i.e. the time interval for playing u media chunks), $Peer - S$ locally computes the current root hash $s(t)$, and submits it to the portal for update. To respond, the portal stores $s(t)$ and $s(t-1)$ for the channel.

$$s(t) = chain(S(t-1)||X[1..u], H, d) \tag{2}$$

Media Chunks' Integrity Verification. Assume that $Peer - D$ wishes to downloads the xth media chunk of the latest updating period from an intermediate sharing peer $Peer - U$ and checks the chunk's integrity locally.

Two types of local verification processes are provided: an eager downloading peer may choose Process 2, while a prudent peer may choose Process 1.

1. **Local verification process 1:** $Peer-D$ checks a chunk's verification path's validity before downloading it and perform the correspondent integrity verification immediately afterwards.
   - **Step1:** $Peer-D$ checks if the local verification vector D contains x's verification path s_x: if so, executes **Step3**; or **Step2**, otherwise.
   - **Step2:** $Peer-D$ asks $Peer-U$ for verification path $s_x = D'[1..u]$, and verifies the validity of the latter's response s_x through the following equation, according to the root hash pair $s(t)$ and $s(t-1)$ from the portal:
   $$s(t) = chain(S(t-1)||D'[1..u], H, d) \qquad (3)$$
   If the equation holds, the path is valid, update the local verification vector $D = s_x$ and executes **Step3**; otherwise, give up $Peer-U$ and choose another intermediate sharing peer for chunk x.
   - **Step3:** After downloading chunk x from $Peer-U$, $Peer-D$ computes its hash checksum h_x and verifies its integrity through the following equation, according to the locally stored verification path s_x:
   $$h_x = s_x[x] \qquad (4)$$
   If the equation holds, chunk x is downloaded successfully; otherwise, give up $Peer-U$ and choose another intermediate sharing peer for downloading x.
2. **Local verification process 2:** $Peer-D$ performs the integrity verification after successfully downloading all the chunks of the latest updating period.
   - **Step1:** After downloading chunk x from $Peer-U$, $Peer-D$ computes its hash checksum h_x and store it to the correspondent element $D[x]$ of the verification vector D.
   - **Step2:** If all the chunks within the latest updating period are downloaded locally, execute **Step3**; otherwise, terminate the current chunk downloading transaction.
   - **Step3:** $Peer-D$ performs the integrity verification for the chunk set locally downloaded for the latest updating period, through using the local verification vector D as well as the root hash pair $s(t)$ and $s(t-1)$ from the portal, according to the following equation:
   $$s(t) = chain(S(t-1)||D[1..u], H, d) \qquad (5)$$
   If the equation holds, terminate the chunk downloading transaction for the latest updating period. Otherwise, starting from the first media chunk of the current period, request the correspondent intermediate sharing peer for the chunk's verification path, until a valid path, satisfying the following equation, is received or no other peer to request:
   $$s(t) = chain(S(t-1)||D'[1..u], H, d) \qquad (6)$$
   If the received D' is valid, $Peer-D$ locates a set of integrity-corrupted chunks S, by comparing h_{x_i} and $D'[i]$. Otherwise, S is set of the complete set of chunks within the latest updating period.
   - **Step4:** $Peer-D$ through the "local verification process 1" for relocating, downloading and verifying each chunk in set S.

Verification Path's Computation and Distribution Assume that $Peer-U$ wishes to share its locally downloaded chunk x, and the correspondent verification path for other downloaders. Intuitively, there are two possibilities:(a) $Peer - U$ downloaded the valid verification path s_x as well as x itself from some other intermediate sharing peer $Peer - X$ through the "Local verification process 1"; or (b)$Peer - U$ downloaded the complete set of media chunks through "Local verification process 2" and acquire correspondent verification paths by local hashing and comparison to the root hash pair. As a result, $Peer - U$ may directly share its locally stored verification vector $D[1..u]$ to another downloading peer as the verification path for any chunk x of the latest updating period. No extra processing is required.

5.2   Tree-Based Scheme for VoD

The root hash of a published VoD file s, is published to the portal by the content source using the hash tree algorithm based on the complete chunk set of the file. The intermediate sharing peer, after downloading a media chunk x, computes its hash checksum h_x and stores the value into the local verification tree D as the xth leaf on the tree. The hash values of the sibling nodes along the tree path from the correspondent leaf node to the root constitute the verification path for chunk x, i.e. $s_x = tree - path(X, d, n, x)$. (See Algorithm 2). The downloading peer needs only to acquire the verification path s_x from any intermediate sharing peer, as well as the root hash s from the portal, to verify chunk x's integrity through Algorithm 3.

Root Hash's Formulation and Publication Assume that a content source $Peer - S$ wishes to publish a VoD file X to the portal. $Peer - S$ locally produces the descriptive information $m(X)$ for the file X, and submits it, as well as the root hash $s(0)$, to the portal for publication. $s(0)$ is computed through the hash tree algorithm where the complete chunk set of the file constitute the leaves of the tree and its degree d is determined locally based on the file's scale and average storage and computation expenditure for verifying the integrity of each chunk. (See Section 6 for reference.)

$$s(0) = tree(X[1..n], H, d) \tag{7}$$

Algorithm 2. $tree - path(T, x)$

$P \leftarrow \emptyset; temp \leftarrow getleaf(T, x)$
while $temp \neq T.root$ **do**
  $AddSibling(T, temp)toP; temp \leftarrow Parent(T, temp)$
end while
return P

Algorithm 3. $tree - verify(P, r, x, d)$

$p \leftarrow Lengthof(P); n \leftarrow (p-1)/(d-1); temp \leftarrow x; k \leftarrow 1$
for $i = 1$ to n **do**
  for $j = 1$ to $d - 1$ **do**
    $temp \leftarrow temp \| P[k]; k \leftarrow k + 1$
  end for
  $temp \leftarrow H(temp)$
end for
if $temp = r$ **then**
  return $TRUE$
else
  return $FALSE$
end if

Media Chunks' Integrity Verification. Assume that $Peer - D$ wishes to downloads a media chunk x of a VoD file X from an intermediate sharing peer $Peer - U$ and checks x's integrity locally.

- **Step1:** $Peer - D$ checks if the local hash tree D contains x's verification path s_x: if so, executes Step3; or Step2, otherwise.
- **Step2:** $Peer - D$ asks $Peer - U$ for verification path s_x, and verifies the validity of the latter's response s_x through the following equation, according to the root hash pair $s(0)$ from the portal:

$$tree - verify(s_x, s(0), x, d) = TRUE \tag{8}$$

If the equation holds, the path is valid, update the local hash tree D with the hash values contained by s_x and executes Step3; otherwise, give up $Peer - U$ and choose another intermediate sharing peer for chunk x.

- **Step3:** After downloading chunk x from $Peer - U$, $Peer - D$ computes its hash checksum h_x and verifies its integrity through the following equation, according to the locally stored verification path s_x:

$$h_x = s_x[1] \tag{9}$$

If the equation holds, chunk x is downloaded successfully; otherwise, give up $Peer - U$ and choose another intermediate sharing peer for downloading x.

Verification Path's Computation and Distribution. Assume that $Peer - U$ wishes to share its locally downloaded chunk x, and the correspondent verification path for other downloaders.

Intuitively, there are two possibilities: (a) $Peer - U$ downloaded the valid verification path s_x as well as x itself from another intermediate sharing peer $Peer - X$, and checked x's integrity; or (b) before downloading x, $Peer - U$ finds that a valid path is available from the current hash tree D stored locally for file X (i.e all the hash values that constitute x's verification path have been computed and stored for some other chunks downloaded earlier). As a result, $Peer - U$ may directly share its locally stored verification path s_x to another downloading peer.

6   Analysis

In Table 2 and 3, variable n stands for the number of chunks contained by an average VoD media file, which is proportional to the length of the file. m represents the average number of chunks downloaded by a participating peer, which is correlated to the coding rate of the media file and the average on-line duration of a typical user. System parameters u and d are constants determined by the system administrator, with the former depicts the number of chunks of an updating period, and the latter the degree of a VoD hash tree. In comparison, System A in the tables refers to a VoD application using a previous method with each chunk's checksum included into a vector as part of the publication data stored at portal/tracker, while system B (live streaming scenario) and C (VoD scenario) employ the proposed chain-based and tree-based schemes, respectively.

Table 2. Requirement-Satisfaction Analysis

Application	Real-Time Verification	Partial Publication	Off-line Verification
A (VoD-vector)	$\checkmark$	$\times$	$\checkmark$
B(Live-chain)	$\checkmark$	$\checkmark$	—
C(VoD-tree)	$\checkmark$	$\checkmark$	$\checkmark$

In system A, a downloading peer may verify the integrity of a newly down-loaded chunk by using the global hash vector from the portal/tracker, therefore fulfilling the requirements of real-time verification and off-line verification. However, it is inapplicable to the live streaming scenario at all since it cannot meet the partial publication requirement as dictated in live streaming scenarios. The content source needs to compute the hash checksums for each media chunk in the form of a hash vector with length n (where n depicts the average number of chunks for a VoD file), resulting in $O(n)$ computation and communication costs for the provider in the content publication procedure, $O(n)$ storage and communication costs for the portal/tracker in checksum distribution procedure, and $O(n)$ computation and/or communication costs for the downloading peers in integrity verification procedure. On the other hand, the proposed two schemes respectively satisfy the requirements in terms of live and VoD scenarios. In both settings, the portal/tracker only needs to store the root hashes for a published program, fulfilling the partial publication requirement and reducing publication expenditures for both the provider and portal/tracker to $O(1)$, with the exception that for live streaming scenario where the root hash of a program is updated periodically (i.e. once every u chunks) by the provider the publication cost is reduced to $O(n/u)$.

In terms of the worse case computational cost for the downloading peer to perform integrity verification of a given chunk, system B and C seem to degrade the $O(1)$ performance respectively to $O(u)$ and $O(log_d(n))$ in order to verify the first chunk of a media content. However, the average verification cost for each chunk will decrease because the internal nodes' values are filled into the chain/tree as the downloaded chunk group grows, and finally approaches $O(1)$.

Table 3. Verification expenditures for each chunk

Application	Publication	Verification	Storage
A(VoD-vetor)	$O(n)$	$O(1)$	$O(n)$
B(Live-chain)	$O(n/u)$	$O(1)$	$O(1)$
C(VoD-tree)	$O(1)$	$O(1)$	$O(1)$

7   Conclusion

This paper proposes an efficient checksum consolidation, distribution and verification scheme to ensure chunk-level integrity protection in P2P streaming systems. The content source publishes only consolidated checksums to the network. A peer securely verifies a downloaded chunk's integrity by combining the checksums from other intermediate peers as well as the one from the content source. Specifically, a tree-based and a chain-based scheme are proposed respectively for VoD and live streaming application scenarios. Preliminary analysis shows that the proposal is highly applicable to the streaming settings and more efficient compared to previous solutions.

References

1. Vu, L., Gupta, I., Liang, J., Nahrstedt, K.: Measurement of a large-scale overlay for multimedia streaming. In: Proceedings of the 16th International Symposium on High Performance Distributed Computing, pp. 241–242. ACM Press (2007)
2. Zhang, Y., Zong, N., Yang, J.S., Problem Statement, R.: of P2P Streaming Protocol (PPSP) (2009)
3. He, P., Wang, J., Deng, H., Sun, P.: Latency Hiding Algorithm for P2P File Integrity Verification. Chinese Journal of Computer Engineering 36 (2010)
4. Merkle, R.: Protocols for Public Key Cryptosystems (1980)
5. Goldreich, O.: Foundations of cryptography: Basic applications. Cambridge Univ. Pr. (2004)

Author Index

Cai, Guoyong   67
Cai, Xingjuan   11
Cai, Yuhao   38
Cao, Cungen   57, 152
Chang, Hongyan   187
Chang, Liang   78, 161, 171, 284
Chen, Jie   47
Chen, Wei   312
Cheng, Wenjun   207
Chu, Hua   113
Cui, Zhihua   11

Deng, Lingli   312
Deng, Ning   20
Deng, Shaobo   141, 152
Diao, Liang   113
Ding, Shifei   133
Dou, Quansheng   216
Du, Chao   278
Duan, Huilong   20
Duan, Xiaodong   312

Fan, Jiancong   38
Feng, Zheng   38

Gao, Yuhui   207
Gu, Tianlong   161, 171, 284
Guan, Yong   178
Guo, Jing   303

Han, Youzhen   133
Hao, Jie   1
He, Hui   303
Herranz, Luis   269
Hou, Fu   122
Hu, Haixiao   197
Huang, Jun   47

Ji, Saiping   230
Ji, Xiang   106
Jia, Meihuizi   90
Jiang, Fei   284
Jiang, Liang   187
Jiang, Ping   216

Jiang, Shuqiang   269
Jing, Quanliang   294

Kang, Qi   11

Li, Fengying   161
Li, Liming   178
Li, Meng   294
Li, Qingshan   113
Li, Shibao   221
Li, Xiangyang   249, 269
Li, Zhijun   29
Li, Zhixin   239, 259
Liang, Yongquan   38
Liu, Jingang   294
Liu, Lizhen   278
Liu, Wenhua   38
Lu, Jingli   278
Lu, Liying   230
Lu, Lu   312
Lu, Xiaobo   230

Ma, Gang   1, 96
Ma, Huifang   90
Mao, Xinjun   122
Meng, Zuqiang   187

Niu, Wenjia   78

Pan, Chao   20
Pan, Xin   221
Peiqing, Wang   90
Peng, Hui   303
Peng, Libin   67
Peng, Wenxian   20

Qi, Baoyuan   96
Qiang, Baohua   47

Shi, Jin   303
Shi, Zhiping   178, 249, 269
Shi, Zhongzhi   1, 96
Shu, Ruo   221
Song, Xinhang   269
Song, Yuanren   29
Sui, Yuefei   141, 152

Sun, Meiying   141, 152
Sun, Weiqiang   78
Sun, Yuexin   90

Tian, Guanhua   303
Tong, Chen   230

Wang, Chaojun   239
Wang, Chenxiao   278
Wang, Hanshi   278
Wang, Lei   11
Wang, Lu   113
Wang, Peng   47
Wang, Wei   96
Wang, Xinsheng   106
Wang, Yaoguang   47, 161
Wang, Yong   67
Wang, Zhuxiao   303
Wei, Hongxing   178
Wu, Minhua   249
Wu, Wei   122

Xia, Fei   57
Xu, Yang   197
Xu, Zhoubo   171, 284
Xu, Ziyao   312

Yao, Zhongjiang   294
Yin, Junwen   122
Yue, Jinpeng   1

Zeng, Peilong   259
Zhang, Bo   1
Zhang, Canlong   239, 259
Zhang, Dapeng   106
Zhang, Jie   178
Zhang, Na   171
Zhang, Ying   303
Zhang, Yuansheng   187
Zhang, Yulin   197
Zhao, Weizhong   78

Printed in the United States
By Bookmasters